D1759159

HPLC COLUMNS

HPLC COLUMNS

Theory, Technology, and Practice

Uwe D. Neue

with a contribution from
M. Zoubair El Fallah

⊛ WILEY-VCH

New York · Chichester · Weinheim · Brisbane · Singapore · Toronto

Dr. Uwe D. Neue
Waters Corporation
34 Maple Street
Milford, MA 01757

Library of Congress Cataloging in Publication Data:
Neue, Uwe D.
    HPLC columns : theory, technology, and practice / Uwe D. Neue ;
  with a contribution from M. Zoubair El Fallah.
      p.   cm.
    Includes index.
    (cloth : alk. paper)
    1. High performance liquid chromatography--Equipment and supplies.
  I. Title.
  QD79.C454N48   1997
  543' . 0894--dc21   97-3394

Printed in the United States of America

ISBN 0-471-19037-3 Wiley-VCH, Inc.

10 9 8 7 6 5 4 3 2

To my science teachers:
Studiendirektor Werner Backes
Oberstudiendirektor Dr. Erwin Distler
Prof. Dr. Istvan Halász
Prof. Dr. Heinz Engelhardt
Dr. Richard M. King

CONTENTS

PREFACE

High-performance liquid chromatography (HPLC) is now nearly 25 years old. It is therefore surprising to find that until now no book has been written that is dedicated to HPLC columns, despite the importance of column technology to the success of HPLC. Hence I was delighted when the idea of writing such a book was proposed to me. It is a commonplace saying that the chromatographic column is the heart of the HPLC system. Consequently, a thorough understanding of the physics and chemistry of the processes that take place in a column is of significant importance to the analytical chemist who uses HPLC.

One of the first steps in writing a book is to define the audience for which it is written. I decided to address the text to the HPLC practitioner working in the laboratory. Therefore the book is largely a practical guide, but it also contains sufficient theoretical background information to give an in-depth understanding of the topics covered. Thus it should be useful for both the novice and the experienced user of chromatography.

Another decision that needs to be made early is which subjects should be covered and what should be omitted. Since the book is intended for the HPLC user, I limited the discussion of packings and columns largely to those that are commercially available. At the same time, I did not include the lengthy tables of commercially available products found in some other books. In my experience, these tables are far from complete and full of erroneous information. Generally, I give sources only for columns or packings, if they are available from only one supplier and if this supplier may not be widely known. Of course, when specific data or examples are shown, a complete reference is given.

The chapters of the book are not intended to be review articles. Therefore, I have been fairly selective with the references. The references given are for the most part key publications on the subject, review articles or sources that are largely unknown. For those who want to explore the subjects in greater depth, I have included recommendations for further reading.

Large portions of the book are based on seminars and lectures that I have put together over the years for both my colleagues at Waters Corporation and Waters customers. However, all material has been updated to incorporate the current state of knowledge.

Many data and examples that I use to illustrate points were provided by my co-workers in the chemical products group at Waters. I want to thank each and every one of them for their contributions.

Throughout the writing of the book, I have asked my colleagues at Waters and Phase Separations for comments and suggestions. Their feedback has strengthened the quality of the book, and I would like to thank Dorothy J. Phillips, Tom H. Walter, Richard M. King, Edouard S. P. Bouvier, Glen E. Knowles, Ray P. Fisk, Peter Myers, Michael S. Young, Jim Krol, Rick Nielson, Yefim Brun, Steve Cohen, Todd Peltonen, Barbara Murphy, Robin Andreotti, and John E. O'Gara for their input.

Also, to further improve the book's content, I have asked M. Zoubair El Fallah to write the chapter on methods development. Zoubair's strong background in chromatographic theory as well as his broad practical experience made him the ideal choice for producing this chapter. Furthermore, we have previously cooperated in the elaboration of method development strategies, and some of the work reported in this chapter originated in this cooperation.

Finally, I want to apologize to my family for spending the better part of my "free" time during the last year writing this book. I promise to make up for it in the future.

I hope that this book will serve you, the reader, well and helps you in the application of HPLC to your analytical or preparative separation problems.

UWE DIETER NEUE
Ashland, Massachusetts

HPLC COLUMNS

1 Introduction

History is not merely a record of the dead past.
— R. A. Raff, *The Shape of Life*

High-performance liquid chromatography (HPLC), together with its derivative techniques, is today the dominant analytical separation tool in many industries. It is used extensively in the pharmaceutical industry in applications ranging from content uniformity assays to pharmacokinetic studies. The chemical industry relies on HPLC in the quality control of raw materials, intermediates, and finished products. In environmental laboratories, HPLC is used for the analysis of pesticides and other contaminants in soil and water. Applications of HPLC in the food industry include the analysis of pesticide residues or the content of nutrients or additives. In clinical analysis, therapeutic drug monitoring is performed by HPLC. These are just a few examples that illustrate the broad scope of HPLC.

1.1 HISTORY

HPLC as we know it today emerged around 1973. During that year, packing technologies were developed that made it possible for the first time to reproducibly prepare high-efficiency columns from 10-μm particles. Older packing techniques had failed to give good results for particles smaller than about 30 μm. Also during 1973, the modification of the silica surface via silanization became commercially feasible. These breakthroughs in the technology lead to the marketing of the first 10-μm reversed-phase columns. An example is μBondapak C_{18}, which became available in August 1973, packed into a 30-cm-long column with an internal diameter of 3.9 mm. Thereafter for the first time sufficient separation power was available to tackle a broad range of real-life problems within a reasonable analysis time. Especially in the

1

pharmaceutical industry HPLC became a nearly instantaneous success, and its impact on the chemical industry in general was significant.

HPLC had now become competitive with gas chromatography, the other major instrumental analytical separation technique of the time. HPLC was able to deal with a broader range of samples than gas chromatography (GC), and had the advantage of simplified sample preparation in those cases that could be solved by either technique. In principle, anything that can be dissolved can be subjected to an HPLC analysis. Consequently the number of application areas of HPLC expanded rapidly.

Between 1973 and 1978 the groundwork was laid for a basic understanding of the fundamentals of the technology. Guiochon and co-workers published a series of papers titled *Study of the Pertinency of Pressure in Liquid Chromatography*, which provided the foundation of rational column design. Halász and co-workers pointed out the "Ultimate Limits of High-Pressure Liquid Chromatography". The research by Horváth and co-workers furnished the first theories of the mechanisms of reversed-phase chromatography, which had quickly become the most popular separation technique. Columns packed with 5-μm particles became commercially available, and column manufacturers wrestled with issues of bed stability and batch-to-batch reproducibility of the packing material. One outcome from these efforts was the stabilization of the packed bed by radial compression.

The next 5 years (1978–1983) were characterized largely by a consolidation of the knowledge and an expansion of the application of chromatography into new areas. Ion chromatography was an important derivative of HPLC and quickly dominated the analysis of inorganic ions. Kirkland and Glajch formed the first framework for rational method development for reversed-phase chromatography based on solvent selectivity. This was soon followed by the development of alternative strategies and the expansion to other parameters to manipulate selectivity. During this time period, the compromises surrounding smaller-diameter columns were explored as well, largely through the work of Novotny, Scott, and Ishii.

The key developments of the subsequent decade were in the realm of bonded-phase technology. The efforts of many researchers gave us a superior understanding of the retention mechanisms, and the structure and preparation of bonded phases, including the pretreatment of the silica surface itself. Important tools were refined chromatographic characterization techniques and the use of sophisticated instruments such as solid-state nuclear magnetic resonance (NMR).

To the casual observer it might appear that HPLC column technology has changed little between 1978 and 1996. We are still using predominantly columns packed with 5-μm particles, and our preferred packing is a C_{18} bonded phase. In reality, much progress has been made, but all advances lie in the details of the column technology. Most major column manufacturers have honed their packing skills significantly to the extent that the reproducibility of the column packing operation and the mechanical stability of the packed bed are no longer an issue. Another area of significant progress is the chemistry of the HPLC packings. Manufacturers have learned to improve the adsorption properties of a silica-based reversed-phase packing and to implement controls over the manufacturing processes that result in better batch-to-batch reproducibility. An example is the drastically reduced tailing of basic compounds on packings that were developed in the late 1980s and the 1990s compared to packings developed in the 1970s.

But the limits of column technology continue to be explored. Areas of developments include smaller particle sizes and smaller column diameters. From the standpoint of column technology, there is no fundamental technological hurdle to overcome to reduce the column diameter. In fact, several advantages could be realized if the column diameter could be pushed to less than 100 μm. However, without a parallel development of an instrument, we will not be able to achieve any gains. In 1973, column technology drove the design of the HPLC instrument. Maybe we will see a similar development again.

1.2 SCOPE OF THIS BOOK

In this book we give a comprehensive overview over the state of the art of column technology and explain the underlying principles. We give practical advice on the use of columns and show application examples. We address the question of which column should be selected and include a section on method development. We also discuss special techniques, including techniques of preparative chromatography, such as continuous chromatography and the simulated moving bed. Finally, this book would not be complete without a troubleshooting section. In summary, we have tried to create a practical guide, designed for the HPLC practitioner working in the laboratory, while supplying sufficient theoretical background information to give an in-depth understanding of the subject.

1.3 WHAT IS HPLC?

Before we delve into the details, let us spend a moment to define high-performance liquid chromatography. Chromatography in general comprises all separation techniques in which analytes partition between different phases that move relative to each other or where the analytes have different migration velocities. The latter part of the definition includes chromatographic techniques in which the analytes are transported via a field, such as in electrokinetic chromatography.

In countercurrent chromatographic techniques, both phases are moving, but in most chromatographic techniques one phase is stationary, while the other one is mobile. In liquid chromatography, the mobile phase is a liquid, while the stationary phase can be a solid or a liquid immobilized on a solid. High-performance liquid chromatography comprises all liquid chromatographic techniques that require the use of elevated pressures to force the liquid through a packed bed of the stationary phase. It is therefore also often called *high-pressure liquid chromatography*. Because sophisticated and therefore expensive instrumentation is required, cynics have also called it *high-priced liquid chromatography*.

HPLC is primarily an analytical separation technique, used to detect and quantitate analytes of interest in more or less complex mixtures and matrices. However, it is also used to isolate and purify compounds. Since the goals of preparative chromatography are substantially different from the goals of analytical HPLC, a separate chapter is dedicated to the special issues associated with this technique.

1.4 SEPARATION MECHANISMS USED IN HPLC

In this section we briefly review the separation mechanisms used in HPLC. The various separation techniques are covered in separate chapters, which describe the mechanisms, the practice, and typical applications in detail.

The traditional form of liquid chromatography employed a polar adsorbent such as silica or alumina, and a nonpolar mobile phase based on hydrocarbons such as petrol ether or chlorinated hydrocarbons such as chloroform. Today, this type of chromatography is known as *normal-phase chromatography*, in contrast to reversed-phase chromatography. It is also often referred to as *adsorption chromatography*. It is based on the interaction of the polar functional groups of the analytes with polar sites on the surface of the packing. Normal-phase chromatography has lost its initial importance and has been eclipsed by reversed-phase chromatography.

In *reversed-phase chromatography*, a nonpolar stationary phase is used in conjunction with polar, largely aqueous mobile phases. Between 70 and 80% of all HPLC applications utilize this technique. Its popularity is based largely on its ease of use; equilibration is fast, retention times are reproducible, and the basic principles of the retention mechanism can be understood easily. Most stationary phases are silica-based bonded phases, but polymeric phases, phases based on inorganic substrates other than silica, and graphitized carbon have found their place as well.

Some applications, such as protein separations, require a surface less hydrophobic than that of reversed-phase packings. Proteins are denatured by the aqueous–organic mobile phases commonly used in reversed-phase chromatography. If the hydrophobicity of the stationary phase is reduced significantly, proteins can be eluted with water or dilute buffer as eluent. In this technique, termed *hydrophobic interaction chromatography*, the analytes are typically adsorbed onto the packing in a buffer with a high salt concentration, and eluted with a buffer of low ionic strength. This technique can be viewed as an extension of reversed-phase chromatography.

In *ion-exchange chromatography*, the interaction of charged analytes with oppositely charged functional groups on the stationary phase is utilized. Elution is effected by either an increase of the ionic strength of the buffer, thus increasing the concentration of competing counterions, or through the change of pH, which can modify the charge of the analyte or of the ion exchanger. Strong and weak cation and anion exchangers are employed. The application range of ion-exchange chromatography is broad, covering organic and inorganic analytes. A large application area is the separation of biopolymers, specifically, proteins and nucleic acids.

In *size-exclusion chromatography*, the separation is based on the partial exclusion of analytes from the pores of the packing, due to the size of the analyte. It is used largely for the analysis or characterization of industrial polymers and biopolymers, but its separation range extends all the way down to oligomers. Known also as *gel-permeation chromatography*, it is one of the parent techniques of today's instrumental HPLC.

Hydrophilic interaction chromatography can be viewed as an extension of normal-phase chromatography to the realm of very polar analytes and aqueous mobile phases. Suitable stationary phases are the same as used in normal-phase chromatography. The most important application is the separation of sugars, oligosaccharides, and complex carbohydrates.

True liquid–liquid *partition chromatography* is seldom used today. In this technique, the pores of a packing are filled with a liquid that is immiscible with the mobile phase. The preparation of the column is complicated, and it is difficult to maintain precise equilibrium conditions. The use of bonded phases has completely displaced classic partition chromatography. One can argue that reversed-phase chromatography or hydrophilic interaction chromatography are really versions of partition chromatography that use a very thin layer of stationary phase. True liquid–liquid partitioning is briefly discussed as a special technique of normal-phase chromatography.

Several special techniques are worth mentioning. An example is *ion-pair chromatography*, which is carried out using ionic surfactants on reversed-phase packings. It is covered in Chapter 10. Another example is *ion chromatography*, which is a special technique of ion-exchange chromatography applied to inorganic ions. The unique requirements of the associated detection technique resulted in distinct stationary phases different from those of standard ion exchangers; these are covered in Chapter 12.

2 Theory of Chromatography

Soll ich dir die Gegend zeigen,
mußt du erst das Dach besteigen.
> —J. W. von Goethe, *Hikmet Nahmeh, Buch der Sprüche*

The theory of chromatography, especially of the hydrodynamics of chromatography, is a subject that chemists may tend to shy away from. Nevertheless, a basic awareness of the theory is useful for the optimization of a method. The choice of the best column dimensions or the best particle size can have a significant impact on the throughput or the sensitivity of a method. Therefore a fundamental grasp of chromatographic theory is important.

In this chapter we will first get a basic understanding of what is going on in a column. We will try to develop the hydrodynamic theory in a rather descriptive form, with a minimal amount of math. We start off with some key definitions, followed by an examination of the parameters that influence column performance. This forms the basis for our discussion in Chapter 3 of the influence of particle size and column dimensions on the quality measures of the separation (resolution, analysis time, backpressure, sensitivity, etc). At the end of these two chapters, readers should be able to select the physical parameters of a column according to the needs of their applications.

2.1 DEFINITIONS

2.1.1 Isocratic and Gradient Chromatography

In isocratic chromatography, all conditions and settings of the separation are held constant. Conversely, in gradient chromatography one or more parameters are varied continuously. In nearly all practical cases, only one parameter is varied. The most typical gradient is a continuous variation of the mobile-phase composition from low elution strength to high elution strength. One may also use a temperature gradient or a flow gradient.

2.1.2 The Height Equivalent to a Theoretical Plate

We will first develop the concept of the *height equivalent to a theoretical plate*. This tongue twister is usually abbreviated as HETP. The origin of the term can be traced back to the early theories of gas chromatography which treated the chromatographic column as a multistage distillation tower. Today, a much simpler concept is used, but the terminology has remained. When a small volume of analyte is injected onto a column, it forms a narrow band on the top of the column. As the analyte migrates through the column, the band becomes broader (Fig. 2.1). In a uniform bed, the peak width increases with the square root of the length that the band has traveled.

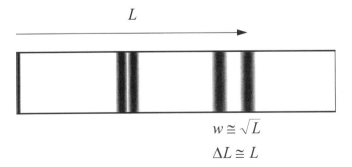

$$w \cong \sqrt{L}$$
$$\Delta L \cong L$$

Figure 2.1 Why does chromatography work? In a uniform bed, the width of the bands of two analytes increases with the square root of the length that they have migrated, while the distance between the centers of the bands increases directly proportionally to the length that the bands have migrated.

If we inject a mixture of two components onto the top of the column, they both form the same narrow band on the top of the column. If they have slightly different affinities to the column, they will migrate at slightly different speeds down the column. As they continue to migrate, the distance between the centers of both bands increases in direct proportion to the length that the bands have traveled.

Since the distance between the bands increases with the length traveled but the width of each band increases only with the square root of this length, these bands will at some point be completely separated. The farther they migrate, the better separated they will be. This is the fundamental reason why chromatography works.

[*Note:* To a first order of approximation (neglecting secondary effects), the width of different peaks (in length units) is roughly the same at any particular point inside the column, including the bottom of the column. When the peak width is measured in length units inside the column, it is roughly independent of the migration velocity of the peak. This is also true for gradient chromatography, again neglecting some second-order effects. Of course, we observe a difference in peak width in an isocratic chromatogram recorded by a data system, which reports the chromatogram, and therefore the width of the peaks, in time units. These differences in peak width measured in time units are due to the differences in migration velocity — and therefore elution velocity — of the peaks.]

A chromatographic band can be considered to be a statistical distribution of molecules. A characteristic parameter of a distribution is its second moment, also called its variance σ^2. Variance is a fundamental characteristic of a distribution independent of the form of the distribution. Therefore it is not important whether the peak is symmetric or skewed or even in a box-shaped distribution. The width of the band is always proportional to the square root of the variance of the distribution, with the proportionality factor depending on the type of distribution. For a typical chromatographic peak, the variance that characterizes this peak is about 25% of the square of the width of the peak at 60.7% of the height of the peak.

We have stated above that the width of the peak increases with the square root of the length that it has traveled. Or, in other words, the variance of the peak σ^2 increases linearly with the distance L that it has traveled. A plot of the variance against length is a straight line (Fig. 2.2). The slope of this relationship is the HETP, sometimes also simply called H.

Most chromatographic beds are uniform from top to bottom. But one can also conceive of beds that are nonuniform. In this case also, the definition of the HETP holds, if we define the slope locally. Mathematically, we formulate this in the form of a differential equation:

$$H \equiv \frac{d\sigma^2}{dL} \tag{2.1}$$

This is the fundamental definition of the HETP.

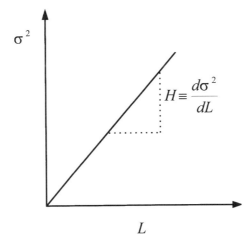

Figure 2.2 The HETP is the slope of a plot of the variance of a peak versus the distance that the peak has migrated.

2.1.3 Resolution

We would like to characterize the amount of the separation between two peaks. An easy measure is the ratio of the distance between the peaks to their peak width. This is called the *resolution Rs*:

$$Rs \equiv \frac{\Delta L}{\frac{1}{2} \cdot (w_{l,1} + w_{l,2})} \tag{2.2}$$

where ΔL is the distance between the peaks measured in length units inside the column and w_l is the width of the peaks measured in the same units. What is the "width" of a peak? As we know, the boundaries between the baseline and the beginning and the end of the peak are fuzzy, depending on the threshold of the peak detection algorithm. For a Gaussian peak, the "width" of the peak has been defined as 4 times the standard deviation of the peak. See Section 2.1.4 and Figure 2.3 for further clarification.

We have seen that the distance between the centers of two peaks increases with the length that they have migrated, while the width of the peaks increases only with the square root of this length. Therefore resolution increases with the square root of the distance traveled by the bands:

$$Rs \propto \frac{L}{\sqrt{L}} = \sqrt{L} \tag{2.3}$$

Resolution is a dimensionless parameter. We can also calculate it from the chromatogram, where the distance between peaks and their width are recorded in time units. This is the most common form in which most of us will calculate resolution:

$$Rs = \frac{\Delta t_R}{\frac{1}{2} \cdot (w_{t,1} + w_{t,2})} \tag{2.4}$$

where Δt_R is the retention time difference between peaks and the w_t values

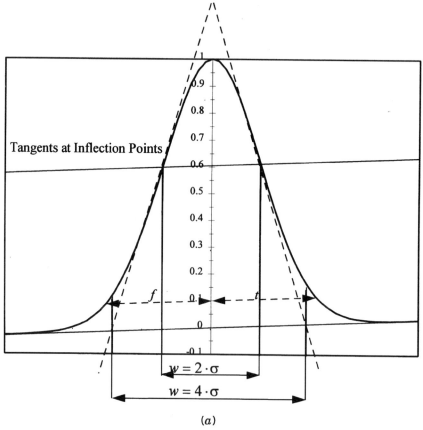

(a)

Figure 2.3 (a,b) Procedures to estimate the variance of a peak. For a truly Gaussian peak, these procedures are accurate. The axis in the middle shows the percentage of the peak height at which the variance is measured. The bottom of the graph shows the width measured by the various methods in units of the standard deviation. In the tangent method tangents are drawn to the inflection points of the peak, and the peak width is determined from the intersection points of the tangents with the baseline. We have intentionally drawn a sloping baseline, since this is commonly encountered in practice, and we wanted to clarify the correct procedure for this case. Also illustrated is the determination of the asymmetry factor, in this case measured at 10% of the height of the peak.

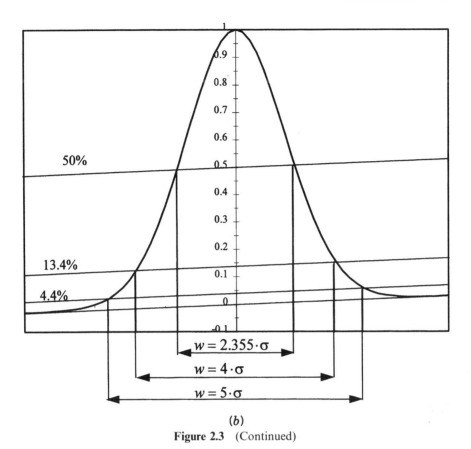

(*b*)

Figure 2.3 (Continued)

designate the peak width, now measured in the same (time) units as the retention-time difference is measured.

(*Note:* It is assumed in this discussion that one peak corresponds to one analyte. This is not necessarily the case, when complex mixtures are separated. In the latter case, a peak may be the composite of several analytes.)

2.1.4 Efficiency, Plate Count

Column efficiency and *plate count* are synonyms. Just like resolution, plate count is a measure of the quality of the separation. While resolution is based on both the distance between two peaks and their dispersion, plate count takes into account only peak dispersion.

The nomenclature also reflects the origin of the concept: the distillation theory of gas chromatography. In this concept, a certain length of the chromatographic column was occupied by a theoretical plate. This length was the height equivalent to the theoretical plate. If we divide the column length by this length, we obtain the number of plates in the column, hence the plate count N:

$$N \equiv \frac{L}{H} = \frac{L^2}{\sigma^2} \qquad (2.5)$$

The quantity N is a measure of the quality of a separation that is based on a single peak. Since the HETP is a function of many different parameters (see Section 2.2), the plate count is not a basic property of a column and therefore not a measure of column quality. Unfortunately, it is frequently used this way, especially in the marketing literature of column manufacturers.

Plate count or column efficiency is conveniently determined from a chromatogram. Since it is also a dimensionless parameter, we can obtain it from the peak width and the retention time, provided the migration velocity of the peak has been constant during the recording of the chromatogram. Therefore, we can determine plate count only from isocratic chromatograms, not from gradient chromatograms.

There are different procedures for determining the plate count. All are different ways to approximate the variance of the peak by using rulers to measure distances and widths. Figure 2.3 shows some of these procedures; see also equation (2.6).

(*Note:* With the computers in use today to record chromatograms, it is easy to determine the true variance of a peak. Nevertheless, the old manual approximations have established themselves and chromatography software programs go to great length to emulate the old manual procedures.)

$$N = \frac{L^2}{\sigma^2} = f \frac{t_R^2}{w_P^2} \qquad (2.6)$$

The peak width is measured in the same time units as retention time. The factor f depends on how the peak width is measured. It is, for example, 16 for the tangent method, 5.545 for the half-height method, and 25 for the 5σ method (see Table 2.1).

TABLE 2.1 Factor f of Equation (2.6) as Function of How Peak Width w_P Is Measured

Method	Position of Peak Width	Factor f
Inflection point	60.3% of peak height	4.000
Half-height	50% of peak height	5.545
4σ method	13.4% of peak height	16.00
5σ method	4.4% of peak height	25.00
Tangent method	Intersection of tangents with baseline	16.00

The details of the measurement are shown in Figure 2.3. We have purposely included a baseline drift in this figure to clarify how the width measurement should be performed in the case of a drifting baseline; while we should draw the lines parallel to the baseline, the width measurement should *not* be performed parallel to the baseline, but horizontally (i.e., in time units). This is not a problem, when the measurement is made with the help of a computer system on the screen display, but needs to be considered when the measurement is made manually on a printed chromatogram or a stripchart.

(*Note:* We can view the plate count also as a measure of the distribution of elution times of the analyte molecules. The relative standard deviation (*RSD*) of these elution times is related to the plate count as follows:

$$\%RSD = \frac{100}{\sqrt{N}} \tag{2.6a}$$

Thus the relative standard deviation of the elution times for a peak with 10,000 plates is 1%. This demonstrates the superb uniformity of HPLC columns.)

2.1.5   Peak Asymmetry

Peak asymmetry is a common practical measure of the quality of a column. As columns age, the peak asymmetry usually deteriorates; thus, one observes tailing peaks. Peak tailing is undesirable, since it can affect the quality of peak integration, especially when the signal-to-noise ratio is low or when peaks are only partially resolved. This tailing can have multiple causes, as will be discussed in later sections.

There are several measures of peak tailing. A common one is the ratio of the width of the tail of the peak to the width of the front of the peak at either 5 or 10% of the height of the peak (see Fig. 2.3 for an illustration of the measurement of the various asymmetry factors):

$$As = \frac{t_P}{f_P} \tag{2.7}$$

where t and f represent the width of the peak at the tail and the front of the peak maximum, respectively, measured at an appropriate height.

A better measure of peak asymmetry is the asymmetry-squared method proposed by King (1). In this method the tail and the front of the peak are measured at 4.4% of the height of the peak and the ratio is squared. The resulting value reflects well the ratio of the sum of the variances of the Gaussian component and the exponential component to the variance of the Gaussian component for an exponentially modified Gaussian peak:

$$As^2 = \frac{t_{5\sigma}^2}{f_{5\sigma}^2} \tag{2.8}$$

A third way to characterize the asymmetry of a peak is the tailing factor used by the U.S. Pharmacopeia (2). It is the ratio of the width of the peak to 2 times the front of the peak, measured at 5% of the height of the peak.

$$T = \frac{f_{5\%} + t_{5\%}}{2f_{5\%}} \qquad (2.9)$$

There is no practical advantage of one method for measuring peak tailing over another.

2.1.6   Linear Velocity

When we set up a chromatographic method, we need to select a flow rate. We know that the choice of flow rate influences the backpressure of the column and how long it takes to complete an analysis. However, the underlying parameter that determines the hydrodynamics of the system is the linear *velocity*. Conceptually, linear velocity is nothing but a way to normalize flow rate by the column cross section.

There are several different linear velocities, which unfortunately in some of the literature of chromatographic theory have been used interchangeably. This occasionally led to confusion. For clarification, we will define all of them and explain their usage.

[*Note:* The *IUPAC Recommendations 1993 for the Nomenclature for Chromatography* (3), which are active at the time of this writing, are proliferating this confusion in Section 3.6.05.1, Mobile-Phase Velocity. There is a need for a differentiation between different linear velocities. Therefore we are forced to depart from the IUPAC (International Union of Pure and Applied Chemistry)-recommended nomenclature.]

The classical definition of the chromatographic linear velocity u is based on the breakthrough time of an "unretained" peak, t_0:

$$u_0 = \frac{L}{t_0} \qquad (2.10)$$

where the subscript 0 indicates that the velocity is based on the unretained peak. Its relationship to the flow rate F is obtained by dividing the flow rate by the cross section of the column that is occupied by the mobile phase. This includes the space between the particles and the pore space. The following equation expresses this relationship:

$$u_0 = \frac{F}{\varepsilon_t \pi r^2} \qquad (2.11)$$

where ε_t is the volume fraction of the column occupied by mobile phase and r is the radius of the column.

In gas chromatography, this is a convenient way to measure a linear velocity. However, this velocity is a fictional velocity from the standpoint of hydrodynamics, since it does not exist anywhere in the column. The mobile phase is moving only between the particles, and is stationary in the pores. Since many of the peak-spreading phenomena in chromatography originate in the difference between the velocity of the mobile phase flowing around the particles and the average migration velocity of the band, it is more accurate and convenient to define a linear velocity that is the average velocity of the mobile phase that moves between the particles. We call this the *interstitial velocity*. It is defined as follows:

$$u_i = \frac{F}{\varepsilon_i \pi r^2} \tag{2.12}$$

where ε_i is the volume fraction of the column between the particles. It can be measured via the elution volume of a high-molecular-weight compound that is excluded from the pores. Or it can be calculated if the fraction of the column occupied by particles is known from other measurements.

2.1.7   Reduced Parameters

The concepts of reduced velocity v and reduced plate height h are powerful ideas that allow us to compare columns to each other under a broad range of mobile-phase conditions and over a range of particle sizes. We use the principle of corresponding states to form dimensionless parameters from the HETP and the linear velocity. The HETP has the dimension of length. To make it dimensionless, we simply divide it by the particle diameter:

$$h = \frac{H}{d_p} \tag{2.13}$$

Similarly, we create a dimensionless parameter from the linear velocity by multiplying it by the particle diameter and dividing it by the diffusion coefficient in the mobile phase:

$$v = \frac{u d_p}{D_M} \tag{2.14}$$

For thorough theoretical work, the type of velocity on which the reduced velocity is based should be specified. In the following, we will follow the common practice to associate the reduced velocity with the mobile-phase velocity on the basis of the breakthrough time of an unretained peak.

2.2   HYDRODYNAMICS OF CHROMATOGRAPHY

Chromatographic peaks have a finite width. This width is an impediment to resolution. It is therefore of utmost interest to understand the parameters that influence peak width. As we have seen above, a good measure for the peak dispersion is the HETP. The theory of the hydrodynamics of chromatography deals with the relationship of the HETP with the flow dynamics, the properties of the stationary phase, and the properties of the sample. A good understanding of these relationships is of vital importance to anybody involved in the design and manufacturing of columns and packing materials. But it is also important to the user of these columns, since a smart choice of, for example, the optimal flow rate or the column length can result in either improved resolution between peaks or a reduced analysis time.

Generally, if we plot the HETP against a linear velocity, we obtain a curved relationship with a minimum and a nearly linear increase of the HETP with linear velocity (Fig. 2.4) at high linear velocity. This relationship can be described by several equations: the van Deemter equation, the Giddings equation, and the Knox equation. We will explore all of them in the following, starting with the van Deemter equation.

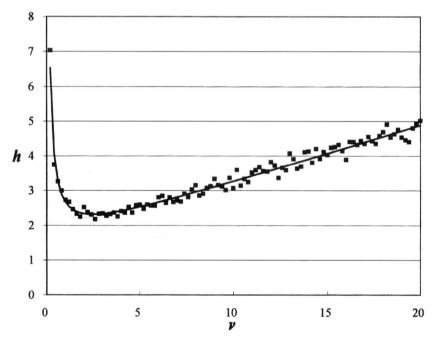

Figure 2.4   Relationship between HETP and linear velocity. The curve has a minimum and the right-hand branch increases usually practically linearly with the linear velocity.

2.2.1 van Deemter Equation

The van Deemter equation (4) is conceptually the simplest among the equations describing the dependence of the HETP on linear velocity. It assumes that the HETP is composed of three different, independent contributions. Since these contributions are independent of each other, they simply add up and thus form the observed curvilinear relationship:

$$H = A + \frac{B}{u} + Cu \qquad (2.15)$$

[*Note:* What is behind this statement is the law of the additivity of variances. It is a basic law of statistics. It states, in simple terms, that if multiple random events contribute to an observed distribution, then the variances of the distribution of the individual events sum up to form the variance of the overall distribution, provided that the individual events are independent of each other. We will encounter this law in various disguises in chromatography. The most obvious form of it is the additivity of the variances of extracolumn and intracolumn effects discussed in the next chapter.)

The velocity-independent term A is a function of the size and distribution of the interparticle channels and other nonuniformities in the packed bed. The second term is inversely proportional to the linear velocity; it describes the molecular diffusion in the axial direction. The third term is directly proportional to the linear velocity; it contains all terms related to mass transfer: from the flowing solvent to the particle through the stationary mobile phase to the retentive phase and back again. It also contains the terms related to sorption kinetics. Let us examine these terms individually. First, let us look at the term that describes the molecular diffusion.

Imagine a tiny droplet of a solute, for example, a dye, located in a vessel of stationary solvent. As time passes, the volume occupied by the solute becomes larger and the boundary more diffuse: the solute is diffusing into the surrounding liquid. If we look at the concentration distribution around the center, we observe a Gaussian distribution. The standard deviation of this distribution increases with time, more specifically, with the square root of time. Or, in other words, the variance of the distribution increases in direct proportion to time:

$$\sigma^2 = 2Dt \qquad (2.16)$$

The proportionality factor D is called the diffusion coefficient. This equation was first developed by Einstein and is therefore called the *Einstein equation*.

Exactly the same phenomenon occurs when we position the tiny droplet of dye inside a packed bed. However, in the presence of the solid walls of the particle skeleton, the rate at which diffusion occurs is smaller than in the neat solvent:

$$\sigma^2 = 2\gamma Dt \qquad (2.17)$$

The factor γ, appropriately called the *obstruction factor*, depends on the amount and the nature of the "obstruction" that is in the way of free movement of the molecule.

Molecular diffusion also takes place when the mobile phase is flowing through the packed bed. If it is flowing very slowly, the analyte will remain in the column for a longer period of time. Therefore the band spreading by diffusion will be larger compared to the case in which the band is leaving the bed quickly. Since the HETP has been defined as the change of the variance of the peak with length, we can write for the contribution of diffusion to the HETP:

$$H = \frac{\sigma^2}{L} = 2\gamma D \frac{t}{L} \qquad (2.18)$$

If the sample molecule diffuses only when it is in the mobile phase (and remains stationary when in or on the stationary phase), we can substitute the time in Equation (2.18) by the time that the molecule spends in the mobile phase. For the same reason, we can define the diffusion coefficient as the diffusion coefficient in the mobile phase, D_M. The result is the desired dependence of the HETP on the linear velocity for longitudinal diffusion:

$$H = 2\gamma \frac{D_M}{u} \qquad (2.19)$$

The coefficient γ is usually between 0.5 and 1, with values centering close to the lower range.

Care must be taken with the definition of the velocity. Some part of the "obstruction factor" can be explained by an inappropriate definition of the velocity. We must keep in mind that it refers to the residence time in the mobile phase of the sample compound whose diffusion is measured. This residence time is not necessarily identical to the residence time of an "unretained" sample compound, which is used to measure the linear velocity. Also, we implicitly assumed that the diffusion coefficient in the pores is the same as the diffusion coefficient in the interstitial mobile phase. This is also not necessarily the case. If the pore size is less than about 10 times larger than the size of the molecule, the diffusion coefficient depends on the ratio of the size of the sample molecule to the pore size of the packing.

Occasionally the measured B term has been much larger than anticipated from an estimation of the diffusion coefficient. This can be explained by a contribution from the diffusion in the stationary phase, which cannot always be excluded. In this case, we need to take the time spent in the stationary phase into account as well, and a more complicated picture emerges than the one painted above. In general, the nature of the diffusion term has not been studied extensively in HPLC, since in most cases it has been negligible in practical chromatography. However, in-depth discussions can be found in References 5 and 6.

Let us now look at the term of the van Deemter equation that describes all the phenomena that result in an increase in HETP with increasing linear velocity. As shown schematically in Figure 2.5, these phenomena include the *mass transfer* from the center of the moving mobile phase to the surface of the particle (A), through the stagnant mobile phase in the pores to the stationary phase on the internal surface of the packing (B), the interaction kinetics with the stationary phase (e.g., adsorption kinetics and desorption kinetics) (C), and then all the way back into the moving mobile phase. One can readily see that this phenomenon is quite complex. But since high plate counts at high linear velocities are desirable, a significant effort has been put into the understanding and subsequent improvement of these effects. To discuss these phenomena in detail is far beyond the scope of this book, but we will sketch out some of the basic principles.

The contributions to the HETP by the resistance to mass transfer depend linearly on the linear velocity, and the different contributions are additive:

$$H = \sum C_i u \qquad (2.20)$$

where the constants C_i depend on the thickness d of the layer over which the mass transfer is happening and the diffusion coefficient D of the analyte in this layer. The general form of the coefficient is

$$C \propto \frac{d^2}{D} \qquad (2.21)$$

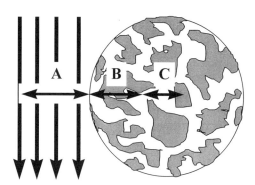

Figure 2.5 Mass-transfer phenomena: the analyte needs to be transported from the moving mobile phase to the surface of the particle (A), through the stagnant mobile phase in the pores to the stationary phase on the internal surface of the packing (B). It then interacts with the stationary phase (C), after which it is transported back into the moving mobile phase.

In principle, two major contributions to the resistance to mass transfer can be distinguished: mass transfer in the mobile phase and mass transfer in the stationary phase. Many equations have been derived describing the terms in a variety of circumstances, including hindered diffusion in the pores, bottleneck pores, shallow pores, etc. We will take a slightly different approach. An easy way to visualize the concepts is to view the packed bed as a system of capillaries. The theory of mass transfer in capillaries has been derived mathematically by Golay (7). Its details have been verified experimentally. Imagine a capillary with a layer of stationary phase immobilized on it (Fig. 2.6). The width of the channel through which the mobile phase is flowing is d_c, and the thickness of the stationary phase is d_s. The diffusion coefficients in the mobile phase and in the stationary phase are D_M and D_S, respectively. There is a parabolic velocity profile in the mobile phase. For this case, Golay obtained for the mobile-phase mass transfer:

$$H_M = \frac{1}{96} \frac{d_c^2}{D_M} \frac{11k^2 + 6 \cdot k + 1}{(k + 1)^2} u \tag{2.22}$$

Note the form of this equation—the mass transfer is proportional to the square of the channel thickness d_c, and inversely proportional to the diffusion coefficient in the mobile phase D_M. Both numerator and denominator contain the retention factor k elevated to the same power. This means that the contribution of the mobile-phase mass-transfer term is smallest but finite when the retention factor is 0 (= the analyte is unretained) and increases as the retention factor increases. This is typical for all contributions to the mass-transfer resistance originating in the mobile phase. The actual form of the coefficient that contains the retention factor is a function of the velocity profile in the mobile phase, which was parabolic in the case of the capillary.

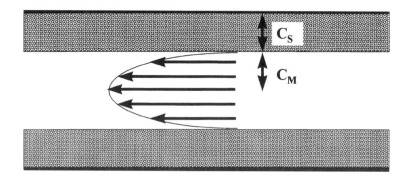

Figure 2.6 Mass transfer in a capillary with an immobilized stationary phase. C_M designates the mass transfer in the mobile phase, and C_S designates the mass transfer in the stationary phase.

For the contribution to the resistance to mass transfer in the stationary phase of a capillary column, one obtains

$$H_S = \frac{2}{3} \frac{d_S^2}{D_S} \frac{k}{(k+1)^2} u \qquad (2.23)$$

Once again the coefficient is proportional to the square of the thickness of the stationary phase layer d_S, and inversely proportional to the diffusion coefficient in this layer D_S. The power of the retention factor is lower in the numerator than in the denominator, and there is no constant term in the numerator. This means that this factor is 0 when the analyte is unretained and approaches 0 when the retention factor is very large. In between it will go through a maximum. This also is typical for all contributions to mass-transfer resistance originating in the stationary phase.

When we sum up the mobile-phase mass-transfer term and the stationary-phase mass-transfer term, we obtain the total mass-transfer term:

$$H = \left(\frac{1}{96} \frac{d_c^2}{D_M} \frac{11 \cdot k^2 + 6 \cdot k + 1}{(k+1)^2} + \frac{2}{3} \frac{d_S^2}{D_S} \frac{k}{(k+1)^2} \right) u \qquad (2.24)$$

Obviously, both contributions increase with linear velocity. If the magnitude of the stationary-phase term and the mobile phase term are similar, the total mass-transfer term will go through a maximum as a function of the retention factor. This is exactly what has been observed in packed beds. Although incorrect in detail, the capillary paradigm provides a good model for the mass transfer in a packed bed.

For the mobile-phase mass-transfer resistance to the stagnant mobile phase in a packed bed composed of particles of different sizes under the assumption of plug flow between the particles, the following equation has been obtained (8):

$$H_M = \frac{1}{30} \frac{\overline{d_p^5}}{\overline{d_p^3}} \frac{1}{\gamma D_M} \frac{\varepsilon_i}{\varepsilon_p} \frac{\left(k + \dfrac{\varepsilon_p}{\varepsilon_i} \right)^2}{(k+1)^2} u_i \qquad (2.25)$$

The coefficient containing the particle diameter d_p describes the volume-averaged surface area of the particles. This is a way of describing the "average" particle size. For all practical purposes, the shape of the curve describing the dependence of this term on the retention factor is indistinguishable from the Golay equation derived for capillaries. Note that the velocity u_i is used. As mentioned earlier, this is the only real velocity that exists in the packed bed. ε_p is the fraction of the column occupied by the pores of the packing. This equation does not apply to nonporous particles, as the term becomes infinite as the pore fraction vanishes.

Because of the complexity of the phenomena, it is difficult to derive a generally valid stationary-phase mass-transfer term for the packed bed. This is the reason why we used the simplified model above to introduce the basic concepts of stationary-phase and mobile-phase mass transfer.

For the user of a chromatographic column, it is usually difficult to distinguish between the different contributions to the resistance to mass transfer. Therefore, it is good to have some rules of thumb on what one can expect from a well-designed packing. Although we have seen that there is a dependence of the mass-transfer terms on the retention factor, it is quite flat in the range of k between 2 and 10, where most retention chromatography is carried out. We can therefore neglect the dependence of the mass-transfer coefficient on the retention factor. For the purposes of a rough estimation, the coefficient of the mass-transfer term in a packed bed with porous particles in the practical range of the retention factor is

$$H = c \frac{d_p^2}{D_M} u \tag{2.26}$$

The coefficient c is between 1/10 and 1/5 in most practical cases.

Finally, let us explore the origin of the flow-independent term A. In a packed bed, the flow path is tortuous. Not all paths are of equal length. If the velocity in all paths were the same, there would be a distribution of breakthrough times of sample molecules that follow different paths. Also, the velocity in some paths may be different from the velocity in other paths. An example is the situation near the wall of a column. Typically, the particles are not packed as densely near the wall as they are in the center of the column. Consequently, the flow near the wall is slightly faster than the flow in the center of the column. The portion of the sample that migrated close to the wall will elute earlier than the portion of the sample that migrated through the center of the bed. This results in a distribution of breakthrough times as well. The ratio of the velocities, however, is independent of the absolute velocity. Consequently, the band spreading is independent of the velocity. In other words, the contribution to the HETP from these microscopic and macroscopic nonuniformities in the packed bed is independent of the linear velocity.

Microscopic nonuniformities originate on one hand from the bare fact that particles are used to form the bed. On the other hand, the pattern in which the particles are arranged relative to each other is also important. As a consequence, the A term of the van Deemter equation is proportional to the particle size and can be influenced by the column packing technique.

The A term of the van Deemter equation is expressed as

$$A = \lambda_i d_p \tag{2.27}$$

For a well-packed bed, the factor λ is between 1.5 and 2. If the factor is significantly larger than 2, this is an indication that the quality of the packed bed is not as good as it could be. Values lower than 1.5, down to 0.5, have been observed in radially compressed beds, which exhibit a high level of structural uniformity.

There are several other contributions to this term. A fairly common one is associated with the distribution of the sample over the cross section of the column and the collection of the band at the column outlet. This may especially play a role with large-diameter columns or more specifically with columns of low aspect ratio (ratio of column length to column diameter). In this case, distributors are needed to distribute the flow and the sample evenly over the entire cross section. We will discuss column design in Chapter 3.

We can now put the whole van Deemter equation together, including the coefficients that we can use as a rule of thumb to estimate the HETP for a normal packed bed of porous particles:

$$H = 1.5d_p + \frac{D_M}{u} + \frac{1}{6}\frac{d_p^2}{D_M}u \qquad (2.28)$$

This equation can be put to use to estimate the plate count of a column, if the diffusion coefficient is known. In a later section of this chapter we show how the diffusion coefficient can be estimated. The reduced form of the van Deemter equation is (using the same estimates of the coefficients)

$$h = 1.5 + \frac{1}{v} + \frac{v}{6} \qquad (2.29)$$

2.2.2 Giddings Equation

The van Deemter equation provides a simple and useful description of the basic phenomena in a packed bed. However, it has two shortcomings. The first is the fact that the A term, describing the nonuniformity of the packed bed, has been introduced in a rather ad hoc manner and does not withstand careful theoretical examination. The second problem arises from the fact that in many practical cases, especially in gas chromatography, a downward curvature is observed on the right-hand branch of the HETP–linear velocity plot, which cannot be accounted for by the van Deemter theory. Something is wrong.

Giddings (9,10) argued that the origin of the problem rests with a basic conjecture of the van Deemter concept: the assumption that the different contributions to the HETP are independent of each other and that therefore the variances of these different contributions can be summed up:

$$\sigma_T^2 = \sum \sigma_i^2 \qquad (2.30)$$

In Giddings' concept, some portions of the packed-bed dispersion and the mass transfer cannot be regarded to be independent. They are coupled to each other. Therefore we are not allowed to simply add up the variances, but we must add the coupled variances harmonically:

$$\sigma_T^2 = \frac{1}{\dfrac{1}{\sigma_1^2} + \dfrac{1}{\sigma_2^2}} \tag{2.31}$$

The simplest justification (10) for this can be derived from the random-walk concept. In this concept, a molecule takes random steps of length l relative to the average position of all molecules. The variance of the distribution is related to this step length and the number of steps n as follows:

$$\sigma^2 = l^2 n \tag{2.32}$$

If we assume that there is a section of length S in the packed bed, in which the velocity of a molecule is preserved, then the step length is related to the velocity difference as follows:

$$l = \frac{\Delta u}{u} S = wS \tag{2.33}$$

where S is called the *persistence-of-velocity span*, and w is the relative velocity difference (or velocity fluctuation) between an individual molecule and the average molecule.

The number of steps taken during the chromatographic process is simply the column length L divided by the persistence-of-velocity span:

$$n = \frac{L}{S} \tag{2.34}$$

We can now express the variance of the peak and then the HETP in terms of the persistence-of-velocity span and the velocity fluctuation:

$$\sigma^2 = w^2 SL \tag{2.35}$$

$$H = \frac{\sigma^2}{L} = w^2 S \tag{2.36}$$

Now comes the important point—both flow and diffusion act together and independently to cut down the persistence-of-velocity span. For example, a flow channel might expand, thus changing the velocity of the molecule. Or the molecule might diffuse into a neighboring flow path that proceeds at a different velocity. Thus each mechanism independently terminates a random-walk-step, and after each terminated step, a new step begins. Therefore, "the only way in which diffusion and flow are additive is in the number of random steps they initiate" (Ref. 10, p. 264):

$$n = n_F + n_D \tag{2.37}$$

where the subscripts indicate the origin of the termination of the step by either a change in flow velocity or diffusion. Consequently, the segment length is

$$S = \frac{L}{n_F + n_D} \tag{2.38}$$

The HETP of these "coupled" mechanisms becomes

$$H_C = \frac{w^2 L}{n_F + n_D} = \frac{1}{\dfrac{n_F}{w^2 L} + \dfrac{n_D}{w^2 L}} \tag{2.39}$$

Since the persistence-of-velocity span for both the flow mechanism and the diffusion mechanism is defined as the column length divided by the number of steps attributable to the respective mechanism,

$$S_F = \frac{L}{n_F} \tag{2.40}$$

$$S_D = \frac{L}{n_D} \tag{2.41}$$

we can write the equation for the HETP in the following form, where the contributions to the HETP from the flow mechanism and the contribution to the HETP originating in a diffusion mechanism add harmonically:

$$H_C = \frac{1}{\dfrac{1}{w^2 S_F} + \dfrac{1}{w^2 S_D}} = \frac{1}{\dfrac{1}{H_F} + \dfrac{1}{H_D}} \tag{2.42}$$

On the other hand, the classic theory simply adds both contributions to the HETP linearly:

$$H = H_F + H_D \tag{2.43}$$

A comparison of the results of the classic concept and the coupling concept is shown in Figure 2.7. Giddings (9) pointed out that the HETP derived from the coupling equation is always less than both H_F and H_D, while the classic H

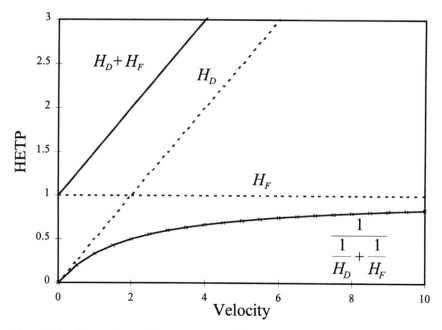

Figure 2.7 Comparison of the treatment of the additivity of the contributions to the HETP from convection and dispersion after the classical concept and Giddings' coupling concept. The HETP derived from the coupling equation is always less than both H_F and H_D, and the classic HETP is always greater than either of these. (Reprinted from Ref. 9, p. 55 by courtesy of Marcel Dekker, Inc.)

is always greater than either of these. With respect to the packed-bed dispersion term and the mass-transfer term in the mobile phase, the equation for the coupled H_c can be written as

$$H_c = \frac{1}{\dfrac{1}{A} + \dfrac{1}{Cu}} \tag{2.44}$$

Since only flow and diffusion effects couple, while longitudinal diffusion and stationary phase mass transfer don't, the complete equation for the dependence of the HETP on the linear velocity should be written as

$$H = \frac{B}{u} + Cu + \sum \frac{1}{\dfrac{1}{A^*} + \dfrac{1}{C^*u}} \tag{2.45}$$

where the star designates the contributions that are coupled. Giddings considered many different contributions to the HETP, including their magnitude and their coupling. Unfortunately, the model is quite complex and "escapes experimental verification" (11).

Essentially the same type of equation was derived by Berdichevsky and Neue (12), but the model was simpler. We considered that the velocities in the packed bed are different from location to location. However, the flow around the particles and molecular diffusion disperse these differences, just as a Galton board disperses the balls that are falling down through it. A rigorous mathematical analysis of this concept results in an equation that is of the same form as the Giddings equation, but assigns a different physical meaning to the coupling principle:

$$H = \frac{B}{u} + Cu + \frac{E}{A + \dfrac{B^*}{u}} \tag{2.46}$$

The denominator of the coupling term describes the radial dispersion. The constant E contains the velocity fluctuations over a characteristic distance.

We are able to show that in a well-packed (i.e., a uniform column), this concept gives rise to a term that is practically constant over the range of interest, that is, in accordance with the van Deemter equation. In a nonuniform bed, on the other hand, it gives rise to the curvature of the increasing branch of the HETP–velocity plot that has been observed experimentally. Consequently, a uniform, well-packed bed can be described by the van Deemter equation, while a poorly packed bed needs to be described by an equation that contains a term incorporating the curvature.

In reduced form, typical coefficients of Berdichevsky's version of the Giddings equation for a well-packed bed and small sample molecules are

$$h = \frac{1.8}{v} + \frac{v}{8} + \frac{v}{1 + 0.4v} \tag{2.47}$$

The value of 0.4 in the denominator of the coupling term is derived from Reference 12, using the reduced velocity based on the migration time of an unretained peak here. The remaining coefficients were chosen from typical curve-fit results.

2.2.3 Knox Equation

Giddings' coupling theory does not result in a practical approach to describe the properties of a packed bed. Therefore Kennedy and Knox (13) introduced an empirical equation that contains a term useful for capturing the observed curvature of a HETP–velocity plot and so overcomes one of the shortcomings of the van Deemter equation. The term introduced by Knox contains the third root of the velocity. It is difficult to associate a physical meaning to this. Therefore the Knox equation is best expressed in its reduced form:

$$h = \frac{B}{v} + Cv + A^*v^{1/3} \tag{2.48}$$

The last term is the empirical correction for Giddings' coupling term. In this form, the equation is stripped of all physical significance, which makes an interpretation of experimental results difficult. However, the Knox equation has become quite popular. The reported coefficients are $\frac{1}{10}$ for the C term and 1–2 for the A^* term. A large value of A^* represents an inferior packed bed.

We choose 1.5 for the B term on the basis of our data. Therefore the normal performance of a typical column should follow the following equation:

$$h = \frac{1.5}{v} + \frac{v}{10} + v \tag{2.49}$$

2.2.4   Comparison of the Different Equations

Which equation is correct? Which should we use to describe the packed bed? The answer to both questions is not necessarily the same. There is little question that the concepts resulting in the coupling equation represent the best description of the packed bed. The observations can be described accurately and interpreted well over a large range of linear velocities. However, the coupling equation also yields coefficients from curve-fitting procedures that are notoriously unreliable. Unfortunately, the Knox equation does not represent much progress, especially if the curvature of the increasing branch of the plot of h versus v is not very pronounced. The van Deemter equation is clearly incorrect in concept, but it describes without difficulty the results obtained with a well-packed column over the usual range of interest in HPLC. For a well-packed column little curvature of the right-hand branch of the h/v plot is observed. Under these circumstances, good reliable results are obtained from curve-fitting procedures to the van Deemter equation. I therefore prefer to use the van Deemter equation in spite of its conceptual weaknesses.

In the last section we have reported what we have found to be typical coefficients for all three equations. Different values for the coefficients have been reported in the literature as well. To some degree, the coefficients depend on the estimation of the diffusion coefficient, which is needed to calculate the reduced velocity. To some degree, the differences in the reported values may also reflect the properties of different packing materials. We should take them clearly as estimates.

Figure 2.8 shows an overlay of the van Deemter equation, Berdichevsky's version of the Giddings equation and the Knox equation using our coefficients. The graph shows the relationships up to a reduced velocity of 20, the typical

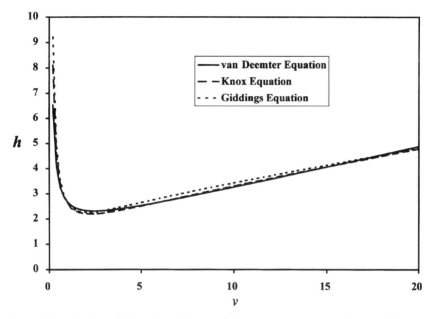

Figure 2.8 Overlay of the reduced forms of the van Deemter equation, Berdichevsky's version of the Giddings equation and the Knox equation. Over the range shown here, up to a reduced velocity of 20, the three curves are indistinguishable from each other.

range of interest to practical HPLC. As one can see, the curves are indistinguishable. As a matter of fact, only at reduced velocities higher than 20 is there a departure between the van Deemter equation and the other curves. The Knox and Giddings equations are virtually identical up to a reduced velocity of 100, and the deviation between both is under 10% up to a reduced velocity of 300. Therefore, for practical estimations of the HETP, the three equations can be used interchangeably over the practical range of HPLC.

In conclusion, the most correct description of the phenomena in a packed bed is a form of the coupling equation like the Berdichevsky equation. The van Deemter equation provides the most practical representation of the packed bed.

2.2.5 Permeability

An important aspect of column performance is its backpressure. Therefore it is important to understand the parameters that influence column backpressure. The relationship between these parameters is described by the equation for permeability. The specific permeability is measured as follows:

$$B_0 = \frac{F\eta L}{\pi r^2 \, \Delta p} \tag{2.50}$$

where F is the flow rate, η is the viscosity, L is the column length, r is the radius of the column, and Δp is the pressure drop across the column. In the older literature, the symbol for the specific permeability was K_F.

As we know from experience, the backpressure of a column packed with small particles is larger than the backpressure of a column packed with larger particles. The specific permeability depends on the particle size d_p and the interstitial porosity ε_i of the packed bed. The relationship is known as the *Kozeny–Carman equation* (16,17):

$$B_0 = \frac{1}{185} \frac{\varepsilon_i^3}{(1 - \varepsilon_i)^2} \, d_p^2 \tag{2.51}$$

Normally, the interstitial porosity of a packed bed is around 0.40. With this value, the constant coefficient is approximately $\frac{1}{1000}$, and the specific permeability simply becomes

$$B_0 = \frac{d_p^2}{1000} \tag{2.52}$$

We can use this relationship to determine the expected pressure drop across the column or to check, if the column backpressure has become excessive. Since this calculation involves the juggling of several conversion factors, which are not necessarily easy to come by, we will show in an example how the specific permeability is calculated.

Example calculation

Column diameter 4 mm; radius 2 mm = 0.2 cm
Column length 10 cm
Particle diameter 5 μm = 5×10^{-4} cm
Flow rate 0.5 mL/min = 0.5/60 cm^3/s
Pressure drop 150 psi (lb/in.^{2}) $\approx$ 10 atm = 10×10^6 dyn/cm^2
Viscosity 0.4 cP = 0.004 (dyn·s)/cm^2

Using Equation (2.50) we can calculate the permeability:

$$B_0 = 2.5 \times 10^{-10} \text{ cm}^2$$

Using Equation (2.52), we can check if this value agrees with the expectation for a column packed with 5-μm particles, and indeed it does.

We also could have calculated the expected pressure directly:

$$\Delta p = 1000 \frac{F\eta L}{\pi r^2 d_p^2} \tag{2.53}$$

In all these calculations we need to know the viscosity of the mobile phase. Table 2.2 contains the data for the solvents commonly used in HPLC. Most mobile phases, however, are mixtures of solvents. For nonpolar solvents, one can estimate the viscosity of a mixture by linear interpolation between the viscosities of the component solvents. This is not true for aqueous mixtures due to the strong association of polar solvents. Since aqueous mixtures are the mobile phases for reversed-phase chromatography, it is important to know their viscosity. Figure 2.9 shows the viscosity of the aqueous mixtures of the polar organic solvents commonly used in reversed-phase chromatography. Note that the water–alcohol mixtures exhibit a pronounced viscosity

TABLE 2.2   Viscosity of Solvents Used in HPLC and GPC (14,15)

Solvent	Viscosity (cP) at 20°C
Acetone	0.32
Acetonitrile	0.37
Cyclohexane	0.98
Diisopropylether	0.37
Diethyl ether	0.23
Dimethyl acetamide	2.1
Dimethyl formamide	0.92
Dimethyl sulfoxide	2.2
Dioxane	1.54
Ethanol	1.2
Ethyl acetate	0.45
Hexafluoroisopropanol	1.0
Isooctane	0.5
Isopropanol	2.5
Methanol	0.6
Methyl acetate	0.37
Methylene chloride	0.44
Methylethyl ketone	0.4
n-Heptane	0.42
n-Hexane	0.33
N-Methyl pyrrolidone	1.67 (25°C)
n-Pentane	0.235
n-Propanol	2.3
o-Dichlorobenzene	1.41
Tetrahydrofuran	0.46
Toluene	0.59
1.2.4-Trichlorobenzene	1.89 (25°C)
Water	1
m-Xylene	0.62
o-Xylene	0.81

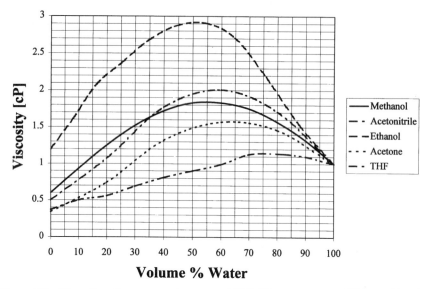

Figure 2.9   Viscosity of aqueous mixtures at 25°C. (Data courtesy of Waters Corp.)

maximum at intermediate compositions. Acetonitrile–water mixtures have the lowest viscosity. From the standpoint of column backpressure, acetonitrile is therefore the preferred mobile-phase modifier in reversed-phase chromatography.

Another measure of column permeability is the flow resistance parameter Φ. It is a dimensionless quantity:

$$\Phi = \frac{d_p^2}{B_0} \tag{2.54}$$

where Φ is about 1000, if the Kozeny–Carman relationship holds. This value is also in good agreement with our experience, but values as low as 500 have been reported in the literature. However, there are several reasons why different values reported by different workers do not represent true differences in the resistance parameter. First, lower values are obtained for porous packings, if the specific permeability is measured on the basis of the breakthrough time of an unretained peak instead of using the flow rate. We follow here the current definition of the permeability by IUPAC; the flow-based permeability has the advantage of being independent of the particle porosity. The second reason is the lack of a universally accepted definition of the average particle size. Nearly all particles used to pack HPLC columns have a size distribution of a finite width, and the designation of the average particle size depends on the way the averaging is done (volume average vs. population average). Also, the measurement technique plays a role. Finally, for irregular-shaped particles the assignment of a particle diameter is quite arbitrary. Our convention assumes a volume-averaged particle diameter.

2.2.6 Estimation of the Diffusion Coefficient in the Mobile Phase

For the calculations of reduced velocities or the prediction of an HETP we need to know the diffusion coefficient of an analyte in the mobile phase. However, this information is seldom available. We therefore have to estimate the diffusion coefficient. There are several different approaches available, but the one most frequently used in the HPLC literature is the Wilke–Chang equation. It is claimed to be accurate for small to medium-sized molecules within 10%, which is sufficient for most purposes (9):

$$D = 7.4 \times 10^{-8} \frac{\sqrt{\Psi_2 M_2 T}}{\eta V_1^{0.6}} \tag{2.55}$$

where T is the temperature (in °K,), M_2 is the molecular weight of the solvent, V_1 is the molar volume of the solute in milliliters, and η is the viscosity in centipoise. Ψ_2 is the association factor for the solvent; it is 1 for nonpolar i.e., nonassociated) solvents, 1.9 for methanol, and 2.6 for water.

It is important to remember that the diffusion coefficient is proportional to temperature and inversely proportional to viscosity. This means that high-viscosity solvents cause slow diffusion, which results in low plate counts (if the HETP is dominated by mass transfer, which it is in most practical cases). It also means that temperature improves the diffusion in two ways: first directly by speeding up diffusion, and then indirectly by reducing the viscosity.

2.2.7 Performance Index and Separation Impedance

Attempts have been made to combine the different parameters of a separation into a single criterion that describes the overall quality of a column or a separation. We would like to combine a measure of the narrowness of the bands, the analysis time, and the pressure needed to drive the separation. Since we want to maximize efficiency while minimizing analysis time and pressure, a possible measure would be the simple ratio of plate count to the product of analysis time and pressure. However, since we end up with two parameters that we want to minimize in the denominator and only one parameter that we want to maximize in the numerator, we end up with a measure that minimizes plate count. To compensate for this, we need to raise the plate count in the numerator to the second power. This measure of overall column performance is the performance index, first developed for gas chromatography by Golay (18):

$$\pi = \frac{N^2}{t_a \Delta p} \tag{2.56}$$

A dimensional analysis of the performance index shows that it has the dimension of an inverse viscosity. With a little algebra, using the equations for permeability [Eq. (2.51)], the definitions of the flow-resistance parameter [Eq. (2.54)], linear velocity [Eqs. (2.10 and 2.11)], the retention factor [Eq. (2.61)] and the reduced plate height [Eq. (2.13)], we can split the performance index into its components:

$$\pi = \frac{1}{\varepsilon_t} \frac{1}{k+1} \frac{1}{h^2} \frac{1}{\Phi} \frac{1}{\eta} \tag{2.57}$$

An inspection of these components (with the exception of the first term) is enlightening. The performance index decreases with increasing retention factor, because of the increase in analysis time associated with increased retention. However, we need retention to perform the analysis. The third term shows that we would like to reduce the reduced plate height as much as possible. This means that we would like to have the best column performance possible. It also implies that we should operate the column at the minimum of the plate height–velocity curve. We will see this in more detail in Chapter 3. The fourth factor contains the flow resistance parameter, which is fairly constant for normal packed beds. The last factor contains the viscosity and gives the performance index its dimension. Choosing a lower viscosity mobile phase results in a higher performance index. A typical value for the performance index is 1/300 (poise^{-1}) or 3000 [plates^2/(s·atm)].

The performance index is a measure of the quality of the separation. Knox (19) was searching for a criterion that is a purer measure of column performance and criticized the performance index for its dependence on the retention factor and the viscosity. Therefore he defined a dimensionless group, the separation impedance E:

$$E = \frac{1}{\pi\eta(k+1)} = \frac{t_0 \, \Delta p}{N^2 \eta} \tag{2.58}$$

From equations (2.54) and (2.53) and the definition of the reduced plate height, we can see that it depends primarily on the reduced plate height and the flow resistance parameter:

$$E = \varepsilon_t h^2 \Phi \tag{2.59}$$

Its value is typically around 5000 for typical HPLC columns, but can be as low 3000.

The separation impedance is a measure of column performance and allows one to compare, for example, capillary columns to packed beds. The performance index is a measure of the quality of the separation and contains a strong dependence on analysis time. I would prefer a criterion that preserves the dependence on viscosity, but eliminates the dependence on the retention factor. One might call this the *quality of the separation medium*, containing both column properties and mobile phase properties. It could be defined as

$$QM \equiv \frac{N^2}{t_0\,\Delta p} = \frac{1}{\varepsilon_t}\frac{1}{h^2}\frac{1}{\Phi}\frac{1}{\eta} \tag{2.60}$$

This criterion does not appear to exist in the literature.

[*Note:* The dependence of the performance index and the separation impedance on the porosity of the column is not found in the older literature, since the specific permeability was defined via the linear velocity u_0 rather than the flow rate (or superficial velocity) $F/\pi r^2$; unfortunately, both permeabilities have been used interchangeably, which is incorrect and causes some confusion. We adhere here to the definition of the specific permeability by IUPAC and are consequently forced to include the porosity factor into the preceding equations.]

2.3 THERMODYNAMICS OF CHROMATOGRAPHY

Most of the specifics of the chemistry of chromatography will be covered in the chapters dedicated to each separation mechanism. Nevertheless, a few topics, including some definitions, are sufficiently generic to be treated separately. In the following chapter, we look at the definition of the retention factor and the relative retention, and then look at the relationship of retention with the free energy of the interaction with the stationary phase.

2.3.1 Definition of the Retention Factor

The retention factor k is measured as the retention time of an analyte minus the retention time of an unretained peak divided by the retention time of the unretained peak. It is a convenient way to normalize retention for comparison of different columns or the same column at different flow rates:

$$k = \frac{t_R - t_0}{t_0} \tag{2.61}$$

However, there is more to it. An unretained peak spends all its time in the mobile phase. Any analyte migrating through the column will spend the same amount of time in the mobile phase as an unretained peak (provided neither is excluded from all or part of the pores). The difference between the total time that an analyte spends in the column and the time that it spends in the mobile phase has to be spent in the stationary phase. Therefore, the retention factor is the ratio of the time t_S that an analyte spends in the stationary phase to the time t_M it spends in the mobile phase:

$$k = \frac{t_S}{t_M} \tag{2.62}$$

Under isocratic conditions, we can also use this factor in another way. Since the retention factor is the ratio of the time that any molecule spends in the stationary phase compared to the time it spends in the mobile phase, it is also the ratio of the number of molecules N that are at any time in (or on) the stationary to the number of molecules that are in the mobile phase:

$$k = \frac{N_S}{N_M} \tag{2.63}$$

This is the basic definition of the retention factor. With this equation, we can now make the link between the retention time of a compound and the chemistry and the thermodynamics of the separation.

2.3.2   Retention Factor and Chemical Equilibrium

The relationship between the retention factor and the chemical equilibria that govern retention is easy to derive. For example, in the case of partition chromatography, we can simply relate the number of molecules in the stationary and in the mobile phase to their respective concentrations c_M and c_S:

$$k = \frac{N_S}{N_M} = \frac{c_S V_S}{c_M V_M} \tag{2.64}$$

where V_M and V_S are the volumes of the mobile and stationary phases, respectively. Their ratio is called the phase ratio β:

$$\beta = \frac{V_S}{V_M} \tag{2.65}$$

The ratio of the concentrations c_S and c_M is the partition coefficient K:

$$K = \frac{c_S}{c_M} \tag{2.66}$$

Therefore the retention factor is the product of the phase ratio and the partition coefficient:

$$k = \beta K \tag{2.67}$$

Similar concepts can be applied to adsorption chromatography or to ion-exchange chromatography. In adsorption chromatography, the concentration of the molecules adsorbed on the stationary is expressed as the number of molecules per surface area:

$$q_S = \frac{N_S}{A_S} \tag{2.68}$$

From Equation (2.63) we know that the retention factor is the ratio of the number of the molecules in the stationary phase to the number of the molecules in the mobile phase. Thus we can express the retention factor in the case of adsorption chromatography as

$$k = \frac{N_S}{N_M} = \frac{q_S A_S}{c_M V_M} \tag{2.69}$$

As in the case of partition chromatography, we can define the phase ratio as

$$\beta = \frac{A_S}{V_M} \tag{2.70}$$

and the adsorption coefficient as

$$K = \frac{q_S}{c_M} \tag{2.71}$$

In the case of ion exchange, there is a chemical equilibrium between the analyte A^{n+} in the mobile phase and the immobilized analyte $A^+E_n^-$ on the fixed ion-exchange sites E_n^-. The counterion in the mobile phase is represented as M^+:

$$A^{n+} + nM^+E^- \Leftrightarrow nM^+ + A^{n+}E_n^- \tag{2.72}$$

The equilibrium constant is

$$K = \frac{[A^{n+}E_n^-][M^+]^n}{[A^{n+}][M^+E^-]^n} \tag{2.73}$$

The first terms in the numerator and the denominator represent the concentration of analyte in the stationary phase and in the mobile phase, respectively. The ratio of the first terms can be related to the number of molecules in the stationary phase and the number of molecules in the mobile phase as follows:

$$\frac{[A^{n+}E_n^-]}{[A^{n+}]} = \frac{N_S/V_S}{N_M/V_M} \tag{2.74}$$

We can now express the retention factor as a function of the phase ratio, the equilibrium constant, and the concentrations of the ion-exchange groups and the competing counterion in the mobile phase:

$$k = \frac{V_S}{V_M} K \frac{[M^+E^-]^n}{[M^+]^n} \tag{2.75}$$

From this equation, we can, for example, see that the retention factor of an analyte is inversely proportional to the concentration of the competing ion in the mobile phase raised to the power of the charge of the analyte (if the competing ion is singly charged, as was assumed in the derivatization above). This can be used to determine the effective charge of an analyte.

2.3.3   Relative Retention α

The relative retention α is defined as the ratio of the retention factors of two compounds:

$$\alpha = \frac{k_2}{k_1} \tag{2.76}$$

Customarily, the retention factor of the later-eluting analyte is in the numerator, forcing the value to be always larger than 1.

The relative retention is purely a chemical or thermodynamic entity that does not depend on the physics of the column any more. This can easily be shown in the case of partition chromatography:

$$\alpha = \frac{k_2}{k_1} = \frac{K_2}{K_1} \tag{2.77}$$

The phase ratio does not appear any more, and we are simply dealing with the ratio of two partition coefficients. One can use this to demonstrate whether the chemistry of a separation remains invariant on transfer of the separation from one column to another.

2.3.4   Relationships to the Free Energy

In Equation (2.67), we established the relationship between the retention factor and a chemical equilibrium constant, in this case the partition coefficient:

$$k = \beta K \tag{2.67}$$

We also know the relationship between the equilibrium constant and the standard Gibbs free-energy difference associated with this chemical equilibrium:

$$\Delta G^\circ = -RT \ln K \tag{2.78}$$

Therefore the logarithm of the retention factor is a function of the free energy:

$$\ln k = \ln \beta - \frac{\Delta G^\circ}{RT} \tag{2.79}$$

We can use this equation for example to explore linear free-energy relationships in chromatography. An example is the relationship between the retention factor of analytes of the same structure but with different aliphatic chains in the molecule and the number n of methylene groups n_c in the chains in reversed-phase chromatography:

$$\ln k = \ln k_0 + Pn_c \tag{2.80}$$

where k_0 is the retention factor of the parent compound. The constant P depends on both the stationary phase and the mobile phase. Linear free-energy relationships are found quite frequently in chromatography. They can be used to predict retention times based on the knowledge of the structure of a molecule.

Equation (2.79) also is the basis for the temperature dependence of retention. Furthermore, it can be used to calculate adsorption energies. We can, for example, calculate the minimum difference in adsorption energy needed for a separation. To obtain baseline resolution between two adjacent peaks at a plate count of 10,000, we need a relative retention α of about 1.1. Combining Equations (2.77) and (2.79), we obtain

$$\Delta\Delta G^\circ = RT \ln \alpha \tag{2.81}$$

for the difference $\Delta\Delta G^\circ$ between the adsorption energies. The numerical calculation yields about 60 cal/mol for a relative retention of 1.1 at room temperature.

REFERENCES

1. R. M. King, private communication.
2. U.S. Pharmacopeia 23, United States Pharmacopeial Convention, Inc., pp. 1776–1777.
3. L. S. Ettre, "Nomenclature for chromatography," Pure Appl. Chem. 65(4), 819–872 (1993).

4. J. J. van Deemter, F. J. Zuiderweg, and A. Klinkenberg, *Chem. Eng. Sci.* **5**, 271 (1956).

5. J. H. Knox, and H. P. Scott, *J. Chromatogr.* **282**, 297–313 (1983).

6. R. W. Stout, J. J. DeStefano, and L. R. Snyder, *J. Chromatogr.* **282**, 263–286 (1983).

7. M. J. E. Golay, *J. Chromatogr.* **186**, 341 (1979).

8. A. Berdichevsky, private communication (1990).

9. J. C. Giddings, *Dynamics of Chromatography*, Part I. *Principles and Theory*, Marcel Dekker, New York, 1965.

10. J. C. Giddings, *Unified Separation Science*, Wiley, New York, 1991.

11. Cs. Horváth and H.-J. Lin, *J. Chromatogr.* **126**, 401–420 (1976).

12. A. L. Berdichevsky and U. D. Neue, *J. Chromatogr.* **535**, 189–198 (1990).

13. G. J. Kennedy and J. H. Knox, *J. Chromatogr. Sci.* **13**, 25 (1975).

14. H. Engelhardt, *High-Performance Liquid Chromatography*, transl. G. Gutnikov, Springer-Verlag, Berlin, 1979, p. 110.

15. J. A. Riddick, W. B. Bunger, and T. K. Sakano, *Organic Solvents*, 4th ed., Wiley, New York, 1986.

16. J. Kozeny, *Sitzungsber. Akad. Wiss. Wien* **136**, 271–306 (1927).

17. P. C. Carman, *Trans. Inst. Chem. Eng.* (London) **15**, 150–156 (1937).

18. M. Golay, *Gas Chromatography 1958*, Desty, ed., Butterworths, London, 1959, p. 36.

19. P. A. Bristow and J. H. Knox, *Chromatographia* **10**(6), 279–288 (1977).

ADDITIONAL READING

For a deeper understanding of the theory of chromatography, References 9 or 10 are highly recommended. There is also an excellent chapter, "The Theory of the Dynamics of Liquid Chromatography," contributed by S. G. Weber and P. W. Carr in *High Performance Liquid Chromatography*, P. R. Brown and R. A. Hartwick, eds., Chemical Analysis Series, Vol. 98, Wiley, New York, 1989, pp. 1–116.

3 Column Design

Whether design is good or bad, it is certainly widespread.
— D. Shasha, *The Puzzling Adventures of Dr. Ecco*

In Section 3.1, we will discuss several aspects of column design. First, we discover the influence of the choice of column length and particle size on resolution and analysis time, and the influence of column diameter on sensitivity. This section should enable the chromatographer to make the best choice of these parameters for the needs of an assay. Then, in Section 3.2, we explore some fundamental limits that constrain the freedom of column choice and design. Section 3.3 we examine the effect of column hardware.

In Section 3.4 we discuss alternatives to the conventional column design. We look into the characteristics of radially and axially compressed columns. The subsequent examination of alternatives to packed beds of particles leads us to a reflection of ways to evaluate the merits of these alternative designs.

3.1 CHOICE OF COLUMN DIMENSIONS AND PARTICLE SIZE

To select a column for a particular application, a chromatographer has to make two basic choices: (1) the selection of the chemistry of the packing that is most appropriate for the separation and (2) the choice of the physical properties of the column, especially particle size and column dimensions. The determination of the correct packing material is the primary task, since it is the interaction of the sample constituents with the packing material that makes the separation possible. However, the choice of column dimensions and particle size is also important, since it influences the speed of the analysis, the resolving power, the column backpressure, the detectability, and the solvent consumption per analysis.

Now that we are armed with a basic understanding of the hydrodynamics of chromatography, we can ask the question of the optimum column choice. We will first look at the interplay between column length and particle size, then we will examine the influence of column diameter.

3.1.1   Influence of Particle Size and Column Length

The foundation of our analysis has been laid by Guiochon and co-workers in three landmark papers published in 1974 and 1975 under the title "Study of the Pertinency of Pressure in Liquid Chromatography" (1–3). These publications represented the first thorough theoretical study of the technology of HPLC, including the subject of column design.

Guiochon's concept starts with the premise that resolution and analysis time are the two major factors that characterize a chromatographic separation. Ideally, one desires to maximize resolution while minimizing analysis time. Although his considerations were limited to isocratic separations, the findings of the study can be generalized to gradient separations as well.

Let us begin by examining the resolution equation:

$$Rs = \frac{\sqrt{N}}{4} \frac{\alpha - 1}{\alpha} \frac{k_2}{k_2 + 1} \tag{3.1}$$

where Rs is the resolution, N is the number of theoretical plates, α is the relative retention, and k_2 is the capacity factor of the second peak of the peak pair, whose resolution is calculated. As we see, resolution is the product of three terms. The first term depends on the number of theoretical plates. The last two terms in this equation describe the influence of column and mobile-phase chemistry on the separation. The first term is the one that is affected by column design and is the subject of this discussion.

Analysis time is related to the speed with which the mobile phase flows through the column:

$$t_a = \frac{L}{u_0} (k_l + 1) \tag{3.2}$$

where L is the column length, u_0 the linear velocity, k_l the capacity factor of the last peak, and t_a the analysis time.

The pressure required to pump the mobile phase through the column is the major obstacle that prevents us from maximizing resolution and minimizing analysis time simultaneously. It depends on flow rate, column length, column diameter, particle size, and viscosity. The relationship is expressed by the Kozeny–Carman equation, which is discussed in detail in Section 2.2.5:

$$B_0 = \frac{F\eta L}{\pi r^2 \, \Delta p} = \frac{1}{185} \frac{\varepsilon_i^3}{(1 - \varepsilon_i)} \, d_p^2 \qquad (3.3)$$

The three parameters resolution, analysis time, and column backpressure are linked to each other, even under optimized conditions. The choice of two of these parameters automatically fixes the third one. Together they determine the particle size and the column length. In the following paragraphs we will show how these relationships operate together.

In order to carry out our analysis, we have to make some choices as to the chemistry parameters of the system that we would like to analyze. Let us choose a reversed-phase column with an aqueous mobile phase. Let us assume that the viscosity of this mobile phase is 1 cP, that the diffusion coefficient of the sample is 1×10^{-5} cm^2/s, that the last peak elutes with a retention factor of 9. We will also imply that all columns are packed equally well and that no extracolumn effects interfere with the performance of the columns.

We will now look at graphs of resolution versus analysis time for various choices of particle size and column length. As a measure for resolution, we simply use the square root of the plate count, as shown in the resolution equation, Equation 3.1. As analysis time, we use 10 times the breakthrough time of an unretained peak, as shown in Equation (3.2). We use the van Deemter equation to calculate the HETP, from which we determine the plate count. From the Kozeny–Carman equation [Eq. (3.3)] we calculate the pressure drop across the column. We set an upper pressure limit of 20 MPa (~ 3000 psi). The curves will stop when this pressure limit is reached.

Let us first examine the influence of column length. Figure 3.1 shows the graphs of resolution (square root of plate count) versus analysis time for three 5-μm columns of 5, 10, and 15 cm length. When examining this diagram, one first observes the obvious; the longer the column is, the higher is the resolution, that is, its plate count. Next, one notices that every column exhibits a resolution maximum. The reasons for this are the fundamental principle of the hydrodynamics of chromatography that we have examined in Chapter 2. The graphs shown here are basically an inverted form of the van Deemter equation. The minimum of the van Deemter equation corresponds to the resolution maximum shown here.

Under ideal conditions, specifically, without slow kinetics or restricted mass transfer, the maximal plate count that can be achieved with a given column is independent of the sample. However, the linear velocity at which this plate-count maximum occurs is a function of the diffusion coefficient of the sample

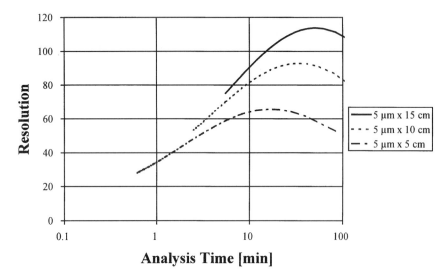

Figure 3.1   Resolution versus analysis time for three 5-μm columns of different lengths. This is an inverted form of the van Deemter equation.

molecule in the mobile phase. Therefore the flow rate and analysis time corresponding to this maximum are a function of the sample and the mobile phase. This is the reason why we had to assume a set of conditions that we deemed typical for reversed-phase chromatography.

At high linear velocities, that is, at short analysis times, the column performance deteriorates as a result of resistance to mass transfer, while at long analysis times longitudinal diffusion is the limiting factor.

On the left side of the graph, the curves stop at some minimal analysis time. This is the point at which the pressure boundary is reached. It does not surprise us that the longer column reaches its pressure limit at longer analysis times than the shorter columns.

For the assumed set of chromatographic conditions, the 15-cm column reaches the resolution maximum at an analysis time of about 50 min; the 10-cm column, around 30 min; and the 5-cm column, at about 15 min. Considering the stipulation of a pressure maximum of 20 MPa, the shortest analysis achievable on the 15-cm column is just under 6 min, about 2.5 min for the 10-cm column, and about 40 s for the 5-cm column. This assumes, of course, that the quality of the separation remains adequate on the shorter columns at the increased speed.

We can also see that the three columns complement each other nicely; there is very little performance overlap between them, and each has its own distinct range of applicability. The 5-cm column should be used for very fast analyses, while the 15-cm column provides sufficient resolving power for even demanding analyses. The 10-cm column provides medium resolution, with the added capability of moderately fast analyses.

To summarize the analysis of the performance of columns of different lengths but constant particle size, resolution increases with the square root of the column length, but analysis time for maximum resolution increases in direct proportion to the column length. How fast an analysis can be achieved depends on the square of the column length. The resolution at the high-speed end drops roughly proportional to column length.

Figures 3.2 and 3.3 demonstrate the capabilities of these columns in the form of chromatograms. The example used is a separation barbiturate standards on a 5-μm reversed-phase material. The column length is varied from 5 cm to 10 cm and 15 cm. The column diameter is 3.9 mm. For each column, two chromatograms are shown to demonstrate the performance range of the columns. One chromatogram was obtained at 1 mL/min. The other chromatogram was obtained at a backpressure of approximately 20 MPa to demonstrate the speed that a particular column dimension can achieve. For the 50-mm column, that translated to a fast flow rate of 5 mL/min. The observations are in agreement with the analysis above. The shortest analysis time is about 0.5 min obtained with the 5-cm column. At such a short analysis time, the resolution between the first two peaks in the chromatogram is lost, and only the last peaks are still adequately resolved.

Let us now examine what is happening if we keep column length constant and change the particle size. We select a 15-cm column. The particle size is varied from 3 μm to 5–10 μm. This covers the range of particle sizes typically used in HPLC.

The results of our analysis are shown in Figure 3.4. Not surprisingly, the maximal resolution obtainable with the 10-μm packing is much lower than what can be achieved with the 3-μm packing. The maximum resolution for the 10-μm packing is 80 at a run time of about 100 min. As we have seen above, the 5-μm packing reaches a resolution of about 115 at 50 min. The 3-μm packing achieves a resolution of nearly 150 at an analysis time of 30 min.

The resolution increases inversely proportionally to the square root of the particle size. What is interesting is that the maximum performance occurs at shorter analysis times as the particle size is decreased. More specifically, the maximum of the resolution shifts to shorter analysis times in direct proportion to the particle size: 100 min for the 10-μm column and 30 min for the 3-μm column. This means in practice that if an analysis is scaled from a 5-μm column to a 3-μm column, the flow rate may need to be increased to take full advantage of the capabilities of the smaller particle diameter. This is counterintuitive, since the backpressure of a 3-μm column is larger than the backpressure of the 5-μm column at equal length and flow rate.

If we look at the part of the graph that depicts the situation at the pressure limit, we observe the following. The 10-μm column can be used down to analysis times of under 1.5 min. At that point the resolution value is about 22 (i.e., only about 500 theoretical plates). As we have seen above, the 5-μm column achieves a resolution value of about 75 (5600 theoretical plates) at an analysis time of just under 6 min. The 3-μm column reaches a resolution of 141

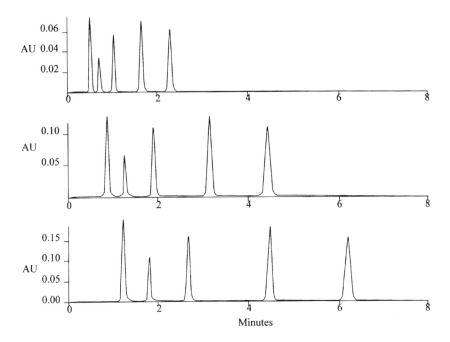

Figures 3.2, 3.3 Comparison of the performance of 5-μm columns of 5, 10, and 15 cm lengths using a separation of barbiturate standards. On the left (Fig. 3.2), the flow rate was kept constant. One observes the increase in resolution with increasing column length at the expense of analysis time. The chromatograms on the right (Fig. 3.3) were obtained at a constant pressure of approximately 20 MPa to demonstrate the speed achievable with each column. (Chromatograms courtesy of D. J. Phillips, Waters Corp.)

(20,000 plates) at an analysis time of about 15 min. Because of the high backpressure, the 3-μm column cannot be used for fast analyses. It is substantially slower than the 5-μm column. The fastest analysis time possible decreases inversely to the square of the particle size.

As a consequence of this, the 3-μm column has only a limited flow-rate range over which it can be used. We can call the ratio of the analysis time at maximum performance to the analysis time at the pressure limit the dynamic range of a column. For the 15-cm 3-μm example discussed here the dynamic range is 2, for the 5-μm 15-cm column it is 9, and for the 10-μm column it is about 70.

It is generally best to select a column with an intermediate dynamic range as a general-purpose column. Columns with a narrow dynamic range should be reserved for special high-resolution analyses. Columns with too broad a dynamic range compromise too much in their performance at both the high-resolution end and at the high-speed end. On the other hand, they may be rugged workhorses for simple analyses, where the need is not performance but reliability.

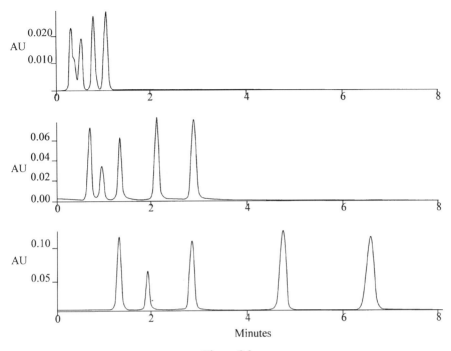

Figure 3.3

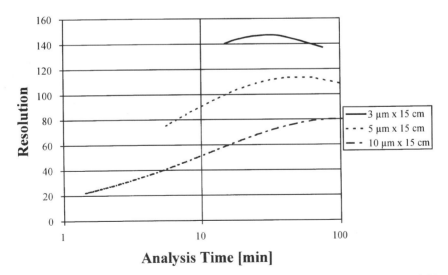

Figure 3.4   Comparison of the performance of columns of constant length but variable particle size. The resolution is plotted against analysis time.

Let us now investigate another case: let us keep the ratio of column length to particle diameter constant. We will study a 30-cm 10-μm column, a 15-cm 5-μm column, a 9-cm 3-μm column, and a 3-cm column packed with 1-μm particles. The results are shown in Figure 3.5. The first observation we make is that the maximum resolution is identical for the first three columns. However, the analysis times corresponding to the maxima are drastically different from each other: 200 min for the 10-μm packing, 50 min for the 5-μm packing, and 18 min for the 3-μm packing. The 1-μm column no longer exhibits a maximum. The reason for this is the pressure limitation. For the rest of the columns, the analysis time at maximum efficiency goes down in proportion to the square of the particle size. Therefore we gain by a factor of about 11 in speed at constant resolving power by going from 10-μm particles to 3-μm particles.

However, something interesting is happening also at the high-speed end of the graph. All columns, including the 1-μm column, reach the pressure limit at exactly the same analysis time. Thus, if raw speed is the only requirement of the separation, all four columns would do equally well. Nevertheless, there are some major differences. They lie in the resolution that can be obtained at analysis times at or close to the pressure limit. At this point, the 10-μm column has only a resolution value of 43 ($=1850$ theoretical plates). The resolution value of the 5-μm column is 75 ($=5700$ plates), and the 3-μm column reaches a resolution of 100, corresponding to 10,000 plates. Since the 1-μm column no

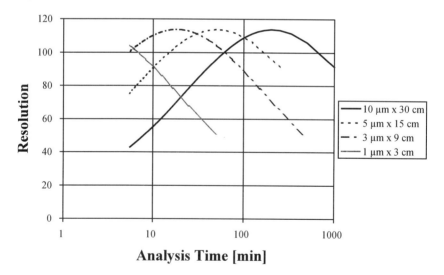

Figure 3.5 Comparison of four columns with constant ratio of column length to particle size. The maximum resolution achievable is the same for all columns, but the maximum occurs at different analysis times. With the 1-μm column, the maximum cannot be reached since it is outside the pressure limit assumed for this graph.

longer reaches a maximum of the resolution, it attains its best performance only at the pressure limit. At this point, it performs merely as well as the 3-μm column. For all the rest of the range, the resolution obtainable with the 1-μm column is inferior to the resolution of the 3-μm column. All other things being equal, the 9-cm column packed with 3-μm particles is the best choice among the four alternatives. The dynamic range of the column (the ratio of the analysis time at the resolution maximum to the analysis time at the pressure limit) is about 3.3, which is still quite acceptable.

What we have learned is that there are two important characteristics of a column: the maximum resolution or plate count and the shortest analysis time that can be obtained with the column. Both are determined exclusively by the ratio of column length to particle size and by nothing else. The particle size itself determines at which analysis time the maximum resolution can be obtained. We should select the smallest particle size possible while proportionally scaling down the column length. However, at some point the resolution maximum moves out of reach, that is, outside the pressure limit. All of these theoretical considerations assumed that all columns can be packed to equal performance. This, however, is not entirely true, since there are some practical limitations that will actually limit the column performance before we reach the particle sizes discussed here. An important factor is the heat generated as a result of the friction of the flow through the packed bed. This subject will be covered in section 3.2.

3.1.2   Influence of Column Diameter

Until now, we have ignored column diameter. Within reasonable constraints, the considerations of column performance are not affected by column diameter. With our current understanding of column technology and packing technology, this is a quite sensible assumption. We will discuss some practical limitations in the subsequent paragraphs, but we can safely say that columns from 1 mm to at least 10 mm i.d. (inner diameter) can be made to perform equally well independent of column diameter. This is true at least as long as we define column performance by HETP versus linear velocity or resolution versus analysis time, as we have done in the previous paragraph. But there are other column performance criteria that are influenced by column diameter. The most obvious one is solvent consumption. If we want to compare two columns of equal length and particle size but different diameter at equal linear velocity, we need to change the flow rate in proportion to the column cross section, that is, proportionally to the square of the column diameter. Under these circumstances, the column performance with respect to analysis time and resolution is identical. But because we use a higher flow rate for the larger-diameter column, the solvent consumption is larger for this column.

A comparison of this type is shown in Figure 3.6. Columns with diameters of 4.6, 3.9, 3, and 2 mm are compared to each other at equal linear velocity. Note that the chromatograms look virtually identical. Also the column

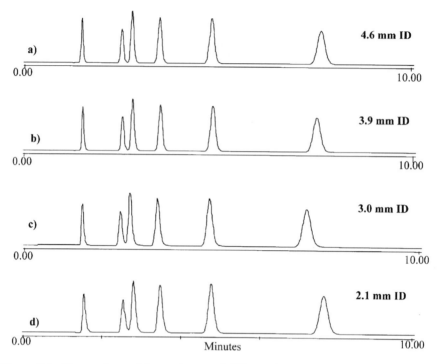

Figure 3.6 Comparison of four columns of different internal diameters but constant column length and particle size. The flow rate and injection volume were scaled in proportion to the column cross section. Identical chromatograms are obtained. (*Chromatographic conditions:* Columns: 5 μm Symmetry C_{18}, 150 mm length, Waters Corp. Mobile phase: water/methanol/glacial acetic acid 79:20:1. Flow rates: (*a*) 1.4 mL/min; (*b*) 1.0 mL/min; (*c*) 0.6 mL/min; (*d*) 0.29 mL/min. Sample: mixture of 6 sulfa drugs. Injection volume: (*a*) 14 μL; (*b*) 10 μL; (*c*) 6 μL; (*d*) 3 μL.) (Chromatograms courtesy of D. J. Phillips, Waters Corp.)

backpressures were nearly identical. The small increase in measured backpressure for the larger-diameter column is due to the increased backpressure in the connection tubing of the HPLC instrument at the higher flow rates used with the larger-diameter columns. It should also be noted, that we changed the sample injection volume in direct proportion to the column volume. As a consequence, the peak heights remained constant as well.

The primary benefit of reduced column diameter is reduced solvent consumption due to the reduced flow rate used with smaller diameter columns. A reduction of flow rate at constant column diameter also results in a reduced consumption of solvent per unit time. But since analysis time increases as well, the solvent consumption per analysis remains constant. It can be reduced only by reducing column diameter.

In principle, solvent consumption is related to column volume. One can get reduced solvent consumption as well by reducing column length while keeping column diameter constant. However, this involves the compromises discussed in the previous section.

Considering the benefit of reduced solvent consumption without compromise in column performance, why would we not choose the smallest column possible? The simple answer is that HPLC instruments were not designed for small column volumes, and the constraints are discussed in Section 3.2.2.

Another very practical aspect of the choice of column diameter is the compromise between sensitivity and resolution that one encounters when varying column diameter at constant volume and amount of sample injected. We should consider this if we have only a limited amount of sample available. A typical example of this type is the analysis of the metabolic fate of a drug in animals or humans. For this kind of analysis, one usually has only a small amount of sample and the concentrations of the analytes of interest are very low.

If we are not limited in the amount of sample available, but would like to achieve high sensitivity, we would simply keep the column constant and inject more sample onto the column. This will work until one of two situations are encountered:

1. *Volume overload* (for a detailed theoretical discussion, see Section 3.2.2). As we increase the injection volume, we will first observe that the peaks become larger (i.e. sensitivity increases) without any effects on the resolution between the peaks in the chromatogram. If we further increase the injection volume, we will encounter a point at which the peaks will become broader, and the resolution deteriorates. At a certain injection volume, which depends on the details of our analysis, a further increase in injection volume is no longer possible because of loss of resolution between the analyte(s) of interest and the remainder of the peaks in the chromatogram.

2. *Mass overload.* This can be found if one or more constituents of the sample are present in much larger amounts than the analytes of interest. An example of this is the analysis of impurities and degradation products in pharmaceuticals, which must be quantitatively analyzed if present at levels down to 0.1% of the parent compound. As one is injecting larger and larger amounts to increase the sensitivity for the contaminants, the peak for the parent compound may become broader as a result of mass overload and start to interfere with the quantitation of the contaminants.

Exactly the same situations are encountered when we decrease column diameter while keeping the amount and volume of injected sample constant. This is something that we would need to do if the amount of sample available for the analysis were limited and if our standard column did not achieve the sensitivity required. As one decreases the column diameter from a standard 4.6-mm-i.d. column to a 2-mm-i.d. column, the sensitivity at constant injection volume increases about 5-fold. It increases about 20-fold if we reduce the column diameter further to 1 mm i.d. However, we may easily encounter a situation in which the smaller-diameter column will be overloaded either by the mass injected or by the volume injected. An example is shown in Figure 3.7. The same volume and amount of a sample of tamoxifen was injected on columns with a diameter of 4.6, 3.9, 3, and 2 mm. The interest was in the quantitation of the impurity peaks that elute just in front of the tamoxifen peak itself. As one can see, the peak height, and therefore the sensitivity, of the

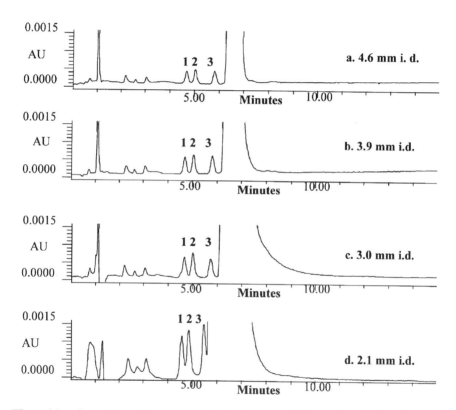

Figure 3.7 Illustration of the compromise between sensitivity and resolution. An equal amount of a tamoxifen sample was injected on all four columns. An increased peak height is observed for the impurities, but for the 2-mm column the resolution between tamoxifen and the impurities is lost due to column overload. *Chromatographic conditions:* Columns: 5-μm Symmetry C$_{18}$, 150 mm length, Waters Corp. Mobile phase: 50 mM potassium phosphate, pH 3.0/acetonitrile 55:45. Flow rates: (*a*) 1.4 mL/min; (*b*) 1.0 mL/min; (*c*) 0.60 mL/min; (*d*) 0.29 mL/min. Detection: UV, 240 nm Sample: tamoxifen, 600 μg/mL. Injection volume: 7 μL.) (Chromatograms courtesy of D. J. Phillips, Waters Corp.)

impurity peaks increases as the column diameter is decreased. However, the tamoxifen peak itself exhibits signs of overloading on the 3-mm-i.d. column, and the resolution between tamoxifen and the closest impurity has suffered too much on the 2-mm column. Therefore, the analysis can not be performed satisfactorily with the 2-mm column, and the 3-mm column is the best compromise given the constraints on the amount and volume of sample that is available for this analysis.

Within the practical range of analytical HPLC, column diameter does not influence column backpressure, plate count, and analysis time. However, parameters such as solvent consumption and sensitivity of an analysis are affected by the choice of column diameter. Smaller diameters lead to reduced solvent consumption, provided the HPLC instrument is designed for the use of small-diameter columns. In the case of sensitivity, a reduction in column diameter involved the identical compromises as an increase in injection volume. If an increase in injection volume is prohibited because of the limited amount of sample available, a correct choice of column diameter can provide a significant increase in sensitivity. Small-diameter columns provide the advantage of increased sensitivity in sample-limited applications.

We should not leave this chapter without at least touching some phenomena that are encountered as the column diameter is decreased to within an order of magnitude of the particle size. If the column diameter is less than about 30 particle diameters, the packed bed becomes less dense than in regular HPLC columns. This is due to the fact that the wall disturbs the regular structure of the packed bed. This is observed in columns of normal diameters as well, but the effects of this wall region are usually negligible. But for columns with very small column:particle diameter ratios, the wall region extends throughout the packed bed. A consequence of the loose bed structure is a decrease in flow resistance. If we go back to the Kozeny–Carman equation (3.3), we see that it contains a term that depends on the interstitial fraction. This factor increases nearly 8-fold if the interstitial fraction changes from 0.4, which is normal for a standard HPLC column, to 0.6 for a loosely packed bed. This means that for a given pressure limitation, the column length can be increased significantly. Also, there is a slight increase in column performance due to increased coupling effects (see Section 2.2.2). As a consequence, columns with a small column diameter:particle diameter ratio have a significant advantage over classic HPLC columns.

3.2 DESIGN LIMITS

In the last paragraph we explored the interplay of particle diameter, column length, and column diameter with only one constraint imposed on the design of the column: the pressure capability of the HPLC instrument. As we will see in this paragraph, there are other constraints on the column design. One is a fundamental one—flow through a packed bed generates heat by friction, which increases the temperature of the mobile phase. The other one is imposed by the need to transport sample into the column and to detect the analytes after the separation. The problems arising from the second constraint are combined under the heading of extracolumn effects.

3.2.1   Heat of Friction

In high-performance liquid chromatography, high pressure is used to force the mobile phase through the packed bed. The mechanical energy needed to maintain the flow of the mobile phase is converted into thermal energy. The thermal energy is dissipated into the mobile phase and the column, and causes an increase in temperature in some parts of the column. This effect has been described in detail by Halász and co-workers in a publication entitled "Ultimate Limits of High-Pressure Liquid Chromatography" (4).

   If the column were an adiabatic system, the increase in temperature would only be a function of pressure. In practice, the problem becomes significant only when short columns packed with small particles ($<5\,\mu$m) are used at high pressures (>20 MPa). We can estimate the effect to a first approximation by calculating the mechanical work under adiabatic conditions and incompressibility of the fluid (5) and equating it to the thermal energy Q:

$$Q = V\,\Delta p = mc_v\,\Delta T \tag{3.4}$$

where V is the volume, Δp is the pressure drop across the column, m is the mass of the solvent, c_v is the heat capacity at constant volume, and ΔT is the temperature difference of the solvent. We can write for the temperature difference

$$\Delta T = \frac{V\,\Delta p}{mc_v} = \frac{\Delta p}{\rho c_v} \tag{3.5}$$

where ρ is the density of the solvent. For an estimate of the order of magnitude of this effect, we can simplify things by assuming a constant density and by using the heat capacity at constant pressure c_p. Endele (5) calculated that a pressure drop of about 15 MPa would result in an adiabatic temperature increase of about 10°C with n-heptane as mobile phase. He was able to verify this by measuring the temperature increase of the solvent when the pressure was released through a needle valve. When a column was used, the temperature increase at the column outlet was between 2 and 3 times less than the theoretical effect due to dissipation of the energy through the column walls.

   One consequence of this effect is the change in retention factor with changing flow rate. This does not represent a problem for the practice of chromatography, since the flow rate is usually not varied. However, if the flow rate is changed to either speed up a separation or to increase efficiency, the change in temperature may influence the chromatogram measurably. Also, shifts may occur when the column length or diameter are changed. Endele measured an increase in temperature of about 10°C and a decrease of the capacity factor by about 20% when the pressure was increased from 1 to 35 MPa.

More bothersome is the fact that a radial temperature gradient develops as well. The heat is generated everywhere throughout the bed, but it is dissipated through the wall. Consequently, solvent near the wall is cooler than solvent in the center of the column. Accordingly, the viscosity in the center of the column is lower, which results in an increased flow in the center of the column. It also results in faster migration of the sample in the center of the column. As a consequence, the chromatographic band becomes distorted and a loss of efficiency results. This means that the theoretically expected gain in column performance through reduction of the particle diameter cannot be fully realized. The radial temperature differences become smaller with decreasing column diameter, but that approach has its own problems, as we will see in the next section. However, packed capillary columns can be used at significantly higher pressures, allowing the use of smaller particles to achieve higher efficiencies in shorter analysis times.

3.2.2   Extracolumn Effects

In current HPLC technology, the sample is injected into the mobile phase outside the column and the mobile phase carries the sample to the column. As the separated analytes exit the column, they are transported by the mobile phase to the detector, which, in turn, senses the analytes and converts the signal received by the sensor to an electrical signal. The electrical signal may be converted further to digital data through an analog-to-digital (A/D) converter. All these steps may broaden or distort the chromatographic band and therefore can be an impediment to the chromatographic separation. Therefore they can impose constraints on the design of the column.

Let us first look at the effects caused by the transport of the analytes through the fluid path outside the column from the injector to the detector. Ideally, we would like to inject the sample in an infinitely small volume, but in reality we have to inject a finite volume of sample onto the column. As the sample is transported through the tubing that connects the injector to the column, it will be subject to a broadening of the band due to differences in the migration velocity of the flow in the tubing between the wall and the center of the tubing. At the wall, the flow is stagnant, while the sample migrates at a certain velocity in the center of the tubing. The sample that ended up near the wall will have to catch up with the sample that is being transported in the center of the tubing. Similarly, there may be sections of the fluid path, for example, in bad connections, where there are pockets of slow-moving or stagnant mobile phase that then will contribute to a spreading of the sample band. These exact same phenomena are encountered at the column exit when the sample is transported through some length of tubing from the column to the detector cell, causing band spreading again. The detector cell itself has a finite volume that is needed to sense the sample. This volume also contributes to band broadening.

These band-broadening effects are all volumetric effects. Each contributes a variance to the width of the band, and the variances of the individual contributions are additive. The variances can be expressed in any unit we want, provided they are measured at the same flow rate. However, it is convenient to declare the variances in volume units. Mathematically the additivity of variances is expressed as follows:

$$\sigma_t^2 = \sigma_i^2 + \sigma_f^2 + \sigma_c^2 + \sigma_d^2 \qquad (3.6)$$

The subscripts have the following meanings: t = total band spreading, i = band spreading in the injector and caused by the injection volume, f = band spreading in the fluid path between injector and detector, c = band broadening inside the columns, and d = band spreading caused by the detector. It is important to remember that the variances are additive, not the broadening itself. This is actually an advantage, since the root of the sum of the variances is always smaller than the sum of the individual contributions to the broadening.

What we would like to know is the influence of the extracolumn effects relative to those inside the column. Let us first calculate the peak width and the variance of a peak inside the column:

$$\sigma^2 = \frac{V_R^2}{N} \qquad (3.7)$$

The variance is proportional to the retention volume V_R and inversely proportional to the plate count N. For most practical purposes, one can assume that the plate count is constant over the chromatogram. The retention volume itself is influenced by two parameters: the column volume itself and the retention factor:

$$V_R = V_C \varepsilon_t (k + 1) = \pi r^2 L \varepsilon_t (k + 1) \qquad (3.8)$$

Let us take a typical column of 15 cm length and packed with 5-μm particles. This column would have a plate count of about 10,000 plates under normal running conditions. Table 3.1 shows the standard deviation of a peak as a function of the retention factor for several internal diameters of the column.

To get a handle on the relative contributions of the different contributions to extracolumn band spreading, we need to make some estimates of the individual variances. Let us first estimate the contribution of the injection

TABLE 3.1   Standard Deviation (μL) for 5-mm Columns of 150 μm Length as a Function of the Retention Factor k and Column Diameter

Retention Factor	Column i.d. (mm)			
	4.6	3.9	2	1
0	17	13	3	1
1	35	25	7	2
2	52	38	10	2
3	70	50	13	3
5	105	75	20	5
8	157	113	30	7
10	192	138	36	9
15	279	201	53	13
20	366	263	69	17

volume. This can be treated as a rectangular signal whose standard deviation is

$$\sigma_i = \frac{V_i}{\sqrt{12}} \tag{3.9}$$

The contribution of the detector volume is of the same nature. The contribution of the connection tubing is usually estimated through the use of the Taylor–Aris (6,7) equation, which assumes that the flow is laminar and that the tubing is long:

$$\sigma_f^2 = L\pi d_t^4 \frac{F}{384 D_M} \tag{3.10}$$

[*Note:* The Golay equation (2.29) becomes the Taylor–Aris equation when the retention factor becomes 0.]

Because of the wall roughness of small-diameter tubing and the fact that the tubing is nearly never straight, the band spreading in the tubing is smaller than predicted by the Taylor–Aris equation, especially at higher flow rates. Also, it has been shown that for the short tubing used as connection tubing, the asymptotic assumption of a long tubing length breaks down as well (8). A better estimate is obtained by assuming a constant HETP of 10 cm for the normal flow-rate range used:

$$\sigma_f = \sqrt{HL}\,\pi\,\frac{d^2}{4} \tag{3.11}$$

This means that the contribution of a 30-cm section of 0.25-mm-i.d. tubing contributes a standard deviation of about 8.5 μL to the extracolumn band spreading. One can see that it pays to reduce the length of connection tubing. A length of only 30 cm is actually quite good for a typical HPLC instrument.

If we assume an injection of 20 μL, a total length of connection tubing (before the column and after the column) of 30 cm, and a detector volume of 10 μL, the total extracolumn band spreading has a standard deviation of nearly 11 μL. If we compare this value to the data in Table 3.1, we see that for normal-diameter columns (4.6 and 3.9 mm) only the early part of the chromatogram close to the unretained peak is affected significantly. For the 2-mm column, however, the influence of extracolumn band spreading remains significant up to a retention factor of about 5. For the 1-mm column the extracolumn effects dominate the peak width over the entire range of chromatographically useful retention factors. Figure 3.8 shows the apparent plate counts obtained with these four columns as a function of the retention factor, given an extracolumn band spreading with a standard deviation of 15 μL, which may be quite typical for an HPLC instrument with long connection tubing and imperfect connections. One can clearly see the strong influence of the extracolumn effects on the performance of the system.

We mentioned above that significant benefits could be realized if the column diameter could be reduced to about $\leqslant 10$ times the particle diameter. For a

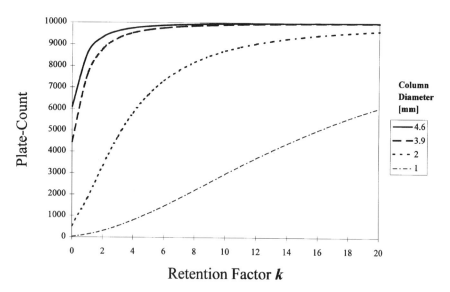

Figure 3.8 Extracolumn effects: apparent plate count obtained on columns with different internal diameters as a function of the retention factor. The extracolumn band spreading was assumed to be 15 μL. A significant deterioration of the performance is observed for the smaller-diameter columns.

5-μm packing, this would mean a column diameter of about 50 μm. To make such a column feasible, one would have to reduce the extracolumn effects, including the detector volume and the injector volume, by a factor of at least 100,000 — quite a significant challenge to designers of HPLC instruments. The largest challenge is the design of the detector, since sensitivity is reduced as the detector cell volume is reduced. But we also need to understand that the injection volume should be reduced in proportion to the reduction in column volume to maintain constant column performance. Therefore we need to ask ourselves the question, at what injection volume a further reduction in column diameter stops making sense. Both the detector volume and the injection volume form the limitation to the design of columns with small diameter. However, some detectors such as the mass spectrometer actually benefit from the use of smaller-diameter columns.

The previous section dealt with extracolumn effects originating in the fluid path of the HPLC instrument. A different type of extracolumn effect is associated with the electronics of the instrument: the time constant of the detector, the sampling rate of the A/D converter, and the digital filtering. All of these factors may contribute to a distortion or broadening of the peak in a similar way as do volumetric effects. The only difference is that they operate in the time domain rather than in the volume domain; therefore it is best to express their influence in time units.

The first effect to consider is the time constant of the detector electronics. If a sharp signal is sent through a simple electronic filter, the output is an exponentially decaying signal. The exponential decay is characterized by a time constant τ. When a Gaussian signal is subjected to the same electronic filter, the output is a peak with a tail, the exponentially modified Gaussian (Fig. 3.9). It has been studied very well and serves in many studies as a model for tailing peaks.

For the exponentially modified Gaussian, the variances of the symmetrical Gaussian distribution σ_G^2 and the tailing exponential decay τ^2 are additive:

$$\sigma_t^2 = \sigma_G^2 + \tau^2 \tag{3.12}$$

A peak with a retention time of 1 min at an efficiency of 10,000 plates has a σ of 0.6 s. This would be the case for an unretained peak eluting from a 5-μm 3.9-mm $\times$ 150-mm column at a flow rate of 1.5 mL/min. If the detector time constant is 1 s, a significant distortion of the peak results, with the total peak width increasing by a factor of nearly 2. Therefore, for normal operating conditions the detector time constant should be significantly less than 1 s, best around 0.1 s. For fast chromatography, special attention should be paid to the detector time constant.

The influence of the sampling rate of the A/D converter can be treated in exactly the same way as the injection volume and the detector volume [Eq. (3.9)], substituting the injection volume with the sampling time $t_{A/D}$ of the A/D

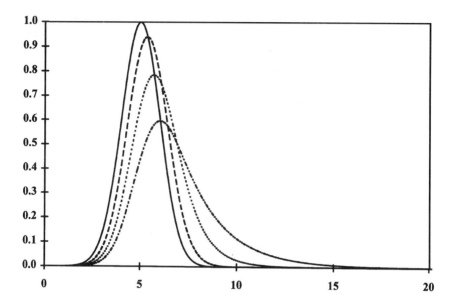

Figure 3.9 Gaussian and exponentially modified Gaussian distributions.

converter:

$$\sigma_{A/D} = \frac{t_{A/D}}{\sqrt{12}} \tag{3.13}$$

This is usually not a problem, since the influence of too low a sampling rate is quite visible: rather that having rounded contours, the peak consists of straight lines connected to each other. This is the best indicator for too low a sampling rate.

In modern HPLC instruments, the classic electronic filters are substituted by digital filters. A detailed discussion of digital filtering is outside the scope of this book. We will only treat the simplest case, in which the filtering is done by "bunching" of the data points. In this case, the same principle applies as in the case of the sampling time of the A/D converter, but the sampling time is multiplied by the bunching factor b:

$$\sigma_b = \frac{bt_{A/D}}{\sqrt{12}} \tag{3.14}$$

A bunching factor of 10 at a sampling rate of $10\,s^{-1}$ results in a standard deviation of about 0.3 s.

The time-based contributions to the extracolumn band spreading play a role only for fast chromatography or early-eluting peaks in normal chromatography. They usually can be chosen by the user through the system operating software of the HPLC instrument. From the standpoint of chromatography, it is desirable to select a fast sampling rate. However, the size of the data file increases with the sampling rate, and the hard drive of the data system fills up more rapidly. Thus a careful compromise needs to be chosen between the needs of the chromatography and the constraints by the data system.

3.3   COLUMN HARDWARE

Today, most columns in use are packed commercially. Consequently, the user has little influence on the column design. Even if the columns are still packed by the user, it is most likely that commercially available column hardware is used. Therefore a discussion of column hardware is of only limited value. Nevertheless, it is an important subject, and a book on column technology can not ignore this subject.

Many different hardware designs are commercially available. It would be inappropriate and quite involved to discuss the advantages or disadvantages of different designs in detail. Therefore we will cover the subject only in broad strokes outlining some basic principles. We will first look at the design of an end fitting, then look at the requirements for the tubing.

The material of construction for most HPLC columns and at least the body of the end fitting is typically 316 stainless steel, which is highly corrosion-resistant. Nevertheless, a small amount of corrosion is possible with chloride salts. Surprisingly, even some organic solvents such as methylene chloride or acetonitrile are capable to leach small amounts of iron from the steel. This is in most practical cases of very little concern, but may play a role for sample compounds that either complex with iron or exhibit structural changes in the presence of small amounts of metal ions. The most common source of metal ions is the frit rather than the column tubing. We will discuss some alternatives in the discussion of the end-fitting design.

3.3.1   End-Fitting Design

The end fitting of an HPLC column has three functions: (1) it should seal the fluid path at the high pressures used in HPLC, (2) it should distribute the mobile-phase flow and the sample evenly over the entire cross-section of the packed bed, and (3) it should retain the packing inside the column. Consequently, most column end fittings comprise three components: the seal, the distributor, and the filter.

The sealing mechanism should be good enough to withstand not only the operating pressure of the column but also the packing pressure, which may be significantly higher than the operating pressure. The pressure used to pack a column may be as high as 70 MPa, while typical operating pressures are around $\leqslant 15$ MPa. The seal must be tight enough to avoid even the tiniest leaks, since they would result in a buildup of buffer-salt crystals on the outside of the seal.

The classic sealing mechanism is through ferrules, which bite into the column tubing and deform the tubing. This kind of seal, which is still common for homemade columns, should not be used at pressures in excess of about 40 MPa. It is quite scary when a ferrule looses its grip on the column at a packing pressure of 60 MPa, and shoots a hole into the floor or the ceiling of the laboratory.

Another sealing mechanism is a metal-to-metal face seal at the ends of the tubing. This seal is highly reliable and can be used in excess of 70 MPa. However, this kind of seal may suffer, if the column is opened frequently, which may be necessary to replace or clean a filter. In practice, this is not a problem over the usual lifetime of a column.

Many proprietary column designs depend on the use of plastic seals, usually made from a fluorocarbon polymer. Although plastic seals can show creep, this is rarely a problem with a well-designed seal.

O-ring seals are less common in HPLC columns, but are quite feasible. The main issue is the compatibility of the O-ring material with the mobile phase. Incompatible O-rings can disintegrate, resulting in a sudden loss of the seal. O-rings made of fluorocarbon-based rubbers such as Kalrez or Chemraz have a sufficiently broad solvent compatibility to be useful with all mobile phases commonly used in HPLC.

Both the mobile-phase flow and the sample should be distributed evenly over the entire cross section of the packed bed. This should be accomplished in a minimal volume to minimize extracolumn effects. How important a good distribution of the sample is depends on both the column performance and its aspect ratio. It is more of a problem with short and wide columns than with long and thin columns. This is why a good sample distribution is more of a concern in preparative columns than in analytical columns. However, measurable effects are observed for high-performance columns if the diameter:length ratio exceeds ~ 0.05. An example would be a 4.6-mm $\times$ 50-mm column packed with 3-μm particles.

Figure 3.10 shows the bands observed near the end fitting in the case of three different distributor designs. In the first case, the sample simply enters the packed bed from a small orifice in the center of the end fitting. We assume that the frit does not have the ability to diffuse the sample. In this case, the sample band at the column inlet is rounded and nearly spherical, with the center of the band advancing over the part of the band next to the column wall. The second example in Figure 3.10 shows the case, when the distributor provides channels to inject the sample onto the column in the shape of a ring. The band is still distorted, but its length in the axial direction is by a factor of ~ 2 shorter than in the case of the point injection in the center. In the third example in this figure, the distribution of the sample is further aided by a diffuser. The diffuser is a layer of filter that has a higher permeability than the bed. In this case the radial spreading of the flow is increased, resulting in a still flatter sample band on the top of the column.

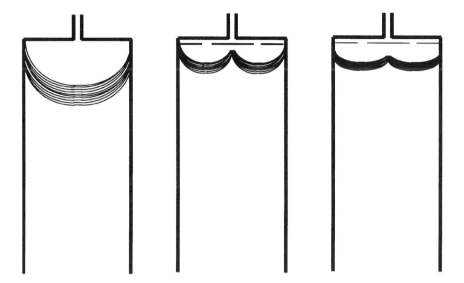

Figure 3.10 Sample bands at the column entrance. In the case on the left, the band resulting from a point source in the center of the column is shown. In the middle, the sample was distributed in a ring. The example on the right shows the band profile resulting from the combination of a circular injection and a diffuser.

Many different practical distributor designs are incorporated into commercial columns. The details of the implementation can be quite different, but they all rely on radial conduits of low flow resistance to distribute the sample over the entire cross section.

The filter itself has several functions. The primary function is the containment of the particles in the column. But as we have seen in the last paragraph, it also contributes to the distribution of the sample. Its third function is the protection of the packed bed from particulate debris, which could come either from the sample or from the wear of pump and injector seals.

Several filter types can be used. The most commonly used one is a frit of sintered-steel particles. It is a depth filter and has the advantage of a large filtering capacity. Its porosity is typically around 30–40%. The manufacturing technique makes it possible to incorporate a flow distributor into the surface of the filter. Also, frits with gradated permeability have been developed for the purpose of flow distribution. While the large porosity of frits results in the advantage of a large capacity, it has the disadvantage of contributing to extracolumn volume and therefore to extracolumn band spreading. The manufacturing process also imposes a lower limit on the thickness of the frit. A frit of 1 mm thickness for a 4.6-mm column has a dead volume of about 5 μL. Since two frits are needed, one can see that this extracolumn volume is not negligible.

An alternative filter design are woven steel-wire screens. They are very thin, with a thickness of about 0.1 mm, and have a low porosity, in the order of 5%.

On one hand, this reduces the dead volume; on the other hand, it limits its filtering capacity. Woven screens tend to clog fairly rapidly and need to be replaced frequently.

The third alternative is a filter design that attempts to combine the advantages of frits with the advantages of the woven screens. The solution is a hybrid of woven screen and pressed-powder frit; metal powder is pressed and sintered into a loose wire mesh. These filters are as thin as woven screens, but have a significantly larger open area and therefore a larger filtering capacity than woven screens.

In some cases, steel filters are undesirable. An analyte may adsorb to the steel or may decompose in the presence of steel. In these cases, plastic frits made from PEEK (polyetheretherketone) or polyethylene can be substituted for the steel filters. They are not as strong as steel frits and are not completely compatible with all mobile phases. Polyethylene swells in solvents such as hexane, methylene chloride, or tetrahydrofuran (THF). PEEK also is incompatible with methylene chloride and THF. In combination with PEEK or glass tubing, these frits can be used for the construction of columns with a metal-free flow path.

3.3.2   Tubing

Even in the early days of column packing it became obvious that the quality of the internal surface of the column tubing can have a significant effect on the quality of the packed bed (9). In general, the packed bed next to the wall has a slightly looser structure than the packed bed in the center. This results in an increased permeability and a faster flow next to the wall. The tubing wall also interacts with the column packing process to create longer-ranging nonuniformities in the bed. The exact nature of the packed bed next to the wall and further away from the wall can be influenced by the nature of the tubing surface. A rough surface results apparently in a looser structure.

The surface of normal steel tubing is usually of insufficient quality for HPLC columns. Therefore, several solutions to this problem have been worked out. In the early 1970s, steel tubing with a high-quality internal surface was available only in a wall thickness that was insufficient for the high pressures of HPLC. As a consequence, a column design was developed by Waters that consisted of a heavy-wall outer tubing for strength and a thin inner tubing with good surface finish. Today, much improved drawing processes have been developed that yield a reasonable surface finish even for heavy-wall tubing.

An alternative approach to an improved surface finish is polishing techniques. Examples are honing, gun drilling, or electropolishing. An interesting method is a patented procedure, in which a silly-putty-like substance is pushed repeatedly through the tubing. Any of these polishing techniques can yield an acceptable surface finish.

Glass or fused-silica tubing has inherently an excellent surface finish. Glass has a sufficient strength only for medium-pressure applications, up to

10–15 MPa for small-diameter columns. It is used in columns for use in biochemical applications. Because the failure of glass is rather unpredictable and catastrophic, column designs with an external shield are recommended for medium-pressure applications.

Fused-quartz tubing has been used successfully for columns of $\leqslant 1$ mm i.d. With these small diameters, it has sufficient strength for high-pressure applications.

PEEK tubing also has been used to prepare HPLC columns. It has intrinsically a very good surface finish. Unsupported heavy-wall PEEK tubing has been used up to 35 MPa; other designs are available in which a PEEK tubing is inserted into a metal tubing that provides the necessary strength.

3.3.3   Columns or Cartridges?

Since about 1977, there have been two basic designs of HPLC columns: one type has a complete end fitting and connects directly to the tubing of the HPLC instrument. The other design, called the *cartridge*, is simply the packed tubing with filters at the end that contain the packed bed. The end fitting consists of separate pieces that are intended to stay permanently connected to the tubing of the HPLC instrument. Connecting the cartridge to the end fittings generally is convenient and does not require any tools. Usually cartridges are cheaper than columns, and after a few purchases this price difference makes up for the additional cost of the end fittings. Which is the better choice?

As always, there are no clearcut answers. If you are planning to use the columns from a single manufacturer exclusively, then you can get the full advantage of the convenience of the permanently attached end fittings and the price advantage as well. However, the cartridges offered by different manufacturers are not compatible with each other's end fittings, so you would have to buy the end fittings from every column manufacturer that you want to use. This means on one hand that you have to disconnect the end fittings frequently from the HPLC instrument and you lose the advantage of the convenience. On the other hand, it will also take longer, until the initial investment into the cartridge end fittings has paid off. Also end fittings that are lying in a drawer have the nasty habit of disappearing, just when you want to use them again.

Today, there are practically no performance differences between a good cartridge design and a good column design, although it is easier to minimize dead volume in a column. Most cartridge designs have permanently incorporated filters. Therefore, if the filter clogs, the cartridge cannot be used any more. On the other hand, if you protect the cartridge with a precolumn, this should not be an issue. Some cartridge designs therefore have precolumns that can be attached directly to the analytical cartridge.

Whether you should choose a column or a cartridge depends on the work that the laboratory is doing and on your personal style of work. In a quality-control (QC) lab, where the same assay is run day-in and day-out using the same column, it makes sense to invest into the end fittings and use the cartridges. In a method development lab, where one is changing columns frequently, it may make more sense to buy the columns.

3.4 ALTERNATIVE COLUMN DESIGNS

Since the beginning of HPLC, alternatives have been sought to the "standard" column, which consists of a bed of particles packed into a piece of rigid tubing. There are several reasons for this search for alternatives. On one hand, column packing technology has always been viewed as "difficult," especially in academia, and in the early times of HPLC, there were good reasons for this view. The reproducibility of the bed formation was poor, and the packed bed often collapsed without any apparent reason. On the other hand, our delight in innovation will drive us to explore alternatives to existing technologies, especially when glory and financial rewards tempt us. HPLC columns and packings are expensive, which makes the search for alternatives financially attractive.

Because of the issues with packed-bed stability, several technologies have been developed that deal with this problem. In radial-compression technology, the bed is contained in a flexible wall, which is radially compressed to achieve a higher packing density and to maintain the position of the particles. In axial compression, the same goal is pursued by pressing a plunger against the top of the packed bed. In a related technique, the end fitting can be moved toward the bed to eliminate any void space that may form during usage due to partial collapse of the bed.

Today, the major manufacturers routinely produce in the order of 100,000 columns per year, and experienced manufacturers have learned to pack columns to a high reproducibility of their performance and stability. Thus the technologies designed to avoid or eliminate column voiding are not needed any more and are found only in niches where the problems have not yet been solved. The primary example is preparative chromatography with large-diameter columns. Because of their importance in preparative HPLC, we will discuss here the principles of radial- and axial-compression technologies.

Many attempts have been made to replace packed beds. Membranes, filters, and fabrics have been stacked or rolled, and fibers have been aligned or wound or otherwise arranged to form structures useful for chromatography. Continuous porous structures akin to open foams have been formed into rods and disks useful for chromatography. After a brief description of some of these very innovative ideas, we will discuss ways to evaluate them from a fundamental standpoint. This should help us to sort good ideas from bad ideas and guide the inventors in their attempts to improve these alternative designs.

3.4.1   Radial Compression

Radial compression technology was developed by workers at Waters as an alternative to classic column technology (10,11). Two weaknesses of classic column technology were addressed by radial compression technology:

1. Inhomogeneities in the packed bed, causing inferior column efficiency and reproducibility problems
2. Low packing density, causing column stability problems during shipment and use

In radial compression technology, the packed bed is contained in a flexible-wall tubing. Controlled hydraulic or mechanical forces are applied uniformly to the outside wall of the column (Fig. 3.11). This compresses the packed bed, resulting in a denser and — when applied properly — more uniform structure. The technology is derived from isostatic compaction, which is frequently used in the ceramics industry, for example, in the mass production of spark plugs.

The compaction of the bed can be carried out both mechanically or hydraulically. In a mechanical scheme, the bed is compressed to a controlled volume, while the hydraulic compression uses a controlled pressure. Most of the initial studies of radial compression were carried out with a controlled compression pressure, applied either to preparative columns packed with irregularly shaped particles with a particle size of about $80\,\mu$m or analytical columns packed with spherical 10-μm particles (12).

The interstitial fraction of an uncompressed bed of hard spherical particles packed by conventional packing technologies is $40 \pm 2\%$. When hydraulic radial compression is applied to such a bed, it readily densifies until an interstitial fraction of $34 \pm 1\%$ is obtained at $3\,$MPa. A further increase in

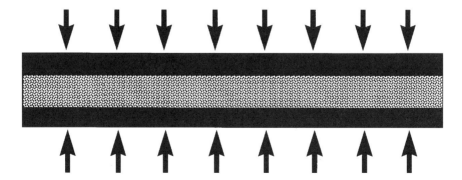

Figure 3.11   Principles of radial compression. The packed bed is contained in a flexible wall. Compression pressure is applied to the outside of the wall to compact the bed uniformly.

compression pressure to 20 MPa barely increases the density. However, above this pressure, even strong particles such as Resolve or Nova-Pak start to show signs of breakage, which imposes an upper limit on the compression pressure that can be applied. But it also indicates that it is very difficult to obtain a still higher density, and beds that have been compressed to this density are extremely unlikely to collapse any further. The theoretical limit for uniformly sized spheres is an interstitial fraction of about 25% (hexagonal or cubic densest packing), but this has never been achieved in praxis for a packed bed of incompressible spheres by any technique other than placing each particle into its proper position. This, of course, can be done only with large beads in relatively small beds and is of no technological interest.

(*Note:* In order to determine the interstititial fraction, one needs to know the interstitial volume and the column volume. The interstitial volume can be measured readily using a marker that is excluded from the pores. For beds in a fixed-volume container, the column volume is known as well, but it is unknown for radially compressed beds. However, the volume occupied by the particles can be determined from the pore volume, which also is easily measurable if the particle porosity is known. The particle porosity, in turn, can be determined in a separate experiment using the same markers for the totally included volume and the interstitial volume in a column of a known column volume. Therefore, after the particle porosity has been determined, the column volume in a radially compressed column can be calculated as the sum of the interstitial volume, the pore volume and the skeleton volume of the particle.)

If the compaction of the bed is carried out below the point where the particles fracture, it results in an improved uniformity of the bed over packed beds obtained by classic packing technologies. In Figure 3.12 the performance of radially compressed columns is compared to the performance of steel columns of the same dimension, packed with the identical batch of particles (12). In this experiment, special care was taken to exclude any extraneous effects. The steel column was equipped with the identical end fittings as the radially compressed column, and the same equipment was used in this comparison. In Figure 3.12, the HETP is plotted against the interstitial velocity u_i. As one can see from the graph, the largest difference between both columns occurs in the range of low velocities and around the minimum of the curve. At high velocity, the performance of the steel column and the radially compressed column is indistinguishable. Fitting the van Deemter equation to the experimental data gave an A term of 1 for the steel columns, which is already excellent, and an A term of 0.5 for the radially compressed columns. The value for the radially compressed columns indicates a significant improvement in bed uniformity.

The homogeneity of the bed of radially compressed columns can also be studied conveniently by injecting dyes into the bed. The walls of a Radial-Pak cartridge are made of polyethylene and can be cut without disturbing the bed. The bands of the dye can then be easily studied. These studies confirm the high uniformity of a properly prepared radially compressed bed.

H/u Curves of a Conventional Column
and a Radially Compressed Column

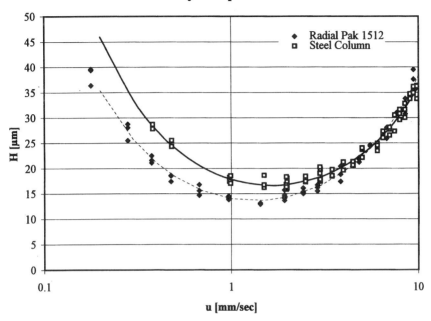

Figure 3.12 Comparison of the performance of a radially compressed column to a column prepared by a conventional slurry-packing technique. The HETP is plotted against the interstitial velocity u_i. Both columns were packed with spherical C_{18}-bonded silica with a pore volume of 0.5 mL/g. The radially compressed column is a 8 mm Radial-Pak cartridge (Waters Corp.). The steel column was a steel tubing inserted into a Radial-Pak cartridge, which allowed the use of the same end fittings.

The improvements in bed uniformity and stability do not come for free. Accompanying the decrease in interstitial fraction is a decrease in permeability. From Equation (2.51) we can calculate that a decrease in interstitial fraction from 40% to 34% should result in a decrease in permeability (= an increase in backpressure) by a factor of 2. This has indeed been observed.

The principle of radial compression has successfully solved the problems of packed-bed uniformity and of bed stability. However, radial compression has its limitations in the strength of the particles. If excessive radial compression is applied, and particles start to fracture, the homogeneity of the bed is lost. This sets an upper limit to the hydraulic compression that can be applied to soft or fragile particles. Since the column backpressure cannot exceed the compression pressure, this limits the applicability of hydraulic compression. Mechanical compression does not suffer from this problem, but it is more difficult to achieve a uniform compression, and therefore a good implementation of mechanical compression is more of a challenge. Classic column packing technology has progressed, and analytical HPLC columns can be packed today with good stability and a high level of reproducibility. Therefore, radial compression is currently used predominantly for laboratory-scale to production-scale preparative chromatography.

Commercial radial-compression cartridges are prepared using polyethylene tubing. This has the side benefit that cartridges that were exposed to biologically or chemically hazardous materials can be incinerated, whereas the disposal of steel columns is more problematic.

3.4.2 Axial Compression

Under the category of axial compression we will discuss several related techniques, ranging from a simple elimination of a void space at the top of the column to active axial compression.

Glass columns used for low-pressure chromatography are nearly always equipped with an adjustable inlet plunger that can be moved up and down depending on the height of the bed. This way, any void space above the packed bed can be eliminated easily. The same principle can be applied to HPLC columns. When the bed collapses, the void space that is formed above the bed contributes significantly to band spreading and peak distortion. However, the bed itself usually has become nonuniform, and this distorts the peaks as well. Thus an adjustable end fitting removes only part of the cause of peak distortion and its effectiveness is limited.

A better approach to the problem is the active compression of the bed through axial compression (13). This technique has not been used for analytical columns, but is widely practiced for large-diameter preparative columns. As is the case for radially compressed column, the permeability of an axially compressed column is lower than the permeability of an uncompressed column (14), but the effect is much smaller because of the lower compression factor. Axial compression maintains column efficiency. In repeated packing trials, reproducible efficiencies can be obtained, but the peak shapes reported do not indicate a good uniformity of the packed bed. There are no studies available yet that allow one to judge whether this is an intrinsic property of axial compression or if this is due to the technical difficulties associated with the packing of large-diameter columns.

3.4.3 Alternatives to Packed Beds

Fundamentally, any arrangement of flow passages through a stationary structure that can carry an interacting phase can be used for chromatography. Therefore, several alternatives to packed beds have been explored. They range from stacked membranes (15,16) to aligned (17,18) or nonaligned (19) fibers to continuous porous structures (20–22) to rolled structures of filter paper (19) or fabric (23). We will discuss some of these approaches in the following.

3.4.3.1 Membranes Membranes are mass-produced for a broad range of industrial filtration applications. For that purpose, the pore size of the membrane needs to be well defined, but the actual construction of the membrane is of secondary importance. Therefore, the structure of membranes can vary widely depending on the type and the fabrication process. Some membranes consist of a network of fibrils, while others have pores that are oriented perpendicularly to the surface and have little radial interconnectivity. In the latter case, the membrane may also be asymmetric, with small pore openings on one side and wider openings on the other side. Such a structure may be advantageous for filtering applications, but is rather a disadvantage in chromatography. Good radial mass transfer is vital for the performance of any chromatographic device, and this kind of a structure is rather an impediment to radial mass transfer.

In packed beds of porous particles, the vast majority of the interacting surface is found inside the pores of the particles. Membranes lack this secondary pore structure and therefore typically have a low surface area. On the other hand, all the free volume in a membrane is considered to be swept by flow, which eliminates the resistance to mass transfer due to diffusion in the pores. Thus the structure of membranes is comparable to packed beds of nonporous particles, and one can expect the tradeoff between capacity and mass-transfer resistances to be similar for membranes and nonporous particles. It should be pointed out that the resistance to mass transfer due to diffusion in the pores is only one part of the overall transport problem. One is still faced with mass transfer from the center of the flowing stream to the interacting surface, which may occur by diffusion or convection, and the kinetics of the interaction of the analyte with the surface. Since membranes have a low surface area per unit volume compared to fully porous particles, the capacity for small molecules is low. On the other hand, particles designed for the chromatography of large molecules need large pore sizes, which result in a low surface area per unit volume. Thus the capacity of membranes and packed beds of particles for macromolecules can be similar when the membrane is properly designed.

Typical membranes are about 100 μm thick. Many types of membranes lack a good connectivity in the radial direction. This is a disadvantage. The strong radial transport in packed beds by the combination of convection and diffusion is one of the reasons for their superb performance in HPLC.

A practical issue in the design of chromatographic devices from stacked membranes is the sealing of the membranes around the edges to prevent preferred flow along the wall. In commercially available arrangements, this flow is prevented by compressing the periphery of the membranes with donut-shaped gaskets (15). This, in turn, seems to cause radial variations of the flow velocity, with a faster flow in the center of the stack compared to the wall region.

Generally, stacked-membrane devices have a large diameter:length ratio, while the opposite is true for packed beds. However, there is no fundamental reason why packed beds of similar aspect ratios as current stacked membrane devices cannot be formed.

3.4.3.2 *Monolithic Structures* Continuous porous structures are the other popular type of alternatives to packed beds. They can be formed in many different forms, such as flat disks or as long rods or as anything in between. Flat disks have occasionally been called "membranes" and have been compared to stacked membranes (21). "Monolithic rods" have been compared to chromatographic columns (22). They are typically prepared in situ (although this is not a necessity) in the cavity that will also form the containment for use.

One can view these monoliths as a single big porous particle. Thus, some of the preparation procedures use similar ingredients as the procedures used to make macroporous particles by suspension polymerization. Consequently, the structures of the monoliths are similar to the pore structure of macroporous particles, as can easily be seen by electron microscopy. Also similar chemistries are available, including styrene–divinylbenzene and methacrylates, which have been proven to form sufficiently rigid structures to be useful in HPLC. But the technology of the formation of the monoliths is less constrained than the suspension polymerization used to form particles, and thus a broader range of chemistries is available. The classic monoliths were based on polyurethanes (20). Recently, silica-based monoliths were formed in a capillary (24).

As is true for the stacked membranes, an unambiguous qualification of the performance of monoliths is not yet available at the time of this writing, although it is unquestionably true that reasonable chromatographic results have been obtained with these devices. When properly prepared, macroscopic uniformity can be achieved and "wall effects" can be avoided. The devices prepared to date do not contain an intentional secondary pore structure; thus they have a low surface area per unit volume. At the same time, the absence of this pore structure also means that resistance to mass transport in these pores is absent. Thus the overall chromatographic performance of the monoliths should be subject to the same rules as stacked membranes or packed beds of nonporous particles. The key advantage of the preparation procedures for monoliths is the ability to design structures with a high porosity, while the porosity of a packed bed of nonporous particles is constrained to be about 40%. The monolith described in Reference 22, for example, has a porosity of 65%. This can result in a significant advantage in permeability if the dynamic chromatographic performance can be maintained.

3.4.3.3 *Fibers* Various structures can be prepared based on fibers. For example, strings of yarn can be lined up in parallel, and heat-shrinkable polyethylene tubing can be shrunk around these strings (19). The resulting structure can be examined just like a radially compressed column, using the same equipment and end fittings. Similarly, the yarn can be wound around a mandrel, and once again a heat-shrinkable tubing can provide the wall. This approach solves the problem of creating a wall around a structure such that a preferential flow near to the wall is avoided. These arrangements were examined isocratically using an unretained sample. Their performance did not match the quality of a packed bed of 10-μm particles. The efficiency of these devices under gradient conditions was not investigated. It is expected to match the performance of stacked membranes and monoliths.

Hartwick (17) aligned uniformly sized fibers into a densely packed hexagonal array. The interstices between the fibers represented the flow channels. There was no transport between the channels. The performance of the device was low relative to its permeability. This is not unexpected. A key property of a packed bed is the radial mass transfer, which evens out flow nonuniformities. This is not possible in a device consisting of parallel independent flow paths. In an array of circular parallel channels, the breakthrough time for an unretained sample is inversely proportional to the square of the diameter of the channel. To obtain a plate count of 10,000 plates, it would be necessary that the relative standard deviation of the channel diameter is under 0.5% (see also the footnote in Section 2.1.4). This is clearly a tall order. For retained peaks, similar demands would need to be placed on the uniformity of the stationary phase from channel to channel.

Czok and Guiochon (18) examined a column prepared from well-aligned fibers made from porous glass obtained from PPG Industries, USA. The dense packing of the fibers leads to a very low interstitial fraction, 9% compared to 40% for a packed bed. If the fibers are 75% porous, one will find a ratio of pore volume to interstitial volume of about 7.5. The same ratio for a packed bed of equally porous particles would yield a ratio of about 1.1. One can see the potential advantage of the bed of fibers. Unfortunately, one can also predict that the low interstitial fraction will result in a low permeability. The efficiency of such a device cannot be predicted a priori, but the results reported were poor. We will discuss this device more in the following section, in which we deal with the evaluation of these alternatives to packed beds.

3.4.4 Evaluating Alternative Designs

How can we evaluate the merits of these vastly different design alternatives relative to each other in an objective way? We will attempt to provide an answer to this question in the following discussion. The basis for the approach presented here are the principles laid out in References 1–5, which we used at the beginning of this chapter.

The purpose of chromatography is to provide maximal resolution in the shortest amount of time possible. In order to eliminate the influence of the retention mechanism from this discussion, we will use plate count as a shorthand description of resolution and linear velocity as a measure for the speed of analysis. What prevents us from maximizing plate count and speed of analysis simultaneously is the pressure requirement that goes along with this. As we have seen in Section 3.1, pressure, speed of analysis, and plate count are inextricably linked.

If we accept this link as a fundamental property of HPLC, we can use it also to compare the merits of different ways to construct chromatographic devices. To do this, we will make use of the concepts of the dimensionless reduced plate height and the dimensionless reduced velocity (see Section 2.1.7). Then we simply treat the column as a black box: we assume that we do not know what is in it, but treat it as if it would be made of a packed bed of particles. Then we can determine whether the quality of this virtual packed bed meets, exceeds, or fails our expectations.

How do we do this? A performance measure of a chromatographic column that is independent of the length is the HETP. The HETP, in turn, depends on the linear velocity with which the mobile phase moves through the column:

$$H = f(u) \tag{3.15}$$

We also know that we can express this relationship in dimensionless parameters, by dividing H by the particle diameter and multiplying the linear velocity with the particle diameter and dividing the product by the diffusion coefficient of the sample in the mobile phase (see Section 2.1.7):

$$h = \frac{H}{d_p} \tag{3.16}$$

$$v = \frac{ud_p}{D_M} \tag{3.17}$$

This allows us now to compare chromatographic columns to each other independent of the column length, column diameter, or particle size. This obviously works well for packed beds, when the particle size is known. But what do we do in the case of chromatographic devices that do not contain particles? We simply treat them as if we did not know their contents, that is, as if they were packed beds made from particles of unknown size. The apparent particle size in such column can be determined from its permeability (see Section 2.2.5):

$$d_p = \sqrt{1000 B_0} \tag{3.18}$$

The permeability of a device can be determined from flow rate, viscosity, column length and pressure as outlined in Section 2.2.5. Now we have accomplished what we have set out to do; we have linked the performance parameter, the speed of analysis and the pressure to each other. All performance parameters can thus be determined from externally measurable observations without reference to the internal content of the chromatographic device:

$$h = \frac{H}{\sqrt{1000 B_0}} \tag{3.19}$$

$$v = \frac{u\sqrt{1000 B_0}}{D_M} \tag{3.20}$$

In order to compare different devices to each other, all we need to do is to plot the relationship of h as defined in Equation (3.19) against v as defined in Equation (3.20). Actually, with the convention that high separation power is associated with a small h, it is more instructive to plot $1/h$ versus v. We have done this in Figure 3.13, using Equation (2.47) as being representative of the performance of a typical HPLC column.

Also shown in Figure 3.13 are the actual results obtained for a steel column and a radially compressed column packed with the same particles. The results for the steel column follow closely the standard curve. The radially compressed column shows inferior results at high velocity. This is due to the convention used here that the apparent particle diameter in the column is determined from the permeability. Radially compressed columns have a denser bed, and thus a higher pressure drop. The particle diameter determined via the measurement of the permeability was 5.7 μm instead of the true diameter of 8 μm. Therefore, the expected performance of a 5.7-μm column is not met at high velocity. On the other hand, the maximum of the curve for the radially compressed column matches the expectation. This is because the increased uniformity of the packed

Non-Dimensional Column Performance vs. Non-Dimensional Speed of Analysis

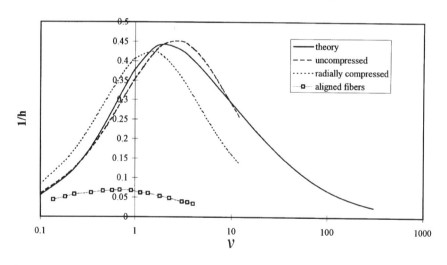

Figure 3.13 Nondimensional column performance versus nondimensional speed of analysis. In this graph the inverse of the reduced plate height is plotted against the reduced velocity. The solid line represents a typical packed bed prepared with an HPLC-grade packing and is the standard of comparison. The dotted line is a radially compressed column; the dashed line is the same material packed into a steel column. At the bottom of the graph the characteristic curve for a bed made from aligned fibers is shown.

bed due to radial compression compensates for the increase in backpressure. This example is a good illustration how the convention used in Equations (3.19) and (3.20) works in practice.

Figure 3.13 also contains the results obtained for a column of aligned porous fibers (PPG Industries, USA) according to the principle outlined above (25). The column consisted of a steel tube packed with controlled-pore glass fibers perfectly aligned in parallel to the column wall. The diameter of the column was 2.36 cm, and the length was 8.84 cm. The diameter of the fibers was 19.5 μm. The permeability of this column was measured to be 1.7×10^{-9} cm^2. Thus the column had the same permeability as a packed bed of 13-μm particles. This value was rather expected. The plate height was measured as a function of the linear velocity. A typical van Deemter curve was found, with the following constants: $A = 109\,\mu$m, $B = 6.82$ cm^2/s, and $C = 246$ ms, using anisaldehyde as sample and 88% hexane/12% ethylacetate as mobile phase. Thus the chromatographic performance of the column was more akin to a bed packed with 50–70-μm particles. Consequently, the performance of the aligned-fiber column was very poor compared to a packed bed of equal pressure drop, as can be seen from Figure 3.13.

The HETP should be measured preferentially using retained compounds that interact with the packing. Figure 3.13 was established using retained peaks. For unretained peaks, one would expect an increase in performance due to the absence of all mass-transfer and kinetics issues associated with the interaction of the analyte with the stationary phase. Therefore a measurement of an unretained peak is an incomplete measure of the performance of the chromatographic device.

The requirement of using a retained peak for the proper judgment of the performance of a chromatographic column causes a problem for devices designed for the retention chromatography of macromolecules, such as proteins. For macromolecules, the retention factor is an extremely sensitive function of the mobile-phase composition, to a degree that isocratic chromatography of macromolecules is completely impractical. Therefore all retention chromatography for macromolecules is carried out using gradients. But an HETP cannot be readily derived from the peaks observed in a gradient chromatogram.

This issue is the primary cause of confusion in the discussion of the advantages and disadvantages of unconventional chromatographic devices such as stacked membranes or monoliths. These devices tend to perform poorly in the isocratic chromatography of small molecules, but appear to give reasonable results in the gradient chromatography of macromolecules. This is due to the fact that a manipulation of the gradient can compensate for a poor hydrodynamic performance of a device. In fact, some novel devices are touted as being breakthroughs in the chromatography of macromolecules. This judgment is often explicitly based on the observation that chromatographic performance does not deteriorate when the flow rate is increased. Unfortunately, this can also be true for a poorly performing device — the chromatography might be so bad to start with that it barely can get worse.

It is therefore desirable to have a tool in hand that can be used to judge the performance of chromatographic devices, when their primary application is the gradient chromatography of macromolecules. Fortunately, a small expansion of the black-box concept outlined above makes this possible.

The whole problem lies in the determination of the HETP from a gradient chromatogram. The HETP can be defined as the variance of the peak (in length units) at the column outlet divided by the column length. The determination of this from the chromatogram is unproblematic in isocratic chromatography, since the migration velocity of the peak is known:

$$\sigma^2 = \frac{\sigma_t^2}{[u(k + 1)]^2} \tag{3.21}$$

where σ is the standard deviation of the peak in length units, as observed inside the column, and σ_t is the standard deviation of the peak in time units, as recorded in the chromatogram. In order to calculate the standard deviation of the peak in length units at the column outlet from the observable standard deviation of the peak in the chromatogram, we need to know the migration velocity of the peak at the moment when it exits the column. How do we do that?

There are actually two solutions to the problem. The first solution is more general, but requires a knowledge of the nature of the dependence of the retention factor on solvent composition. The second solution is simpler, but depends on the (verifiable) assumption that the retention factor of the analyte at the column outlet is 0.

Let us examine the first case in more detail for reversed-phase chromatography. We can generally assume a linear dependence with a slope s of the logarithm of the retention factor on the volume % percent (vol%) of the organic modifier φ:

$$\ln(k) = \ln(k_0) - s\varphi \tag{3.22}$$

Equipped with this knowledge, we can solve the gradient equation for reversed-phase chromatography and obtain an equation that describes the elution time t_g of an analyte as a function of the gradient slope g:

$$t_g = t_0 + t_d + \frac{1}{gs} \ln[gst_0 k(\varphi_0) + 1] \tag{3.23}$$

There are basically two unknowns in this equation: the slope s and the retention factor at the start of the gradient $k(\varphi_0)$. If we run two gradients with different slopes g, we obtain two results and can calculate the two unknowns. We now know the dependence of the retention factor on solvent composition. Since we know the composition of the mobile phase at the moment when the peak of interest exits the column, we can determine the retention factor of the analyte at the column exit. This, in turn, allows us to calculate the variance of the peak at the column exit using Equation (3.21).

Experimentally, this approach is rather simple, since it involves only two gradients with different gradient slopes but the same starting solvent composition. The same principle can be used for ion-exchange chromatography, or other types of chromatography.

Since we are working with large molecules, it is highly likely that the result of the experiment above tells us that the slope s is very steep (which is why we need to use this approach to start with), and that the retention factor of our analyte at the column exit is essentially 0. This is the assumption that we use in our second, simpler approach.

If we know that the retention factor of our analyte at the column outlet is 0, all we need to do is measure its migration velocity at a mobile-phase composition, where the analyte is unretained. Thus we would select a mobile phase that is slightly stronger than the composition at which the analyte elutes in the gradient and measure its retention time isocratically. This knowledge then allows us to determine the variance of the peak in length units inside the column:

$$\sigma^2 = \sigma_t^2 \frac{L^2}{t_R^2} \tag{3.24}$$

Whether we have used the first or the second approach, we can now proceed to determine the true HETP as a function of the velocity. An example of results obtained (26) using this procedure is shown in Figure 3.14. The HETP was measured for several proteins eluted under gradient conditions from a 5-cm column packed with 8-μm particles and a 10-cm column packed with 20-μm particles. The packing was a 300 Å C_{18} packing. In this case, we did not need to estimate the particle size from the permeability of the column, since the particle size was known. The results for both particle sizes line up quite nicely. This demonstrates that this procedure works well and that true reduced HETP values data can be obtained even under gradient conditions.

In summary, the procedures outlined here enable us to unambiguously compare chromatographic devices to each other independent of the nature of the construction of the device. Of course, one should also use other measures that are more specific to the type of chromatography for which the device is used. For example, one should measure the dynamic protein-binding capacity for a device that is used for the preparative chromatography of proteins. Specific procedures like this are well established and can be found in the literature. But the procedures outlined here establish objective quality criteria for a chromatographic device. These quality criteria are of a general nature and are independent of the specific mode of chromatography. They are especially valuable when the quality of unusual devices, such as stacked membranes or aligned fibers, must be assessed unambiguously.

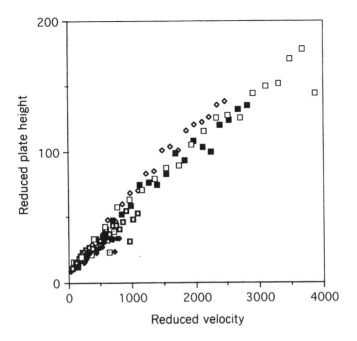

Figure 3.14 Reduced plate height–reduced velocity plot for protein samples on a reversed-phase column. The data were derived from gradient chromatography.

REFERENCES

1. M. Martin, C. Eon, and G. Guiochon, *J. Chromatogr.* **99**, 357 (1974).

2. M. Martin, C. Eon, and G. Guiochon, *J. Chromatogr.* **108**, 229 (1975).

3. M. Martin, C. Eon, and G. Guiochon, *J. Chromatogr.* **110**, 213 (1975).

4. I. Halász, R. Endele, and J. Asshauer, *J. Chromatogr.* **112**, 37 (1975).

5. R. Endele, dissertation, Universität des Saarlandes, Saarbrücken (1974).

6. G. I. Taylor, *Proc. Roy. Soc.* (London) **A219**, 196 (1953).

7. R. Aris, *Proc. Roy. Soc.* (London) **A235**, 67 (1965).

8. M. J. E. Golay and J. G. Atwood, *J. Chromatogr.* **186**, 353–370 (1979).

9. J. Asshauer and I. Halász, *J. Chromatogr. Sci.* **12**, 139–147 (1974).

10. P. D. McDonald and C. W. Rausch, US Patent 4,211,658 (1981).

11. C. W. Rausch, Y. Tuvin, and U. D. Neue, US Patent 4,228,007 (1980).

12. U. D. Neue, unpublished results (1977).

13. E. Godbille and P. Devaux, *J. Chromatogr.* **122**, 317 (1976).

14. M. Sarker and G. Guiochon, *J. Chromatogr.* **A702**, 27–44 (1995).

15. D. K. Roper and E. N. Lightfoot, *J. Chromatogr.* **A702**, 3–26 (1995).

16. J. A. Gerstner, R. Hamilton, and S. J. Cramer, *J. Chromatogr.* **596**, 173–180 (1992).

17. R. F. Meyer, P. B. Champlin, and R. A. Hartwick, *J. Chromatogr. Sci.* **21**, 433–438 (1983).
18. M. Czok and G. Guiochon, *J. Chromatogr.* **506**, 303–317 (1990).
19. U. D. Neue, unpublished results (1979).
20. L. C. Hansen and R. E. Sievers, *J. Chromatogr.* **99**, 123–133 (1974).
21. T. B. Tennikova, B. G. Belenkii, and F. Svec, *J. Liq. Chromatogr.* **13**, 63–70 (1990).
22. M. Petro, F. Svec, I. Gitsov, and J. M. J. Fréchet, *Anal. Chem.* **68**, 315–321 (1996).
23. M. Ladish, presentation at the 1996 PrepTech Conference, East Rutherford, NJ (USA) (1996).
24. Steven M. Fields, *Anal. Chem.* **68**, 2709–2712 (1996).
25. U. D. Neue, B. Alden, unpublished results (1990).
26. U. D. Neue, B. Alden, and B. R. San Souci, poster presentation at HPLC'91, Basel (1991).

4 Physical Properties of HPLC Packings

Von der Stirne heiß
rinnen muß der Schweiß,
soll das Werk den Meister loben.

— F. Schiller, *Das Lied von der Glocke*

There are two fundamental aspects of HPLC packings: (1) the surface chemistry of the packing which determines the interaction of the analytes with the packing; and (2) the physical properties of the packing. Physical properties include the particle size and shape, the pore size and porosity, the specific surface area, and last, but not least, the particle strength. These properties determine, among other factors, the column efficiency, the retentivity, and whether the particle is suitable for HPLC at all. We will address these issues in this chapter.

4.1 PARTICLE SIZE, PARTICLE SHAPE, AND PARTICLE SIZE DISTRIBUTION

The particle size of a packing is of utmost importance. As we have seen in Chapters 2 and 3, it determines the permeability of the column as well as the column efficiency.

Most packings used in HPLC have a distribution of particle sizes. This is inherent in the manufacturing processes. For example, irregular-shaped particles, the oldest type used in HPLC, are made from chunks of silica that are milled and subsequently classified. For large particle sizes, (above ~30 μm), classification of the particles can be accomplished by sieving, using a stack of sieves of progressively smaller screen sizes. The particles in a sieve are larger than the mesh size of the screen, but smaller than the mesh size of the next larger sieve. The resulting particle size fractions are designated by the screen sizes used to generate them. An example is the designation of 55–105 μm for a preparative packing.

The sieving of particles smaller than $\sim 30\,\mu$m is inefficient on an industrial scale. Therefore smaller particles are usually prepared by air classification. The technique is used for both irregular-shaped and spherical particles. A typical particle size distribution for an irregular-shaped packing is shown in Figure 4.1.

Spherical particles are synthesized specifically for HPLC. Most common manufacturing techniques for spherical particles also yield a particle size distribution that is too broad for direct use, and therefore a classification is necessary.

A particle size distribution can be displayed in several different ways. The simplest view is a plot of the number of particles in a narrow particle diameter window versus the average particle diameter in this window. From this distribution, called the *population distribution*, the number-averaged particle size $d_{p,n}$ can be calculated:

$$d_{p,n} = \frac{\sum d_{p,i} n_i}{\sum n_i} \qquad (4.1)$$

More, commonly, the volume or weight of the particles in this narrow-particle-diameter window is computed and plotted against the particle size. From this distribution, the volume- or weight-averaged particle size $d_{p,v}$ is determined:

$$d_{p,v} = \frac{\sum d_{p,i} d_{p,i}^3 n_i}{\sum d_{p,i}^3 n_i} \qquad (4.2)$$

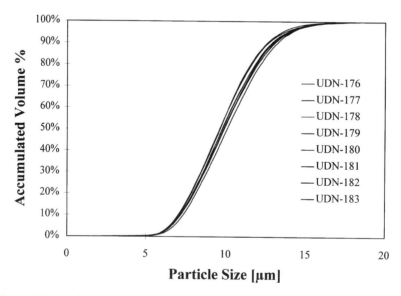

Figure 4.1 Particle-size distribution of several batches of Waters μBondapak C_{18}.

The product of the third power of the particle size and the number of particles is proportional to the volume of the particles in the channel. It is also proportional to the weight of the particles in this channel, assuming that the particles have a uniform density. The volume-averaged particle size is always larger than the number-averaged particle size, unless the particles are monodisperse (they all have the same diameter).

The volume-averaged particle size is most often reported in the HPLC literature, but not everybody abides by this convention. Therefore, the "5-μm" particles of one manufacturer are not necessarily the "5-μm" particles of another. To complicate issues, other averages can be determined as well.

The average particle size determines both the column backpressure and its efficiency. But what average? For narrow particle size distributions, this matters little, since the difference between the averages is small. But for broad particle size distributions, one can apply the rule that the column backpressure is determined by the population-averaged particle size, while the column performance, the C term of the van Deemter equation, is determined by the volume-averaged particle size (1). This is demonstrated in Figures 4.2a, b. For a range of particle size distributions generated by blending 4, 6, and 12-μm Nova-Pak silica, the C terms of the van Deemter equation and the permeabilities were measured using Radial-Pak cartridges. In Figure 4.2a both parameters are plotted against the square of the population average of the particle size distributions. In Figure 4.2b, the same parameters were plotted against the square of the volume average of the particle size distributions. As one can see, a straight line is obtained only for the plots of the C term versus the volume average and the permeability versus the population average. The particle size distributions of most commercially available HPLC packings are quite narrow. Early studies in HPLC (2) have found that as long as the particle size distribution is not wider than about $\pm 40\%$ around the mean, acceptable column performance and pressure drops can be achieved. Nevertheless, narrower distributions still give better results (3), and the distributions tend to be no wider than about $\pm 25\%$ around the mean for today's high-performance packings.

From the standpoint of the user, reproducible averages are desirable. For example, the packed-bed permeability depends on the square of the particle size. Therefore a 10% shift in the average particle size would result in a 20% change in column backpressure. Such a large difference is unacceptable. Figure 4.3 shows the particle size distribution of several batches of Symmetry silica. One can see that today's sizing techniques can give quite reproducible results.

Multiple techniques can be used to measure the particle size distribution, for example electrozone sensing, sedimentation, laser diffraction, and microscopy. With the exception of microscopy, they all require calibration and the results depend on the technique. For example, in a round-robin study reported in Reference 4, the commonly used electrical sensing zone technique (Coulter Counter) was compared to microscopy and sedimentation. The average particle size determined by the electrical sensing zone method was by about 25%

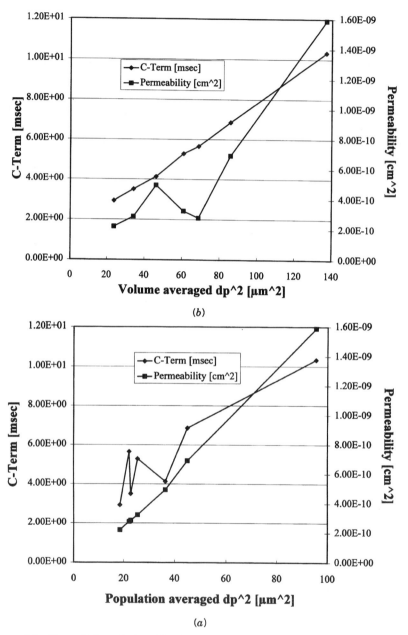

(b)

(a)

Figure 4.2 Test of the relationships between particle size distribution and column characteristics. In both plots the van Deemter C term and the permeability are plotted against the square of the average particle size. In Figure 4.2a the x axis is derived from the population-averaged particle size, and in Figure 4.2b the x axis is derived from the volume-averaged particle size.

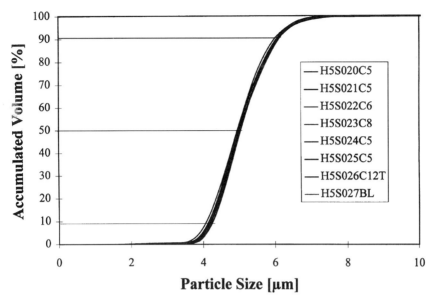

Figure 4.3 Reproducibility of the particle size distribution of Waters Symmetry silica. The graph shows the accumulative distributions of eight separately sized batches.

lower than the averages determined by the other techniques. The electrical sensing zone technique is also sensitive to the porosity of the packing (4). Thus, different results are obtained for a C_{18}-bonded phase and its parent silica. This dependence on the measurement technique and on the properties of the particle represents another obstacle to the comparison of data published in the literature that depend on a determination of the "particle size," as, for instance, the reduced plate height, the reduced velocity and the flow-resistance parameter. The best absolute technique for determining the true particle size distribution and the true average particle size is still microscopy, but it is too tedious to be of practical value. For manufacturers of HPLC packings, an absolute measure is less important than a measure that yields reproducible results with a high resolution. In this respect, the electrical sensing zone technique is superior to microscopy.

While the designation of a particle size by microscopy is unproblematic for spherical particles, it represents a significant challenge for irregular-shaped particles. This is obviously a problem, if we try to determine whether irregular or spherical particles make better columns. Halász (5) pointed to a simple and practical solution to this dilemma; he suggested that the average particle size should be determined from the backpressure of the column. We would then use the Kozeny–Carman equation to calculate the particle size, which then can be fed into the equations for reduced plate height and reduced velocity. Thus the definition of the chromatographic particle size is

$$d_p = \sqrt{1000\,B_0} \qquad (4.3)$$

If we apply this technique, little performance difference can be found between irregular and spherical particles.

(*Note:* One might argue that this suggestion itself is problematic. The packed beds obtained with irregular particles tend to be looser, i.e., have a higher interstitial fraction than beds prepared from spherical particles. Thus the inferior performance of irregular packings is compensated for by a superior permeability. From the standpoint of practical chromatography, i.e., looking at the column as a black box, this argument becomes irrelevant.)

Using special synthetic techniques (6), polymeric and silica particles can be prepared with a uniform particle size. Despite occasional unsubstantiated statements to the contrary in the literature, no intrinsic advantages can be found for monodisperse particles. Neither do they form a closest packing structure (which would actually be a detriment since the pressure drop would rise about 6-fold), nor do they yield an increased plate count, nor is there any theoretical reason why they should do such things.

4.2   POROSITY, PORE SIZE, AND SPECIFIC SURFACE AREA

Most particles used in HPLC are fully porous. In size-exclusion chromatography, the pores are needed to effect the separation. In retention chromatography, the walls of the pores provide the large surface area needed for retention. The external surface of a 5-μm particle is only $0.02\,\text{m}^2$ per 1 mL of packed bed, while a fully porous HPLC packing provides a surface area of $\sim 150\,\text{m}^2$ per 1 mL of bed volume. One can see that the contribution of the external surface of a porous particle to the total surface area is negligible.

The external surface of a packing increases in inverse proportion to the particle size. For a normal packed bed with an interstitial porosity of 40%, the formula for the surface area per volume is $3.6/d_p$. Similarly, the internal surface of a particle increases with decreasing pore size. Typical surface areas are $100\,\text{m}^2/\text{g}$ for a pore size of 30 nm, $300\,\text{m}^2/\text{g}$ for a pore size of 10 nm and $500\,\text{m}^2/\text{g}$ for a 6-nm pore size.

It is instructive to calculate the wall thickness of a typical HPLC silica with a surface area of $330\,\text{m}^2/\text{g}$ with a skeleton density of $2.2\,\text{g/mL}$. If this surface were spread out over a two-sided sheet, the thickness of the sheet would be about 2.75 nm. Similarly, if we view a silica particle as being composed of little spheres, the diameter of these spheres would be about 8 nm. The latter is the more realistic view of the structure of an HPLC particle.

If a large surface area is desirable, why wouldn't we choose the particle with the largest surface area? Unfortunately, the diffusion of a molecule in the pores of a packing slows down measurably as the pore size becomes smaller than about 10 times the size of the analyte molecule. As a result, the efficiency drops sharply. As a consequence, the best choice for general-purpose HPLC packings is a nominal pore size of approximately 10 nm. Particles designed for the chromatography of large molecules such as proteins have a nominal pore size of at least 30 nm, and often 100 nm.

All packings have a finite pore size distribution that often spans a full decade (Fig. 4.4). There is no consensus as to what is the "average" pore size. Therefore the designation of the nominal values are quite arbitrary. Two techniques are commonly used to measure the pore size distribution: mercury porosimetry and nitrogen adsorption. Their ranges overlap; mercury porosimetry can measure the pore size from the interstitial space between the particles down to 3 nm, and nitrogen adsorption is good for pore sizes from ~ 0.2 to ~ 30 nm. Mercury porosimetry is not suitable for fragile or compressive materials, like polymeric packings.

To interpret the results from pore size distribution measurements, it is usually assumed that the pores are tubular. However, this is far from the truth. The nature of the pores depends to some degree on the method used to synthesize the particles, but in all cases they form an interconnected network of irregular-shaped channels. For most preparations the nature of the pore network is much like the interstitial space formed in a packed bed. Electron microscopy is used to examine the internal structure of HPLC packings. Except for some special cases, the internal structure is homogeneous throughout the particle.

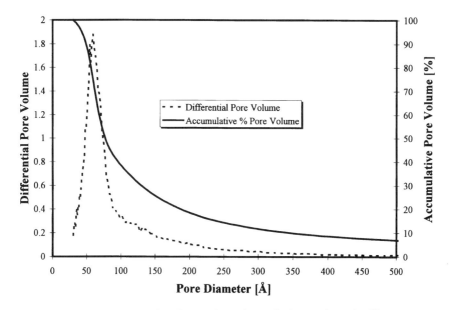

Figure 4.4   Pore size distribution of a typical HPLC-grade silica.

For size-exclusion chromatography the specific pore volume is very important, since this is where the separation takes place. If all other things are equal, a larger pore volume results in a better separation. Manufacturers usually give the specific pore volume in milliliters per gram. This is convenient, since the measurement techniques used to measure pore volume generates this number. However, the specific pore volume does not give an adequate representation of the usefulness of a packing. A material with a specific pore volume of 1.5 mL/g is not twice as useful for size-exclusion chromatography as a material with a specific pore volume of 0.75 mL/g. Also the specific pore volume does not allow for a comparison of packings made from different materials, such as polymers versus silica.

A better measure of pore volume is the particle porosity ε_{pp}. It is the fraction of the particle volume that is occupied by pores.

$$\varepsilon_{pp} = \frac{V_{pp}}{V_p} = \frac{V_{pp}}{V_{pp} + V_{sk}} = \frac{V_{sp}}{V_{sp} + \frac{1}{\rho_{sk}}} \qquad (4.4)$$

where V_{pp} designates the pore volume, V_p is the volume of the particle, V_{sk} is the volume occupied by the skeleton of the particle, V_{sp} is the specific pore volume (in mL/g), and ρ_{sk} is the skeleton density of the particle. For pure, underivatized silica, the skeleton density is about 2.2 g/mL.

The particle porosity for a silica-based particle with a specific pore volume of 0.75 mL/g is 0.62, and it is 0.77 for a particle with a specific pore volume of 1.5 mL/g. As one can see, the improvement is not as drastic as the numerical value of the specific pore volume indicated. Also, a specific pore volume of 1.5 mL/g for a styrene–divinylbenzene-based particle with a skeleton density of about 1 g/mL is not very impressive, since it translates into a particle porosity of only 0.6.

For retention chromatography, the pore volume is only a necessary evil, providing access to the internal surface. The surface area within the pores of the particle is responsible for retention, and the surface area per column volume is a measure of the retentivity of the material. The surface area per mobile-phase volume is called the *phase ratio*. As we have seen for the specific pore volume, the specific surface area, given in square meters per gram, is also a misleading indicator for either the retentivity or the capacity of a packing. In order to obtain a meaningful indicator, we can either use the ratio of the specific surface area to the specific pore volume or calculate the phase ratio from the specific pore volume, the specific surface area, and the skeleton density of the packing, assuming a normal packed bed with an interstitial fraction of 40%. The first indicator can be called the *particle phase ratio*:

$$\beta_p = \frac{A_{sp}}{V_{sp}} \qquad (4.5)$$

The true phase ratio can be calculated as follows:

$$\beta = (1 - \varepsilon_i) \frac{A_{sp}}{V_{sp} + \dfrac{1}{\rho_{sk}}} \qquad (4.6)$$

Let us calculate the phase ratio and the particle phase ratio for two silicas with significantly different properties. Nova-Pak silica has a specific surface area of about $120 \, m^2/g$ and a specific pore volume of about $0.3 \, mL/g$. Symmetry silica has a specific surface area of $330 \, m^2/g$ and a specific pore volume of $0.9 \, mL/g$. The particle phase ratio for Nova-Pak silica is $400 \, m^2/mL$; it is $366 \, m^2/g$ for Symmetry silica. The true phase ratio is about $100 \, m^2/mL$ for Nova-Pak silica and about $150 \, m^2/mL$ for Symmetry silica. Therefore the retentivity of a column packed with Symmetry silica is only about 50% larger than that of a column packed with Nova-Pak silica, not at all 3-fold as the specific surface areas would have indicated.

Also, for comparison, a styrene–divinylbenzene packing with a specific surface area of $500 \, m^2/g$ and a specific pore volume of $2.3 \, mL/g$ has a phase ratio of $90 \, m^2/mL$, if the porosity of the bed is similar to that of the silica-based particles.

In summary, to get a true impression of the capabilities of a packing, one should calculate the particle porosity in size-exclusion chromatography or the phase ratio in retention chromatography. The commonly used specifications of a packing, such as the specific surface area and the specific pore volume, are misleading measures of the performance of a packing.

4.3   POROUS VERSUS NONPOROUS PARTICLES

As we have seen at the beginning of the previous section, there is a difference of nearly 4 orders of magnitude between the external and the internal surface of a $5 \, \mu m$ fully porous particle with a pore size of around 10 nm. If we compare a 1-μm nonporous particle to a fully porous particle, the difference in surface area per column volume is still about 3 orders of magnitude. This means that the mass loadability of a packed bed with a 1-μm nonporous particle is 3 orders of magnitude lower than that of the fully porous particle with 10-nm pores.

This reasoning, however, assumes that a pore size of 10 nm is adequate for the size of the analyte. This is the case only for small molecules. For macromolecules such as proteins or nucleic acids, larger pore sizes are necessary. To have access to most of the surface of a particle, the pore size should be at least 3 times the size of the analyte, and for unrestricted diffusion, it should be at least 10 times its size. For a large protein, the pore size should therefore be around 100 nm ($= 0.1 \, \mu m$) or larger. For a particle with such a large pore size, the surface area per column volume is in the same order of magnitude as the surface area of the nonporous 1-μm particle, and the loadability advantage of the porous particle becomes small or even vanishes.

On the other hand, we have seen in Chapter 3 that for small molecules a reduction of the particle diameter below about 1 μm stops making sense because of the pressure requirements and the high linear velocities associated with the optimum of the plate height–velocity curve. However, for larger molecules with lower diffusion coefficients, the optimal linear velocity moves to smaller values in direct proportion to the diffusion coefficient. This means that the pressure is also reduced by the same factor, if we keep column size and particle size constant. Since the pressure constraints are lowered, we can reduce the particle size further, beyond that which made sense for small molecules. So while a good choice of a particle size for a small molecule was around 3 μm, it may be around 0.3 μm (300 nm) for a macromolecule.

The need for a large pore size drives us to a pore size of >0.1 μm, the desire for maximal efficiency and short analysis times drives us to a particle size of <1 μm. Both effects cooperate to make nonporous particles of a size $\sim 1 \mu$m or slightly less an attractive option for the chromatography of macromolecules. The exact point where both effects meet is a question of both the size and the shape of the macromolecule. Indeed, careful studies (7–10) have demonstrated that small nonporous particles perform well for protein and nucleic acid separations.

What about nonporous particles for the separation of small molecules? In current HPLC practice, the columns are used at or close to the minimum of the plate height–velocity curve. This is also the point of the lowest pressure for a particular combination of analysis time and plate count. In this range of the curve, the HETP is dominated by the flow-dispersion term and to a lesser degree by the longitudinal diffusion. Nonporous particles eliminate the term associated with the mass-transfer resistance in the stagnant mobile phase in the pores, which is only part of the total resistance to mass transfer. Therefore, for small molecules, the gain in efficiency of nonporous particles over porous particles is marginal. At the same time, as we have seen above, the surface area per column volume and therefore the loadability is about 3 orders of magnitudes better for porous particles. When using nonporous particles, we would need to give up this advantage for only a marginal gain in efficiency. Consequently the use of nonporous particles for small molecules does not make sense.

4.4   PARTICLE STRENGTH

High-performance liquid chromatography is high-pressure liquid chromatography. The pressure is absorbed by the particles. Therefore, they must be strong enough to withstand the pressure necessary for using the column without deformation or breakage, which would result in a collapse of the packed bed.

In a longitudinally uniform packed bed, the pressure decreases linearly with the length. This means that the forces on the particles are the same in every part of the column. We can therefore use the pressure drop per unit length to measure the strength of the particles that form the bed.

There is a fundamental difference in the behavior of polymeric particles compared to particles based on inorganic oxides such as silica and alumina. Polymeric particles will exhibit a deformation of the bed already at low pressures, in the order of 0.2–0.5 MPa/cm. When the particles are examined afterward, no damage is observed. Beds prepared from silica particles, on the other hand, do not collapse until pressures in excess of 10 MPa/cm are reached. In contrast to the polymeric particles, inorganic particles are found to have shattered, when the bed collapsed.

Column collapse by breakage of particles has been studied in detail by Groh (11). He slowly increased the flow through a column packed with silica particles until the bed collapsed. He found that for a given particle size the point of collapse could be described by either the product of linear velocity and viscosity or by the pressure divided by column length. The dependence on particle size was not as clearcut because of experimental noise, but in a good approximation the critical value was proportional to the particle size:

$$C_{\text{crit}} = \frac{u_i \eta}{d_p} = \frac{\Delta p}{L/d_p} \qquad (4.7)$$

Thus, this critical value can be viewed either as the shear on a particle or the pressure drop per layer of particles.

When the behavior of inorganic packings is studied in detail, it is found that the primary parameter that determines a particle's strength is the specific pore volume. Particles with a high pore volume break more easily than do particles with a small pore volume. However, since the pressure per unit length in most chromatographic applications rarely exceeds 2 MPa/cm, the commonly used particles with a pore volume of 1 mL/g are of sufficient strength.

This is not true for the strength of polymeric particles. While the strength of polymeric particles presents no problem in applications of size-exclusion chromatography or retention chromatography of large molecules, where perforce low linear velocities are used, it does present a limitation in retention chromatography. Newer polymeric particles exhibit improved properties, reportedly due to an increase in crosslinking, but there remains a significant gap between the inorganic and the polymeric particles.

REFERENCES

1. A. Berdichevsky, B. Alden, and U. D. Neue, unpublished data (1990).
2. I. Halász and M. Naefe, *Anal. Chem.* **44**, 76 (1972).

3. C. Dewaele and M. Verzele, *J. Chromatogr.* **260**, 13–21 (1983).

4. U. M. Schön, dissertation, Universität des Saarlandes (1984).

5. R. Endele, I. Halász, and K. Unger, *J. Chromatogr.* **99**, 377 (1974).

6. J. Ugelstad, L. Söderberg, A. Berge, and J. Bergström, *Nature* (London) **303**, 95 (1983).

7. D. C. Lommen and L. R. Snyder, *LC-GC* **11**, 222–232 (1993).

8. G. Jilge, K. K. Unger, U. Esser, H. J. Schäfer, G. Rathgeber, and W. Müller, *J. Chromatogr.* **476**, 37–48 (1989).

9. J. J. Kirkland, *Anal. Chem.* **64**, 1239–1245 (1992).

10. K. Kaghatgi and Cs. Horváth, *J. Chromatogr.* **398**, 335–339 (1987).

11. R. Groh, Diplomarbeit, Universität des Saarlandes (1975).

5 Column Packing and Testing

It must be precisely thus, and we know all about it.

— E. A. Abbott, *Flatland*

The ability to pack columns with particles of a diameter of 10 μm or less was a major milestone in the history of HPLC. Prior to the development of this key enabling technology, HPLC columns were packed with $\geqslant 30$-μm particles using vibration and tapping procedures that had originally been developed for gas-chromatography columns. This technique failed for particles under 30-μm because of the increased interaction of particles with each other. Efficiencies were low, reproducibility was poor, and the columns often exhibited distorted peak shapes. Although slurry-packing techniques using high-pressure filtration were not unknown, there was concern over the settling of the dense silica particles ($\rho = 2.2$ g/mL) in normal solvents. Kirkland (1) explored first the suspension of particles in a mixture of high-density solvents for the high-pressure packing of controlled surface porosity supports with a particle size of ~ 40 μm. The same principle was then employed by Majors (2) and Strubert (3) for 5–10-μm porous silicagel particles using solvent mixtures that had the same density as the silica itself. Good column efficiencies were achieved reproducibly, and high-performance liquid chromatography was born.

A necessary component of the balanced-density slurry solvent was tetrabromoethane, which has a density higher than that of silica. However, it decomposes easily and its toxicity is a major drawback, especially for commercial column production. Therefore alternatives were sought quickly. In 1972, Kirkland (4) described a technique in which silica particles were suspended in an aqueous ammonium hydroxide solution, and in 1974 Asshauer and Halász (5) reported success with slurries based on high-viscosity solvents.

5.1 COLUMN PACKING

5.1.1 Principles

High-pressure filtration has remained the technology of choice for the packing of HPLC columns, but the details of the column packing techniques used today are significantly different from the ones reported in the early literature. Neither the balanced-density technique nor the high-viscosity technique were ever compatible with routine column manufacturing, and the industry immediately developed other approaches. The industrial packing techniques, however, were closely guarded secrets, and the knowledge of the underlying principles remained proprietary. Meanwhile the technology was viewed as a mysterious "art" in academic circles (6). In this section we will examine some of the basic principles of column packing technology, taking into account the latest publications on the subject as of the time of this writing.

Column packing by any "slurry packing" technique is a high-pressure filtration technique: the solvent carrying the particles is pushed through the already formed bed, resulting in a further buildup of the bed. Therefore the properties of the slurry, the properties of the packed bed and the filtration regime all should have an influence on the result.

The early investigators of the technology were concerned about the settling of the particles in the slurry. To get an initial understanding of the packing process, let us therefore calculate the settling velocity of a typical HPLC packing and compare it to the filtration velocity through a fully formed bed. This will give us a handle on the relevance of the concern about the settling of particles during column packing.

The sedimentation velocity of a porous particle in a solvent can be estimated by the Stokes equation:

$$u_s = \frac{1}{18} \frac{d_p^2}{\eta} \varepsilon_{sk}(\rho_{sk} - \rho_l) \tag{5.1}$$

where d_p is the particle diameter, η is the viscosity, ε_{sk} is the fraction of the particle occupied by the particle skeleton, ρ_{sk} is the density of the skeleton, and ρ_l is the density of the liquid.

For a 10-μm porous silica particle with a particle porosity of 75%, the calculated settling velocity in water is 1 mm/min. From Equations (2.47) and (2.48) we can calculate that at a pressure of 40 MPa, the linear velocity of water through a 30-cm column packed with the same 10-μm particles is 1.33 cm/s. This is about 800 times larger than the settling velocity. Thus the cake formation velocity is some 3 orders of magnitude faster than the sedimentation velocity of the particles. This shows that the concern over the settling of particles during column packing is unwarranted. Therefore it is not necessary to try to match the density of the particle with the density of the slurry solvent.

Let us now examine the influence of the viscosity of the slurry solvent. The settling velocity of a particle decreases with the viscosity of the solvent. But also, the filtration rate through an already formed bed decreases with increasing velocity. Moreover, the ratio of the settling velocity to the filtration velocity is independent of the viscosity of the slurry solvent. Consequently, if particle settling were of concern in column packing, an increase in the viscosity of the slurry solvent would do us no good. Packing with a high-viscosity solvent only increases packing time.

If neither solvent density nor viscosity are important in column packing, then what is? The missing factor is the interaction of the particles with each other, which is mediated by the slurry solvent. This is the factor that had been recognized by Kirkland (4) when he developed the packing technique using a stabilized aqueous suspension of silica particles. The technique as implemented by Kirkland, however, works only for unmodified silica, and cannot be used for bonded phases without modification of the procedure. But the underlying principle is correct.

If the particle–particle interaction is stronger than the particle–solvent interaction, the particles will flocculate and particle aggregates will settle rapidly. On the other hand, if there is no attraction or even a repulsion between particles, the particles will settle as individuals and the slurry is more stable. This phenomenon can be studied conveniently by particle size analysis by sedimentation (7). If particles flocculate, they settle rapidly and form a loose sediment. If there is no flocculation, settling is slow and the sediment is dense. Thus, we can determine whether a slurry solvent flocculates a packing by either comparing the density of the sediment obtained with different solvents (8–10) or by measuring the sedimentation velocity and comparing it to the theoretically derived settling velocity (11).

The importance of flocculation is debated in the literature. Shelly and Edkins (9) propose that flocculation of the slurry should be avoided and that slurry solvents with minimal agglomeration of particles yield the best columns. Vissers et al. (11) suggest that the slurry solvent is not important, at least for the packing of microcolumns. However, we will follow the reasoning of Shelly and Edkins.

Another point of discussion in the literature is how the filtration is carried out: by application of constant pressure or by using a constant rate. For silica-based particles, good results have been reported with both techniques. Thus the filtration technique can be selected freely depending on the available equipment. It should be pointed out, though, that the filtration rate should not exceed the strength of the particles as discussed in Chapter 4.

While the basic filtration technique does not seem to matter, the details of the technique matter a lot. The study by Meyer and Hartwick (12), for example, demonstrated that the slurry concentration, the flow rate (at constant rate filtration), and the pressure (at constant pressure filtration) all influence the plate count of the column. In many cases, an optimal setting was found for one parameter, while the other parameters were held constant.

The column packing studies found in the HPLC literature are usually concerned with the packing of hard, silica-based particles. Little attention is given to the packing of polymeric particles, which are more compressible and also can swell and shrink. All the considerations mentioned above apply to polymeric particles as well. Because of the lower density of polymeric particles, settling can be eliminated fairly easily. But one does have to pay attention to the swelling and shrinking of the particles. Most polymer-based HPLC packings are highly crosslinked. Consequently swelling and shrinking is minimal, maybe around 10%. This is still sufficient to cause the collapse of a bed upon shrinking or an undue decrease in permeability upon swelling. This behavior needs to be taken into consideration when choosing a slurry solvent for polymeric packings.

5.1.2   Apparatus

The apparatus for slurry packing consists basically of a slurry reservoir attached to a high-pressure pump. All parts of the equipment need to be compatible with the intended pressure for column packing. In many cases, the pressures used are larger than those normally encountered in HPLC. Therefore, safety is a primary concern. You should check with the manufacturer of your column hardware and the fittings used for building the equipment as to the pressure limitations. Ferrules can be a weak link. The ferrule that is used to attach the column to the slurry reservoir is a point of concern. What is the pressure to which the manufacturer recommends the use of this ferrule? If you are repacking used columns, you should check whether the ferrule is in a good shape or is worn from frequent use. Also, if small-diameter tubing is used to connect the pump to the slurry reservoir, you want to make sure that the tubing is not kinked or otherwise worn, resulting in the danger of rupture. A fluid stream shooting at high velocity out of a 0.5-mm tubing can easily penetrate the skin and inject a toxic solvent into the body.

One can use an HPLC pump for column packing. These pumps are designed to deliver a constant flow at pressures of up to $\sim 40\,\mathrm{MPa}$. Also, the flow rate can be programmed, if one so chooses. In small labs where columns are packed only occasionally , this may be the most cost-effective approach.

An alternative frequently described in the literature is the use of air-driven amplifier pumps, like the ones supplied by Haskel (13,14). They supply a much higher flow than HPLC pumps and are also capable of much higher pressures.

Many different designs for the slurry reservoirs have been proposed. Several publications deal with reservoirs that allow for a stirring of the slurry during packing to prevent settling of the slurry (7,15). As I have pointed out before, settling is of little concern. On the other hand, the stirring of the slurry in the reservoir has some disadvantages. Because the slurry is continuously diluted with the solvent entering the slurry reservoir, the packing process takes longer and becomes more complicated. Therefore a straightforward design of the slurry reservoir is best. Also, for simplicity, I advocate simple downward packing rather than upward packing as proposed in some publications (e.g., 6).

In the construction of the slurry reservoir and in the connections of the slurry reservoir to the column it is important that there be no horizontal surfaces where the packing can accumulate. A very good design of a packing chamber and connections is described in Reference 13. Both the body and the bottom of the chamber are modular, making it possible to pack columns with different end fittings and different volumes. This reference also includes good information on column preparation, including honing, polishing, and degreasing. Another interesting design can be found in Reference 16 — the cap of the reservoir can be opened and closed fairly rapidly, which allows for the use of slurries based on low-viscosity solvents.

The connection of the column to the slurry chamber should be made through a precolumn, that has the same diameter as the column to be packed. There are several reasons for this; two practical ones are immediately obvious:

1. It is difficult to accurately determine the amount of material necessary to fill a column; therefore we always use an excess of packing.
2. The presence of the packed precolumn prevents the contents of the slurry chamber from pouring over the operator's hands while disconnecting the packed column. (This is one point where upward packing has an advantage.)

An inexpensive portable system is described in Reference 14. The system can be built from easily available components, but is limited in its use. One should also review carefully the pressure up to which this system can be used safely.

Several systems are described in the literature that allow for the parallel packing of several columns. The design of such a system is, however, more of interest for commercial column packing than for the occasional column packing needed by chromatographers. Therefore a discussion of this kind of apparatus is outside the scope of this book.

5.1.3   Slurry Preparation

The first step in slurry preparation is the selection of the slurry solvent. If the manufacturer of the packing suggests a particular solvent or solvent mixture, you can save yourself a lot of work by following the manufacturer's recommendations. If this information is not available, you can follow the thought process developed in this section to select an appropriate slurry solvent yourself.

Consideration should first be given to the viscosity of the slurry solvent. If the handling time of the slurry — including the residence time in the packing chamber before the start of the filtration — is long (i.e., minutes rather than seconds), you should select a slurry solvent of intermediate viscosity, between 1 and 5 cP. While we have pointed out above that the viscosity of the slurry solvent is irrelevant for the packing process from a fundamental standpoint, an intermediate viscosity can still prevent undue settling of the slurry before the actual packing process starts. If, on the other hand your equipment allows for a rapid start of the packing process, you should select a low-viscosity solvent ($\leqslant 1\,cP$), since the time required for packing is directly proportional to the viscosity of the slurry.

You also need to consider the chemistry of the packing and its compatibility with the solvent. For example, since primary amines react easily with ketones to form imines, you should not use acetone or methyl ethyl ketone as a slurry solvent for bonded phases containing primary amino groups.

If you are dealing with a polymeric packing, you need to consider the swelling and the shrinking of the packing in the slurry solvent relative to the mobile phase in which the column will be used. For example, when preparing size-exclusion columns based on polymeric packings, it may be convenient to pack the column in the mobile phase. Polymer-based ion exchangers exhibit swelling in deionized water and are therefore best slurried up in a salt or buffer solution.

Next you should select among the remaining choices a solvent or solvent mixture that prevents the agglomeration of the particles. That means that you should determine which solvents are flocculating and which are deflocculating. As we have pointed out above, there are two tests that can be applied. A slurry prepared in a flocculating solvent will settle faster than in a nonflocculating solvent of equal density and viscosity. It also will form a looser cake than the one obtained in a nonflocculating solvent. Therefore we can compare either the settling velocity of a slurry to the theoretical settling velocity without floccula-tion or the height of the settled cake in different solvents.

Let us look at the settling velocity first. The settling velocity of a particle in a slurry is different from the velocity calculated by the Stokes equation. Because of the counterflow of the solvent, it depends on the volume fraction ϕ of the particles in the slurry (11):

$$u_s = \frac{(1 - \phi)^{-\kappa_2} d_p^2 \varepsilon_{sk} (\rho_{sk} - \rho_l)}{18 \eta} \qquad (5.2)$$

where the coefficient κ_2 is an empirical factor. With the exception of the first term, this equation is identical to Equation (5.1). If the pores are not filled with liquid, but contain some amount of air, further corrections need to be introduced (11):

$$u_s = \frac{(1 - \phi)^{-\kappa_2} d_p^2 [\rho_{sk} \varepsilon_{sk} + \rho_l (\varepsilon_f - 1)]}{18 \eta} \qquad (5.3)$$

where ε_f is the fraction of the particle filled with liquid. With the help of these equations and the knowledge of the porosity and the density of the particles, one can compare the actual settling velocity with the settling velocity calculated by Equation (5.2) or (5.3).

Alternatively, the volumes of the settled cake obtained in different solvents can be compared. For that purpose, a fixed amount of packing is slurried up in a fixed amount of solvent and allowed to settle completely. A detailed procedure is given in Reference 10. The solvent that exhibits the smallest cake forms the slurry with the least amount of particle agglomeration.

Either technique results in a determination of which slurry solvents tend to agglomerate particles and which ones don't. This is different for every type of packing, depending on the surface chemistry of the packing. For the ODS material studied by Shelly and Edkins (9), the most deflocculating solvents were acetone, methyl ethyl ketone, and tetrahydrofuran, while methanol flocculated the packing. The results by Vissers et al. (11) are similar. Nevertheless, the optimal solvent may be slightly different for different brands of the same type of packing due to subtle differences in the surface chemistry of different brands.

After the solvent is chosen, the slurry concentration needs to be selected. As shown in Reference 11, there is an optimal value for the slurry concentration that depends on the other parameters of the packing procedure. Therefore, the slurry concentration that results in the best column needs to be determined empirically. The concentration that you start with might be dictated by the packing equipment, but if you are free to choose, a concentration of 10% v/v is a good starting point.

The actual preparation of the slurry is straightforward: the packing is combined with the solvent and stirred vigorously. Some investigators recommend putting the slurry into an ultrasonic bath to remove air from the pores. Considering the pressures involved in slurry packing, this reasoning is questionable. On the other hand, sonication does no harm as long as it is not excessive: particle breakage may occur during lengthy sonication.

We should always use a slight excess of packing over what is needed to fill the column. A 10% excess is usually adequate. This is insurance that the column is really filled. It also packs the precolumn, which prevents a spill of the contents of the slurry chamber when the column is disconnected from the equipment.

In the case of polymeric packings, one might want to allow some time for the equilibration and swelling of the packing in the slurry solvent. In our experience, a few minutes are sufficient.

5.1.4 Packing Techniques

We have discussed the equipment and the preparation of the slurry. In this section we will discuss the filtration technique itself. There are two basic types of packing techniques: under constant pressure and using constant flow. Both techniques have been used successfully for hard particles, such as silica-based packings. However, in a constant-pressure technique, the initial flow rate is very high and the load on the particles may exceed their strength. For this reason, constant-flow packing techniques are preferred for at least soft or fragile particles. For hard particles, constant-pressure techniques are feasible and give results that are as good as constant-flow techniques. Most packing procedures for silica-based packings found in the literature are based on constant pressure.

The goal of the packing process is to create a packed bed of maximal radial uniformity. To work out an optimal packing procedure is an empirical exercise. Slurry concentration, slurry viscosity, and flow rate interact with each other. Usually, the choice of the slurry solvent is fixed on the basis of the flocculation behavior of the slurry. We can then use the simultaneous variation of both the slurry concentration and the flow rate to optimize column performance. Often, variation of one of the two parameters results in an optimum, when the other is held constant (12). Often, more than one optimum can be found, where equal column performance is achieved. To find an optimum, one parameter is varied and the response to the variation is recorded. Both column efficiency and peak asymmetry should be measured. Normally, peak asymmetry will change from fronting to tailing or vice versa, while column efficiency will go through a maximum. Peak asymmetry is the prime indicator of bed uniformity. The parameter space of column packing can also be explored using a statistically designed experimental matrix.

The packing process itself should be continuous. An interruption of the process usually results in inferior columns. Although some early publications report a stressing of the column by rapidly increasing and releasing the pressure, this practice does not yield reproducible results.

Up to now, we have entirely ignored the solvent that is used to displace the slurry in the slurry reservoir. Ideally, the solvent used to prepare the slurry should also be used to displace the slurry. However, there are some practical limitations; for instance, the materials of construction in the pump may not be compatible with the slurry solvent. This is not an issue if HPLC pumps are used, but is of serious concern for air-driven pumps that may use elastomeric seals, which usually have a limited solvent compatibility. In this case a different solvent needs to be chosen to push the slurry into the column. This solvent has several different requirements to fulfill:

- It should be compatible with the material of construction in the fluid path of the pump.
- To prevent mixing with the slurry, it should have a lower density than the slurry solvent.
- If we let the push solvent enter the packed column, it should be compatible with the packing and miscible with the slurry solvent.

In some packing experiments reported in the literature, not all these criteria were fulfilled. For example, push solvents with a higher density than the packing solvent have been used, which then obviously mixed with the slurry and influenced the properties of the slurry, for example, by changing its flocculation characteristics. Also, some studies reported that the shape of the packing chamber has an influence on column performance. This is most likely due to the interaction of the push solvent with the slurry, the degree of which can be influenced by the geometry of the chamber. The mixing behavior of the push solvent and the slurry can easily be observed in a glass graduate.

5.2 COLUMN TESTING

The primary purpose of a column testing procedure in the context of column packing is to verify the quality of the packed bed. This information can be used to optimize a packing procedure or to verify the quality and reproducibility of an already established packing procedure. Therefore we need a procedure that is sensitive to differences in the packed bed, but insensitive to other parameters such as extracolumn effects or the chemistry of the packing. Also the test procedure should provide results fairly quickly. We can therefore write down a few requirements for such a testing procedure:

1. The equilibration with the mobile phase should be fast. This means that we should choose an analyte as test marker, whose interaction with the stationary phase is simple and does not depend on the concentration of small impurities in the mobile phase. For example, we would choose an aromatic hydrocarbon as a test compound for reversed-phase packings. For silica columns, one should use a mobile phase with a controlled water content to speed up equilibration.
2. The largest influence of the column packing process on the quality of the column is observed at the minimum of the plate height–velocity curve. Therefore we should test the column close to this minimum. The flow rate at which this minimum occurs is a function of the viscosity of the mobile phase; a lower viscosity allows for a higher flow rate and therefore a faster column test. Therefore, acetonitrile/water mixtures are preferred over methanol/water mixtures in reversed-phase chromatography.
3. The peak volume should be large enough to be free from distortions by extracolumn effects. This is best accomplished by an adjustment of the retention factor of the test compound.

Suitable samples for reversed-phase columns packed with C_{18}, C_8, phenyl, or hexyl packings are aromatic hydrocarbons such as toluene, xylene, cumene, naphthalene, acenaphthene, and anthracene and long-chain phthalates such as dihexyl phthalate or dioctyl phthalate. These samples give large retention factors and symmetric peaks with acetonitrile/water mobile phases of high acetonitrile content. Typical mobile phases contain 50–70% acetonitrile. The optimal linear velocity depends on the particle size, the mobile phase composition, and the sample chosen, as well as on the quality of the packing. For a 5-μm column, 1 mm/s is a good starting point. For CN columns used in reversed-phase mode, the same samples can be used, but more polar mobile phases are required, containing 30–50% acetonitrile. The same holds true for other short-chain reversed-phase packings such as C_4, C_2, or C_1.

Suitable samples for silica columns are acetanilide, short-chain phthalates such as dimethyl phthalate or diethyl phthalate and anisaldehyde. They exhibit weak interaction with the silica surface that prevents tailing. Good retention factors are obtained with methylene chloride-based or hexane-based mobile phases with small amounts of methanol, isopropanol, or acetonitrile. Examples are hexane with 0.5% acetonitrile, methylene chloride with 0.5% isopropanol, or methylene chloride with 0.1% methanol. To ensure rapid equilibration with the mobile phase, the mobile phase should be half-saturated with water. For a 5-μm column, a good linear velocity in these solvents is between 2 and 3 mm/s. Nitroaromatics are used sometimes as samples for the testing of silica columns, but they may exhibit peak tailing depending on the pretreatment of the silica.

Polar bonded phases such as propylamino-, cyano-, or diol packings used in normal-phase chromatography can be tested exactly like silica columns. Amino-bonded phases used for hydrophilic interaction chromatography are best tested with a carbohydrate sample in 70% acetonitrile 30% water.

Silica-based ion-exchange packings designed for small molecule separations can be tested with benzoic acid or related compounds for anion exchangers and phenylethylamine and related samples for cation exchangers. Columns for ion chromatography should be tested with the ions for which they were designed such as sulfate or nitrate for an anion-analysis column. Ion-exchange columns designed for proteins are best tested with a neutral unretained marker that does not interact with the packing (e.g., dihydroxyacetone).

Size-exclusion packings are usually tested in the target mobile phase using a totally permeating marker. Since this mode of chromatography is often carried out using a refractive-index detector, a broader choice of samples is available. For aqueous size-exclusion chromatography, a common marker is D_2O. If a UV detector is used, a good neutral, polar, noninteracting UV-absorbing test compound is dihydroxyacetone. However, this sample has a limited shelf life and is best prepared fresh. For organic size-exclusion chromatography, typical markers are toluene or o-dichlorobenzene, which are both readily available in the polymer-characterization laboratory. Specialty packings like chiral and other packings are best tested following the manufacturer's recommendations.

The concentration of the test sample and the amount injected should be chosen such that the column is not overloaded. For the samples proposed for the testing of reversed-phase and normal-phase columns, which are all good UV absorbers, a concentration of about 100 μg/mL and an injection volume of 1–10 μL are recommended for a standard HPLC column. Another indicator for a correct choice of injected mass is a peak height that does not exceed 0.1

AUFS. The retention of the sample should be adjusted to minimize extracolumn influences on the measurement. If we would like less than a 5% influence of extracolumn band spreading on the plate-count measurement, the standard deviation of the peak should be about 5 times larger than the standard deviation of the extracolumn band spreading (see Section 3.2.2).

These tests are used to calculate the HETP, the column plate count, and the peak asymmetry. A detailed description of the various calculation procedures is given in Section 2.1. If we measure the HETP close to the optimum, we should expect a value smaller than four particle diameters for an acceptable column and between two and three particle diameters for a good column. A full characterization of the column by a complete van Deemter plot is too much work for simple column testing, but should be considered for future reference once a packing procedure has been optimized. This allows for a check of the chromatographic properties of a new batch of packing against the properties of the one that was used to establish the packing procedure.

Every column test should include a recording of the column backpressure and a determination of column permeability (see Section 2.2.5). The column permeability should be compared to the expected value for the particle size ($\sim d_p^2/1000$). A permeability that is lower than expected indicates that the particles are compressed or crushed during the packing procedure, or that the packing contains excessive amounts of fines. Section 2.2.5 contains viscosity data for most solvents of interest in HPLC, which are needed for calculating permeability.

5.3 SUMMARY

As we have seen, many different elements need to be taken into account when developing a packing procedure. Also, the testing procedure plays an important role. Ultimately, the results of column packing can only be as good as the test procedure allows us to measure.

If we do everything right, it is not difficult to achieve reduced plate heights of <3. If we are dealing with a high-quality packing, reduced plate heights of ~ 2 are possible. Also, the reproducibility of a well-designed packing procedure is high. In commercial production, a high-quality manufacturer can achieve relative standard deviations of $<5\%$, in some cases $<3\%$, for plate count and peak asymmetry. The permeability is reproducible within the measurement error of the test instrument. Of course, this level of quality and reproducibility may be more difficult to achieve, when only a limited number of columns are packed. Considering this, one can question whether the development of a packing procedure in a user's lab makes sense at all, if the same column is available commercially. While column packing by the user was popular in the early times of HPLC, the chromatographers today prefer professionally packed columns.

REFERENCES

1. J. J. Kirkland, *J. Chromatogr. Sci.* **9**, 206 (1971).
2. R. E. Majors, *Anal. Chem.* **44**, 1722 (1972).
3. W. Strubert, *Chromatographia* **6**, 50 (1973).
4. J. J. Kirkland, *J. Chromatogr. Sci.* **10**, 593 (1972).
5. J. Asshauer and I. Halász, *J. Chromatogr. Sci.* **12**, 139 (1974).
6. M. Verzele and C. Dewaele, *LC-GC* **4** 614–618 (1986).
7. M. Broquaire, *J. Chromatogr.* **170**, 43–52 (1979).
8. U. D. Neue, unpublished results (1975).
9. D. C. Shelly and T. J. Edkins, *J. Chromatogr.* **411**, 185–199 (1987).
10. D. R. Absolom and R. A. Barford, *Anal. Chem.* **60**, 210–212 (1988).
11. J. P. C. Vissers, H. A. Claessens, J. Laven, and C. A. Cramers, *Anal. Chem.* **67**, 2103–2109 (1995).
12. R. F. Meyer and R. A. Hartwick, *Anal. Chem.* **56**, 2211–2214 (1984).
13. T. J. N. Webber and E. H. McKerrel, *J. Chromatogr.* **122**, 243–258 (1976).
14. E. J. Kikta, *J. Liquid Chromatogr.* **2**, 129–144 (1979).
15. H. R. Linder, H. P. Keller, and R. W. Frei, *J. Chromatogr. Sci.* **14**, 234 (1976).
16. G. E. Berendsen, R. Regouw, and L. de Galan, *Anal. Chem.* **51**, 1091–1093 (1979).

ADDITIONAL READING

Kaminski, M., J. Klawiter, and J. S. Kowalcsyk, "Characteristics of the Tamping Method for Packing Preparative Columns; Part II," *J. Chromatogr.* **243**, 225–244 (1982). (characteristics of the slurry method of packing chromatographic columns).

Klawiter, J., M. Kaminski, and J. S. Kowalczyk, "Investigations of the Relationship between Packing Methods and Efficiency of Preparative Columns, Part I," *J. Chromatogr.* **243**, 207–224 (1982).

OTHER LITERATURE RELATED TO COLUMN PACKING

Cassidy, R. M., D. S. Le Gay, and R. W. Frei, "Study of the Packing Techniques for Small-Particle Silica Gels in High-Speed Liquid Chromatography," *Anal. Chem.* **46**, 340–344. (1974).

Coq, B., C. Gonnet, and J. L. Rocca, "Techniques de remplissage de colonnes de chromatographie liquide à haute performance avec des microparticles poreuses," *J. Chromatogr.* **106**, 249 (1975).

Keller, H. P., F. Erni, H. R. Linder, and R. W. Frei, "Dynamic Slurry Packing Technique for Liquid Chromatography Columns," *Anal. Chem.* **49**, 1958–1963 (1977).

Bristow, P. A., and J. H. Knox, "Standardization of Test Conditions for High-Performance Liquid Chromatography Columns," *Chromatographia* **10**, 279–289 (1977).

Elgass, H., H. Engelhardt, and I. Halász, "Reproduzierbare Methode zum Packen von Trennsäulen für die Chromatographie mit Kieselgel ($5-10\,\mu$m)," *Fresenius Z. Anal. Chem.* **294**, 97–106 (1979). (includes a table describing the details of published packing methods including the reported results from 1972 to 1978).

Kuwata, K., M. Uebori, and Y. Yamazaki, "Rapid Method for Packing Microparticulate Columns Packed with a Chemically Bonded Stationary Phase for High-Performance Liquid Chromatography," *J. Chromatogr.* **211**, 378–382 (1981).

Dewaele, C., and M. Verzele, "Influence of the Particle Size Distribution of the Packing Material in Reversed-Phase High-Performance Liquid Chromatography," *J. Chromatogr.* **260**, 13–21 (1983).

Gluckman, J. C., A. Hirose, V. L. McGuffin, and M. Novotny, "Performance Evaluation of Slurry-Packed Capillary Columns for Liquid Chromatography," *Chromatographia* **17**, 303–309 (1983).

Davis, R. H., and A. Acrivos, "Sedimentation of Noncolloid Particles at Low Reynolds Numbers," *Ann. Rev. Fluid Mech.* **17**, 91–118 (1985).

Shelly, D. C., V. L. Antonucci, T. J. Edkins, and T. J. Dalton, "Insights into the Slurry Packing and Bed Structure of Capillary Liquid Chromatograhy Columns," *J. Chromatogr.* **458**, 267–279 (1989).

Wang, T., R. Hartwick, N. T. Miller, and D. C. Shelly, "Packing of Preparative High-Performance Liquid Chromatography Columns by Sedimentation," *J. Chromatogr.* **523**, 23–34 (1990).

Farkas, T., J. O. Chambers, and G. Guiochon, "Column Efficiency and Radial Homogeneity in Liquid Chromatography," *J. Chromatogr.* **A679**, 231–245 (1994).

Sarker, M., and G. Guiochon, "Study of the Packing Behaviour of Radial Compression Columns for Preparative Chromatography," *J. Chromatogr.* **A683**, 293–309 (1994).

Zimina, T. M., R. M. Smith, P. Myers, and B. W. King, "Effects of Kinematic Viscosity of the Slurry on the Packing Efficiency of PEEK Microbore Columns for Liquid Chromatography," *Chromatographia* **40**, 662–668 (1995).

Baumeister, E., U. Klose, K. Albert, E. Bayer, and G. Guiochon, "Determination of the Apparent Transverse and Axial Dispersion Coefficients in a Chromatographic Column by Pulsed Field Gradient Nuclear Magnetic Resonance," *J. Chromatogr.* **A694**, 321–331 (1995).

6 Column Chemistry

Hear me, all ye hallowed beings,
both high and low, of Heimdall's children:
thou wilt, Valfather, that I well set forth
the fates of the world which as first I recall.

         — Voluspa, Edda, *translated by Lee M. Hollander*

In chapter 4 we discussed the physical properties of chromatographic adsorbents. In this chapter we will discuss their chemical properties. The most important aspect of the chemistry of a packing is the character of the adsorbing surface. But the chemistry of the packing also plays a role with respect to its hydrolytic stability or whether and to what degree it shrinks and swells in various solvents. In addition, the chemistry of the packing influences its physical strength. The chemistry of the surface influences adsorption kinetics and mass transfer.

Currently, we can distinguish between two major types of packings: those based on inorganic supports and those based on organic polymers. We will discuss the properties of both classes as they present themselves today. We will then explore in detail the properties of silica-based bonded phases, which represent by far the most widely used HPLC packings. This discussion will also include other bonding techniques that have been derived from classic bonded phases. Then we will examine the chemistry of polymeric packings.

In Section 6.4, we will investigate other surface modification techniques that have been developed for inorganic packings and do not rely on the attachment of the interacting layer to the surface of the packing. This area of HPLC technology is still in flux, and new technical developments can be anticipated.

6.1 INORGANIC VERSUS POLYMERIC PACKINGS

Classic HPLC packings are based on silica. Silica-based packings are very hard (= mechanically strong). They can easily withstand the high pressures encountered in HPLC without measurable compression or breakage. They also do not swell or shrink when exposed to different solvents. In fact, the hardness of the silica was an important enabling factor in the development of HPLC. The other important factor was the availability of a surface modification technology that allowed the design of chemically stable surfaces with a wide range of adsorptive properties. The interacting layers on these surfaces are very thin, on the order of a monomolecular layer. This provides for fast mass transfer of analytes to and into the adsorbing layer, which results in narrow peaks at short analysis times.

HPLC packings are generally porous, to maximize the surface area available for interaction with the analyte. The surface inside the pores must be accessible to the analyte without excessive restriction of diffusion. The technology to manipulate the pore size, surface area, and pore volume of silica was developed quickly and is well understood today. Thus the pore size may be tailored to the size of the analyte, and packings can be created and optimized for small molecules as well as for polymers.

Silica would be the dream packing for HPLC, were it not for one major disadvantage — silica dissolves at alkaline pH values. How rapidly this occurs depends on the nature of the stationary phase, the pH and the type of the buffer, the nature and content of the organic modifier in the mobile phase, and the temperature. Also, the time that it takes for the column to collapse may depend on the specific pore volume of the silica; packings with a small pore volume (and therefore more massive skeleton) appear to be affected less than packings with a high pore volume. Furthermore, the type of bonded phase and the surface coverage play an important role in the hydrolytic stability of a packing. So while column manufacturers usually recommend against using bonded-phase columns beyond pH 8, this is a very soft limit. Columns have been run successfully at pH 9 for extended periods of time at room temperature in mobile phases containing a large percentage of organic modifier in which the solubility of silica is reduced. On the other hand, columns have been found to collapse after only weeks, when phosphate buffers at pH 8 were used at elevated temperature.

Nevertheless, chromatographers would like to be able to expand the use of chromatographic packings into the alkaline pH range. Variation of the pH is one of the most powerful tools in the manipulation of chromatographic selectivity. Polymeric packings such as those based on vinyl monomers are stable over a broader range of pH values. Styrene–divinylbenzene polymers, especially, are stable over the entire pH range from pH 0 to pH 14. Styrene–divinyl–benzene-based packings have been used since the 1960s for the size-exclusion chromatography of polymers and are therefore available in a broad range of pore sizes. They are also manufactured in the same particle-size range as silica-based packings. The same is true for other polymeric packings based on methacrylates or polyvinylalcohol. Chemical modification of the surface of these polymeric particles are possible, too. Then why are they not used more frequently in HPLC?

The first issue is that polymeric packings are softer than inorganic ones and are not able to withstand pressures quite as high. Nevertheless, we rarely need the full capability of the inorganic packings, and fast separations are possible at pressures under 15 MPa, which is quite compatible with some of the better polymeric packings. The second issue is that in one-for-one comparisons of polymeric and bonded-phase silica-based packings, polymeric packings always exhibit significantly lower efficiency (in the retention chromatography of small molecules) than do the silica-based packings. This is due to hindered mass transfer into the stationary phase, which plagues all currently available polymeric packings. To date, no solution has been found to this problem, despite the fact that polymeric materials of quite different makeup have been explored.

A third drawback of polymeric packings is that they imbibe solvent and therefore swell and shrink depending on the mobile phase in which they are used. Since the HPLC-grade packings are usually highly crosslinked, the swelling and shrinking is limited, on the order of $\leqslant 10\%$ of the particle size. Although this change is not much, it is still enough to cause the collapse of a column when the mobile phase is changed from a swelling solvent to a nonswelling solvent. In actuality, this problem can be handled by restricting the use of a column to a range of mobile phases and by selecting the optimal packing solvent for this range. Swelling and shrinking is only a minor problem in practice.

The dual disadvantage of polymeric packings — reduced pressure capability and hindered mass transfer for small molecules — has limited their use to applications where either these issues are not important, or the selectivity of the separation is sufficiently unique to compensate for these disadvantages. For large molecules such as proteins, the hindered mass transfer is not observed. Also, because to the slow diffusion of large molecules, the velocities (and therefore pressures) used are much smaller than for the chromatography of small molecules. Therefore, polymer-based packings dominate the field of biopolymer separations and size-exclusion chromatography. Examples of specialty applications that successfully employ polymeric packings are ion-chromatography applications, the analysis of oligosaccharides by size-exclusion chromatography, and the separation of carboxylic acids by ion-exclusion chromatography on cation exchangers.

Attempts have been made to design packings with an expanded pH compatibility compared to silica, but with a hardness comparable to silica. Other inorganic carriers such as alumina, titania, and zirconia have been explored. Indeed, their hardness matches that of silica, and being impervious to small molecules, they also exhibit the same advantageous mass-transfer properties as silica. However, no simple surface modification techniques are available as yet that match the silanization chemistry used for silica. Therefore, polymeric coatings have been used, which then in turn exhibit inferior mass-transfer behavior.

A packing with all the advantages of silica, but with an expanded pH range is still the holy grail of HPLC.

6.2   SILICA-BASED BONDED PHASES

The surface of silica is composed of silanols, $\equiv$Si—OH, and siloxane bridges, $\equiv$Si—O—Si$\equiv$. Several different types of silanols can be distinguished spectroscopically: single silanols versus geminal silanols (silanediols), and bridged silanols versus lone silanols:

Single   versus   geminal   bridged   versus   lone

Silanols

Bridged silanols are also called *vicinal silanols*. Single silanols that are not bridged are called *lone silanols*.

These silanols are the adsorption centers in normal-phase chromatography. They are very hydrophilic, that is, they interact with polar functional groups of mobile-phase and analyte molecules. They are also the reactive centers that are used in the surface modification of silica. The siloxane bridges, on the other hand, are rather hydrophobic and largely unreactive.

Silanols are acidic. The surface of a fully hydroxylated silica without metal impurities is uncharged around pH 3. At higher pH values, the silanols dissociate, leaving a negative charge on the surface. This makes silica suitable for ion-exchange chromatography (it is a weak cation exchanger), but in bonded phases the interaction of dissociated silanols with positively charged analytes may result in tailing peaks. The acidity (or activity) of the silanols depends on their chemical environment. While there has been much debate on which type of silanol is the most acidic, the most recent evidence suggests that lone silanols are more acidic than hydrogen-bridged silanols. Furthermore, metal impurities incorporated into the silica matrix, especially trivalent iron and aluminum, contribute to the acidity of surface silanols. These impurities give rise to acidic sites as shown schematically in the following diagram:

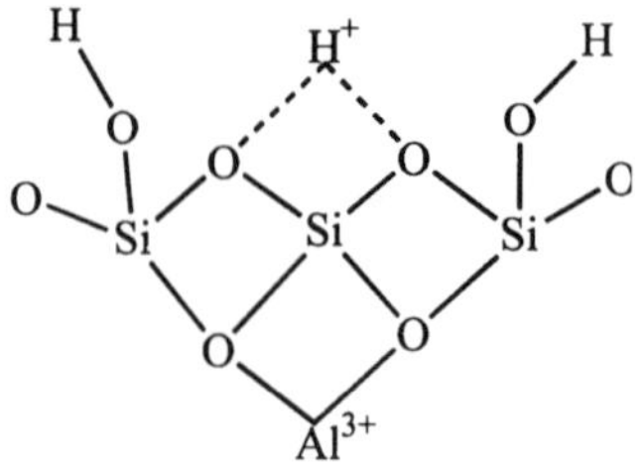

The weakly bound hydrogen on the surface (Brönsted acid) compensates for the lower formal charge of the aluminum atom $(3+)$ compared to the silicon atom $(4+)$ in the matrix.

The generally accepted value for the surface concentration of silanols on a typical silica is about 8 μmol/m^2 (4.6 groups/nm^2). Higher values, 14 μmol/m^2 (8 groups/nm^2) have been reported as well and are still plausible. The actual value depends on the pretreatment of the silica, but it is also a function of the measurement technique. If we accept the value of about 5 groups/nm^2 or more for a highly hydroxylated silica, the average distance between silanols is less than 0.5 nm. At that high a level of hydroxylation, a large fraction of the silanols are geminal and bridged. The ratio of single to geminal silanols can be measured by ^{29}Si-CPMAS-NMR. One can get a measure of the number of bridged versus lone silanols from infrared spectroscopic measurements.

When silica is heated, neighboring silanols condense to form siloxane bridges, leaving lone silanols. Lone silanols are more acidic than bridged silanols and represent strong adsorption centers. Above a temperature of $\sim$800°C, more than 90% of the silanols have been removed, and the remaining silica contains mostly hydrophobic siloxane bridges. Silica treated at high temperature is therefore hydrophobic. However, the reaction is reversible, and the surface can be rehydroxylated through extended exposure to water.

Manufacturers of silica-based HPLC packings have spent much effort learning to understand and mitigate the influence of silanols, and much improvement has been made with newer bonded-phase packings. For a more detailed discussion, see Chapter 10.

6.2.1   Bonding Technology

The silanols on the surface of silica can be reacted with many different agents to prepare the so-called bonded phases. For example, they condense with alcohols to form silicon esters. This reaction is reversible, however; that is, the bond hydrolyses readily in aqueous solvents. Therefore it is unsuitable for the preparation of HPLC phases, but it is used extensively in the synthesis of phases for gas chromatography.

Silica can be chlorinated, and the resulting product reacted with amines (1,2). Stationary phases prepared by this procedure contain a silicon–nitrogen bond, which is more stable than the silicon ester, but is still sensitive to hydrolysis. A better choice is the silicon–carbon bond, which is hydrolytically very stable. Chromatographic phases based on this bond can be prepared by reacting the silica with reactive silanes. Today's bonded phases are all based on silicon–carbon bonds.

In a typical reaction, a chlorosilane is reacted with the silanols on the silica surface in the presence of a basic scavenger of the hydrochloric acid (3,4):

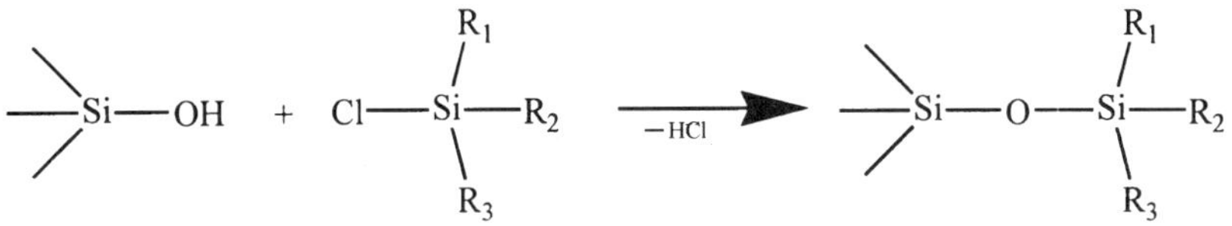

Instead of the chlorosilanes, other reactive silanes can be used as well. Alkoxysilanes are less reactive than chlorosilanes, but will form bonded phases in the presence of acidic or basic catalysts. Silazanes, on the other hand, are as reactive or more reactive than chlorosilanes. A well-known example is the reaction of the silica surface with hexamethyldisilazane:

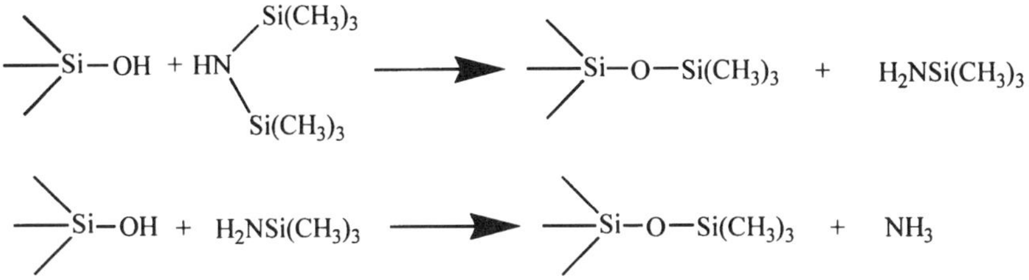

This reaction is frequently used for endcapping. As a result of steric hindrance or limits in the reactivity of a silane, only a part of the silanol groups on the surface are reacted. Therefore often a second reaction is added, in which a small, highly reactive silane is used to remove additional silanols. This second reaction is called "endcapping."

The silane may have one, two or three groups that can react with the surface. Depending on this number, it is called monofunctional, difunctional, or trifunctional:

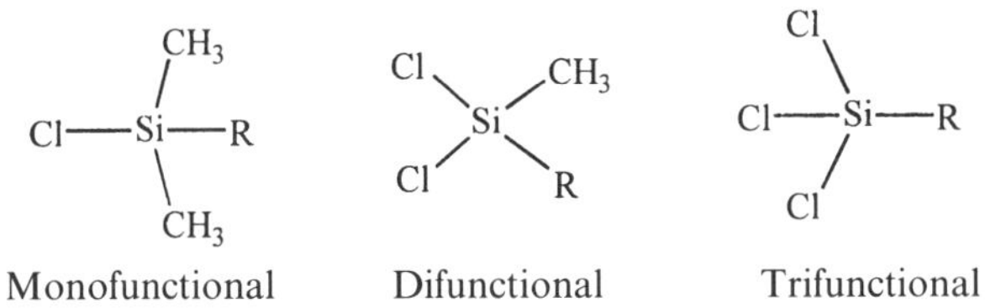

Monofunctional   Difunctional   Trifunctional

How much is bonded to the surface can be determined by elemental analysis. The result is expressed in micromoles per square meter of bonded phase, and is called the surface coverage χ. The surface coverage is calculated by determining the number of moles of ligand from elemental analysis and dividing it by the surface area of the parent silica. The percentage of parent silica in the bonded sample is obtained by subtracting the total amount of weight gain due the bonded ligand from the weight of the silica:

$$\chi = \frac{\%C}{100 \cdot SA \left(1 - \frac{\%C}{100} \frac{MW-1}{nC \cdot 12}\right)} \tag{6.1}$$

where $\%C$ is the carbon percentage from elemental analysis, SA is the specific surface area of the parent silica, MW is the molecular weight of the attached ligand, and nC is the number of carbon atoms in the ligand.

When a monofunctional silane is reacted with the silica surface, even under the best bonding conditions only about 4 μmol/m^2 can be bonded to the surface. The steric hindrance of the other groups surrounding the silicon atom of the reagent (usually methyl groups) prevents a higher coverage. As stated above, a fully hydroxylated silica has a silanol content of 8 μmol/m^2. Therefore at least half of the original silanols are still present after bonding. This applies to all bonded phases prepared by this technique, even those that are "fully endcapped," or have "maximal coverage," or whatever other expression can be found in the marketing literature of packing manufacturers. Although these "residual" silanols are not available for additional bonding, they may still interact with analytes and contribute to the adsorptive properties of a bonded phase. A typical "fully endcapped" bonded phase is schematically depicted in Figure 6.1. We see that the total amount of bonded-phase groups and endcapping groups equals the amount of residual silanols, and that the amount of bonded endcapping reagent is small compared to the amount of bonded-phase groups. The details of the structure of a bonded-phase surface were studied by Berendsen and de Galan (5), using phenyldimethylsilane or trimethylsilane bonded phases as models. Their model of a trimethylsilyl bonded phase is shown in Figure 6.2. Clearly, it is impossible to obtain a perfect match between the space requirement of the bonded phase and the distribution of silanols on the surface. As a consequence, even for bonded phases with maximum coverage, residual silanols are still available for interaction with analytes and need to be taken into account when examining the retention mechanism for an analyte. They are also the points of entry for agents that cause the hydrolysis of the bonded phase or the silica itself.

6.2.2 Monomeric versus Polymeric Bonding

In the previous paragraph we mentioned di- and trifunctional silanes, but we discussed only the surface reaction with monofunctional silanes. When a

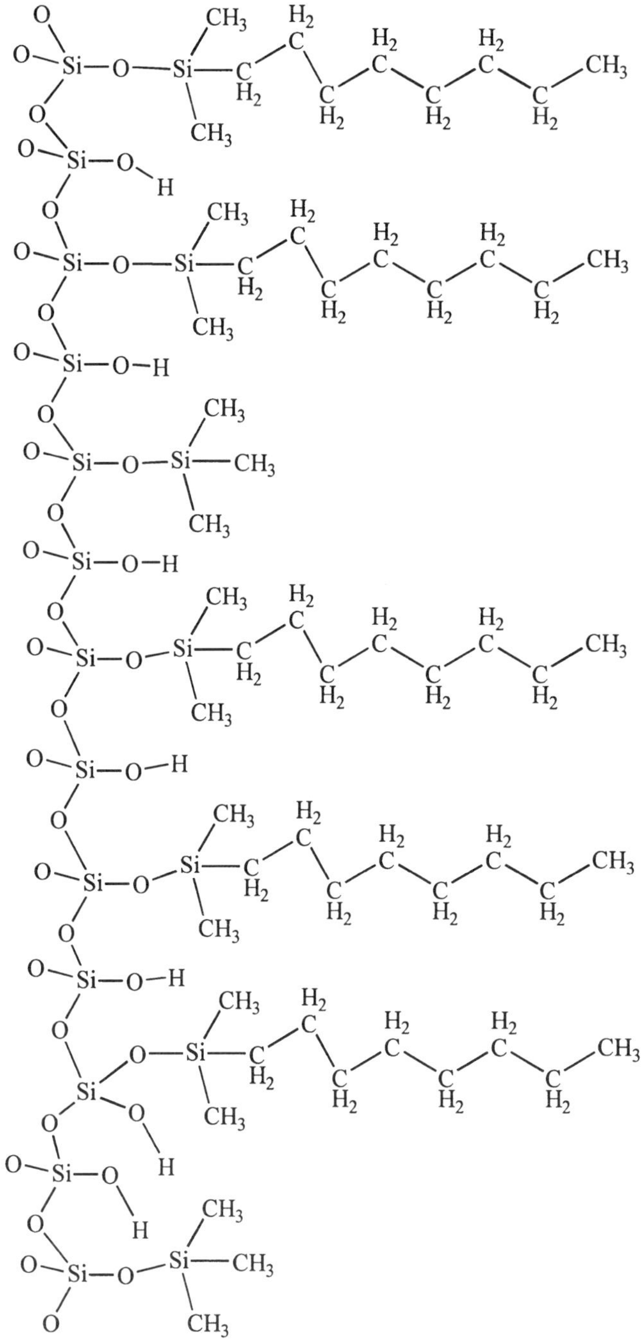

Figure 6.1 Typical "fully endcapped" C_8 bonded phase. There are about as many silanols left on the surface as there are bonded ligands.

Figure 6.2 Surface of a trimethylsilyl bonded phase after Berendsen and de Galan. (Reprinted from Ref. 5, p. 419, by courtesy of Marcel Dekker, Inc.)

difunctional silane is reacted with the surface, more than one silanol group can react. Once again, a complete removal of all surface silanols is not possible because of the randomness of the bonding process. On average, roughly half of the bonded-phase molecules bind to the surface via two siloxane bonds, while the other half can bond only via a single bond. The singly bonded difunctional silane contributes one new silanol to the surface, after the second functional group is hydrolyzed. Therefore, the amount of unreacted silanols in a bonded phase based on a difunctional silane is barely different from the silanol content of bonded phases based on monofunctional silanes.

The situation does not improve with the use of trifunctional silanes. It appears that a tridentate reaction with the surface is impossible. Consequently, a large number of new silanols is formed on hydrolysis of the unreacted functional groups. Some of these groups, however, condense with other newly formed silanols to form siloxane bridges. The overall result of both effects is that the total content of silanols in a bonded phase based on a trifunctional silane is at least as high as the silanol content of bonded phases based on monofunctional or difunctional silanes.

While the surface coverage obtainable with monofunctional silanes is limited by steric hindrance, the reaction of multifunctional silanes can be carried out under conditions that result in a significantly higher content of bonded phase than is possible with monofunctional silanes. The addition of water to the reaction results in the formation of additional silanols, which then can react further and add more bonded phase. Traditionally, phases with a bonded-phase content of more than about $4 \, \mu\text{mol/m}^2$ have been called *polymeric phases*, although the polymeric nature of such a phase is debatable. After all, we know from the study of surfactant monolayers that the space requirement of a hydrocarbon chain is under $0.20 \, \text{nm}^2$. Consequently, a coverage of $< 8 \, \mu\text{mol/m}^2$ can still be considered to be a monomolecular layer. However, the selectivity, especially the selectivity for molecular shape, is a delicate function of the chain spacing. Sander and Wise (6) have carefully studied the properties of C_{18}-type bonded phases with a coverage between 4 and $8 \, \mu\text{mol/m}^2$ for the selectivity of polynuclear aromatic hydrocarbons.

The majority of the commercially available bonded phases are today based on monofunctional silanes. The main reason for this is the increased reproducibility of the preparation. With di- and trifunctional silanes, even small amounts of water present in the reagents, either on the surface of the silica or in the reaction solvent, can increase the amount of bonded phase attached to the surface substantially. This problem does not exist with monofunctional silanes. Also, hindered mass transfer has been observed when multifunctional silanes were reacted with small-pore silicas to a high coverage level. The latter issue should, however, not be a fundamental problem, since the pore size can be matched to the coverage level to avoid this phenomenon.

6.2.3 Types of Bonded Phases

Until now we have referred to bonded phases in generic terms without specifying the chemical entity that is attached to the surface. For the majority of the bonded phases in use, the bonded entity is simply a hydrocarbon. Table 6.1 lists the most popular bonded phases in use today and their designations:

The other groups attached to the silicon atom have been given the general designation R_1 and R_2. As we have seen in the previous section, R_1 and R_2 could be either additional siloxane bonds to the silica surface or silanols from hydrolysis of the functional groups, or they could be other hydrocarbon groups. The most typical side substituent is the methyl group, exhibiting the least steric hindrance and allowing the highest coverage. However, bulky isopropyl groups have been demonstrated to improve the hydrolytic stability of the bonded phase at acidic pH (7).

The hydrocarbon bonded phases are used in reversed-phase chromatography. For a given substrate and a given bonding technology, retention increases with increasing chain length. There is seldom much of a selectivity difference between different straight-chain hydrocarbons (C_6, C_8, C_{12}, C_{18}), but some selectivity difference can be observed between a phenyl bonded phase and a C_8 or C_6 bonded phase. The nomenclature in the shorter chain length is ambiguous; a trimethylsilane bonded phase which for consistency with the nomenclature of the longer chains should be called C_1 is sometimes referred to as a "C_3 phase."

The bonded ion exchangers are used for ion-exchange chromatography, but can also be used for hydrophilic interaction chromatography. The most

TABLE 6.1   Commonly Used Bonded Phases

C_{18}

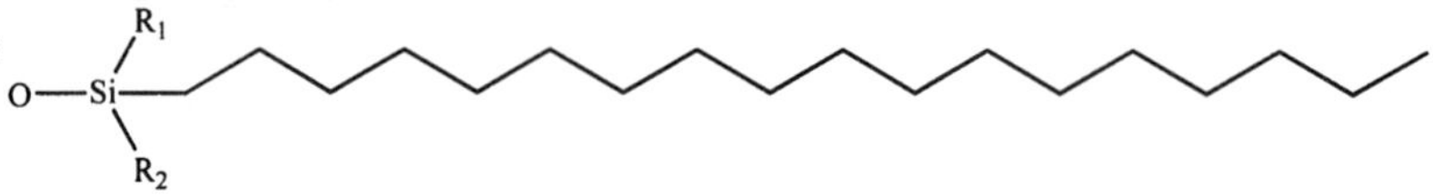

C_{12}

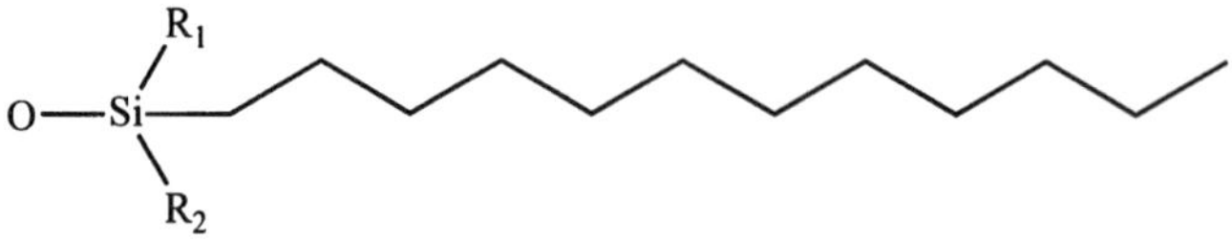

C_8

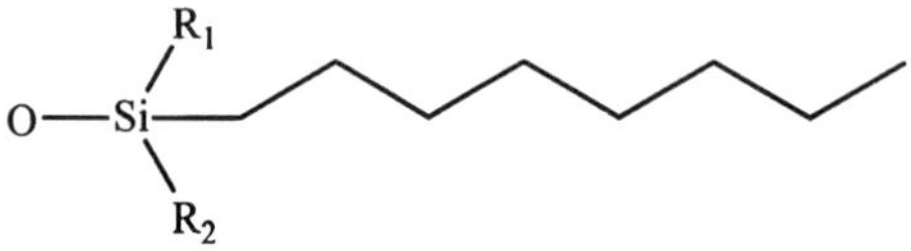

C_6, hexyl

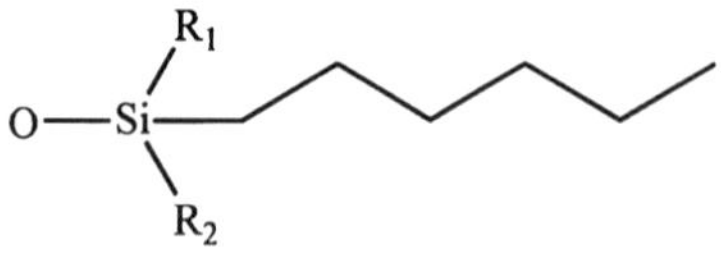

Cyclohexyl

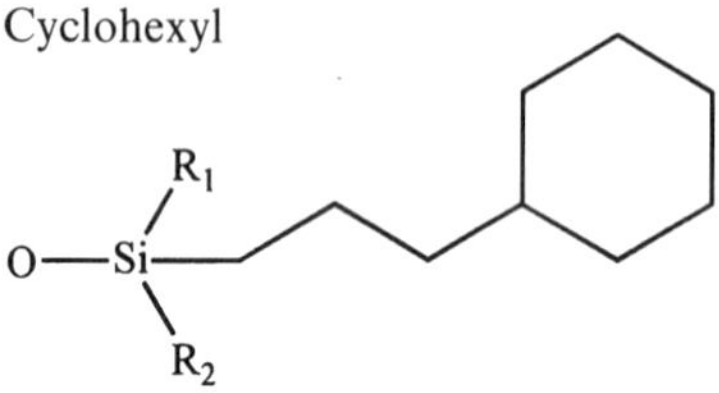

Phenyl

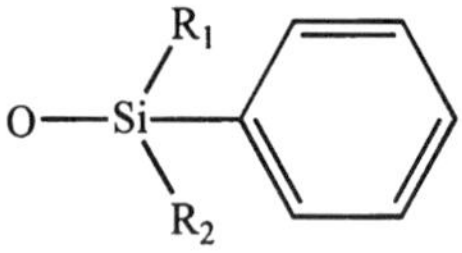

Alkylphenyl

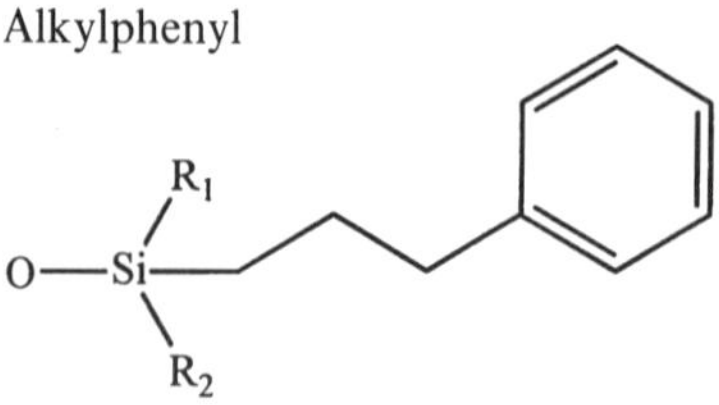

TABLE 6.1   Commonly Used Bonded Phases (Continued)

C$_4$

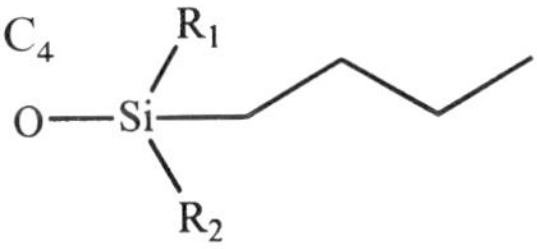

C$_3$

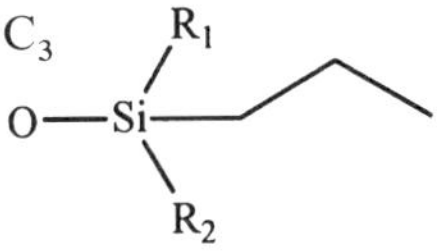

C$_2$

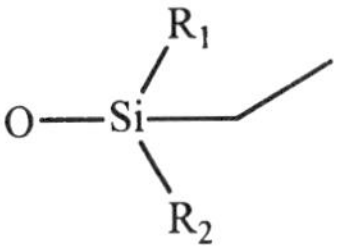

C$_1$

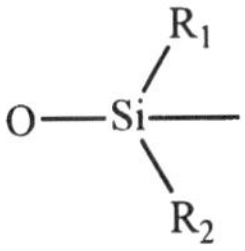

Cyano, nitrile

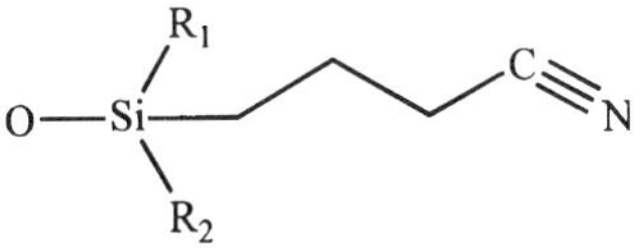

Amino

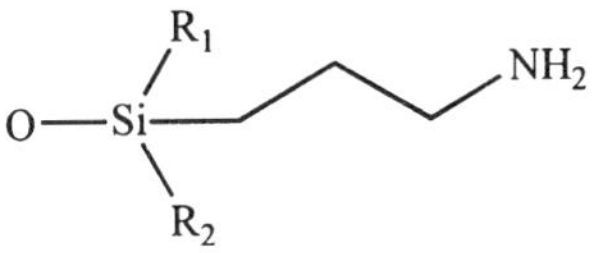

Diol

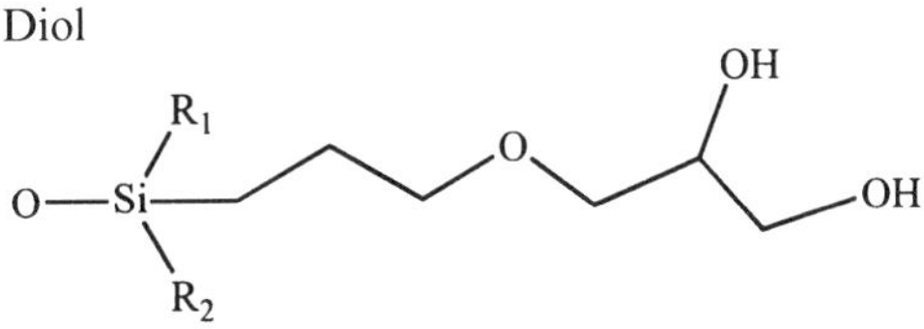

SAX

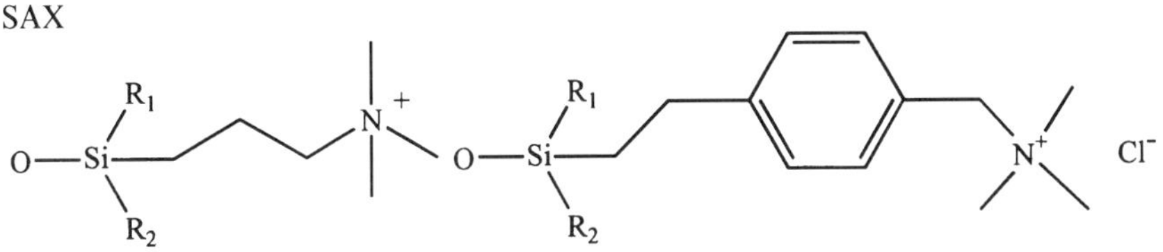

TABLE 6.1   Commonly Used Bonded Phases (Continued)

SCX

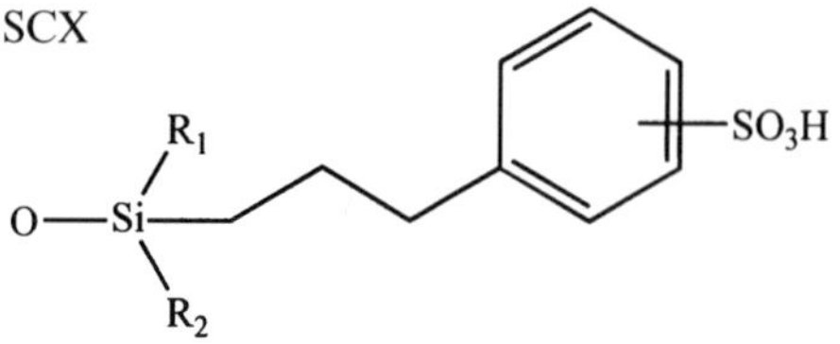

Nitro

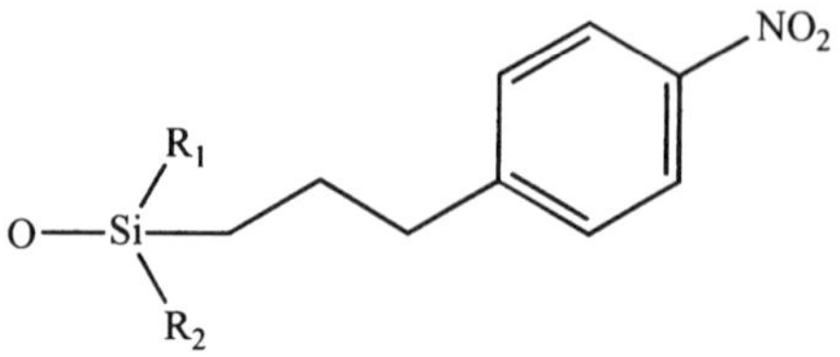

Phases with embedded polar groups

   SymmetryShield RP$_8$

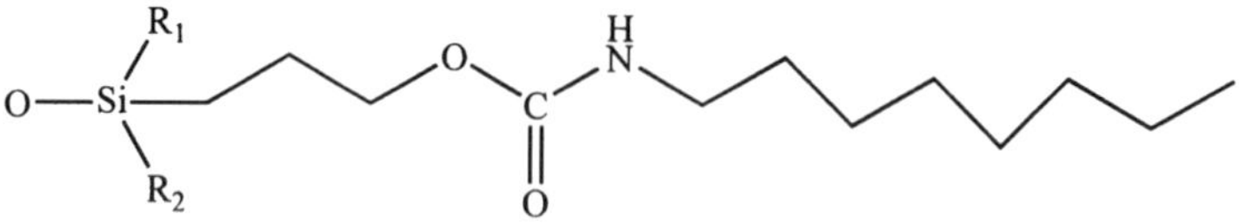

   Supelcosil ABZ

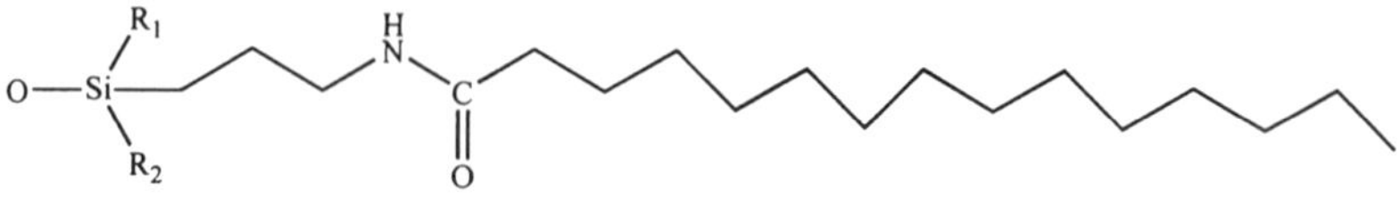

important phase for hydrophilic interaction chromatography is the amino bonded phase, which, in turn, can serve as an ion exchanger. Diol bonded phases are utilized predominantly for aqueous size-exclusion chromatography, but can also be used in hydrophilic-interaction and in normal-phase chromatography. Cyano columns are employed in both reversed-phase and normal-phase chromatography, while nitro columns exhibit specific selectivity for aromatic compounds in normal-phase chromatography.

The special phases with an embedded polar functional group are designed for reversed-phase chromatography. The polar function shields the silica surface, preventing the interaction of analytes with the acidic silanols on the silica surface. For many samples, they exhibit a significant difference in selectivity compared to simple hydrocarbon bonded phases. Also, the tailing of basic analytes is reduced.

A few specialty phases are available as well, and many more bonded phases have been described in the literature. A discussion of all these options is beyond the scope of this book. Furthermore, such a variety really is not necessary in liquid chromatography, where the variation of the mobile-phase composition gives us such a powerful tool to change the selectivity of the separation. This is the reverse of gas chromatography, where the lack of selectivity of the mobile phase requires the existence of a myriad of stationary phases.

6.2.4 Reproducibility of Bonded Phases

The selectivity of a separation is not only a function of the type of the bonded phase but can also depend on the spacing of the bonded-phase chains and the availability and activity of residual silanols. If we want to achieve highly reproducible separations, it is therefore very important that the composition of the stationary phase be tightly controlled. Luckily, the technology for preparation of silica-based bonded phases is now over 20 years old, and manufacturers have had ample time to learn to measure and control the important properties of bonded phases.

In chapter 4 we discussed the physical properties of a packing: the pore size, specific pore volume, specific surface area, and particle size. In retention chromatography, the first three parameters may affect retention, while a change in the fourth influences column efficiency and backpressure. However, none of these parameters will change the relative position of peaks in the chromatogram, which is what we call the "selectivity" of the separation. On the contrary, the chemical properties of the packing are responsible for the selectivity. Therefore we need to control tightly the population density (= surface coverage) of the bonded phase and the residual silanols. The first step in this is the adjustment of the concentration, type, and acidity of silanols on the surface of the parent silica. To prevent the formation of acidic silanols, the silica must be free from metal contaminants. This is accomplished through the quality of the raw materials used in the synthesis of the silica. The pretreatment of the silica is the second most important factor determining the nature of the silanol population. The next step is the bonding reaction, usually followed by an endcapping step. With today's bonding technology using monofunctional silanes, it is possible to control the surface coverage to under $\pm 3\%$. This results in an equally tight control over the retention times and selectivities of a separation.

How closely can a packing be reproduced today? Table 6.2 shows the results of a study of the reproducibility of a wide range of separations, using a recently developed packing material. Two columns were prepared from the same batch of a packing. They represent the control of the experiment. The third column was prepared from a different batch. All experiments were carried out in sequence, using the identical mobile phase under controlled temperature to minimize external influences. The difference in retention times between the two columns prepared from the same batch of packing was typically around 1%.

TABLE 6.2 Reproducibility of Column-to-Column and from Batch-to-Batch Retention Times for Several Different Assays[a]

Compound	Retention Time (min)			Difference (%)	
	Column 1	Column 2	Batch 2	Between Columns	Between Batches
Acetaminophen	1.87	1.87	1.89	0.0	1.1
Amitriptyline	25.34	25.06	24.63	1.1	2.3
Barbital	2.05	2.07	2.01	1.0	2.5
Cefixim	1.77	1.77	1.79	0.0	1.1
Cefuroxim	3.4	3.4	3.52	0.0	3.5
Codein	9.84	9.62	9.8	2.3	0.7
Dextromethorphan	3.09	3.12	3.13	1.0	0.8
Imipramin	19.21	18.97	18.68	1.3	2.2
Mephobarbital	7.74	7.89	7.8	1.9	0.2
Protriptylin	5.06	5.00	4.98	1.2	1.0
Secobarbital	9.93	10.1	10.0	1.7	0.1
Succinylsulfathiazol	8.83	8.64	8.56	2.2	2.0
Sulfadiazine	3.19	3.15	3.14	1.3	1.0
Sulfamethazin	6.24	6.12	5.98	1.9	3.3

[a]*Columns*: Waters Symmetry C_8, 3.9 mm × 150 mm.

Source: Data courtesy of D. J. Phillips and M. Capparella, Waters Corporation.

The difference between batches was not much worse, averaging 1.5%. Consequently, the reproducibility between batches was of the same order of magnitude as the experimental error in this experiment, where all attempts were made to minimize the experimental error. In general, one should expect a modern packing obtained from a reputable manufacturer to yield a batch-to-batch standard deviation of around 2% for the reproducibility of retention times and around 1% for the reproducibility of the relative retention between similar analytes.

6.2.5 Alternative Bonding Technologies

There are several different alternative techniques for the surface modification of silica. In the following, we will describe two approaches that are closely related to classic bonded phases. They are both characterized by the facts that the surface modification agent is (1) anchored to the surface via $\equiv$Si—O—Si—C— bonds and (2) applied in a very thin, approximately monomolecular layer. Alternative coating techniques, which do not depend on the attachment of the adsorptive layer to the surface, are described in Section 6.4.

In the approach discussed here, a silane is first bonded to the surface that serves as an anchor for the subsequent attachment of a polymerizable monomer carrying the desired functional groups. An example of this is the technique developed by Engelhardt et al. (8), in which the first step consists of the derivatization of the silica with a vinylsilane. The vinyl-modified silica is then coated with a vinyl monomer, typically an acrylate or methacrylate, and a radical initiator. Then the polymerization is carried out along the surface by increasing the temperature to the decomposition temperature of the radical initiator. The average thickness of the surface layer can be controlled via the amount of monomer. A wide range of surface chemistries, from ion exchange to reversed phase, is available through the choice of the monomer. Phases obtained this way are supposed to be more pH-stable than conventional bonded phases, but under standardized test conditions this could not be verified for a commercial version of this type of sorbent (9). The quality of the mass transfer of the packings can be optimized through the thickness of the layer. A recent study (10) also indicated that the choice of the polymerization conditions influence the quality of the mass-transfer, and that wet polymerizations can yield C_{18}-type reversed-phase packings with a mass-transfer behavior that is competitive to classical bonded phases.

Naturally, other silanes can be used as anchors for similar polymerization reactions on the surface. For example, in the related technique developed by Alpert (11–13) the silica is first derivatized with an aminopropylsilane. In the second step, the aminopropyl silica is reacted with a pre-polymerized polysuccinimide (obtained from self-condensation of aspartic acid). The resulting material contains excess succinimide groups, which can be further reacted to yield an poly(aspartic acid) derivative (11) as a weak cation exchanger, poly(aspartic alkylamide) derivatives (12) for hydrophobic interaction chromatography and poly(2-sulfoethyl aspartamide) derivatives (13) as strong cation exchangers. These stationary phases are useful for the chromatography of proteins and peptides and are available from PolyLC (USA).

In an alternative technique, a siloxane polymer is attached to the surface and the polymer is subsequently modified with other reagents to create the desired stationary phase. Ohtsu and co-workers (14,15) describe a procedure in which the silica is first reacted with 1,3,5,7-tetramethylcyclotetrasiloxane. To create a reversed-phase packing of the C_{18} type, the hydrosiloxane-coated silica is then derivatized with octadecene in the presence of a soluble platinum catalyst. The resulting material reportedly exhibits a pH stability much improved over that of conventional bonded phases without an appreciable compromise in efficiency. To date, this technology has only been applied to reversed-phase packings. A commercial version is Capcellpak C_{18} available from Shiseido (Japan).

6.3   THE CHEMISTRY OF POLYMERIC PACKINGS

As pointed out above, a key requirement for an HPLC packing is a minimum mechanical strength that makes the packing compatible with the typical pressure drops required for using a column packed with 5–10-μm particles. To date, three types of polymeric packings have a sufficient strength and are commercially available: divinylbenzene-based packings, methacrylate-based packings, and vinylalcohol-based packings. All of them also have an increased range of pH stability compared to silica, which makes them attractive as an alternative to silica. Other polymeric packings with an appropriate combination of desirable properties have not yet been developed. Packings based on polysaccharides, which are so popular and successful in low-performance chromatography, are much too weak for HPLC conditions. Occasional publications of polymeric packings based on other chemistries have not yet found a commercial implementation.

The skeleton of an inorganic packing is impervious to solvents, even to gases. This is not true for the skeleton of a polymeric packing, which still imbibes a large enough amount of solvent to exhibit some swelling and shrinking of the matrix as a function of the solvent. This effect may very well be a fundamental property of polymeric packings, since the size of a cell in the crosslinked network of a polymer is fairly large compared to the tight three-dimensional network of an inorganic packing. The same cells that allow the penetration of solvent molecules are also accessible to small analyte molecules. However, because of the tightess of the network, the diffusion of the analyte in this network is slow. A consequence of this is the poor mass transfer of analytes into and out of the surface layer of the polymeric packing, which, in turn, results in broadening of the chromatographic peaks. Therefore, the lower performance of polymeric packings compared with silica-based packings may also very well be a fundamental property of polymeric packings. However, much progress has been made recently in the design of polymeric packings for the chromatography of small analytes, and the final word may not yet have been spoken.

The slow mass transfer of polymeric packings is limited to small molecules. High-molecular-weight analytes such as proteins or synthetic polymers do not suffer from this problem, which is one of the reasons why the predominant application of polymeric packings is in the chromatography of macromolecules.

In the following sections we will discuss the properties of the three polymeric packings currently used in HPLC.

6.3.1   Divinylbenzene-Based Packings

Divinylbenzene- or styrene–divinylbenzene-based packings were originally developed for the size-exclusion chromatography of industrial polymers (16). They are prepared via radical polymerization of the monomers. The monomers used are, in principle, styrene and divinylbenzene isomers, but industrial-grade divinylbenzene also contains ethylvinylbenzene isomers. Thus the picture of what forms the network of a divinylbenzene particle is complex.

Styrene   *o*-Divinylbenzene   *m*-Divinylbenzene   *p*-Divinylbenzene

There are two types of styrene–divinylbenzene packings: microreticulate packings and macroreticulate packings. Microreticulate packings are less crosslinked. They are nonporous in the dry state and acquire a pore volume only by imbibing a solvent. Thus the pores are formed by the holes in the crosslinked polymer network [reticulum (Latin) = net]. These packings are designed largely for the size-exclusion chromatography of small molecules and oligomeric analytes.

The opposite of microreticulate is macroreticulate. However, this word is a misnomer, since the pores of a macroreticulate packing are not formed by a network, but by the spaces between spherules of "solid" polymer (17). In this aspect they are identical to inorganic packings and it is more appropriate to categorize these packings as macroporous. Macroporous packings have a permanent pore structure and internal surface that can be explored and measured by the same techniques as used for inorganic packings. Packings with a permanent pore structure are used in size-exclusion chromatography of larger molecules, but they also form the basis for packings used in retention chromatography (18). The pore size of the packing is determined by the size of the spherules that form its skeleton, and the size of the spherules, in turn, is determined by the preparation conditions of the polymer. A broad range of pore sizes can be synthesized, which is a necessity for size-exclusion chromatography.

The resulting matrix consists nearly entirely of aliphatic and aromatic hydrocarbons (some amount of radical initiator is always incorporated into the matrix). Therefore it is highly hydrophobic. A dry powder is not wetted by water. Also, water does not swell the packing at all. However, organic solvents, even polar ones such as methanol, wet the packing and swell the matrix.

The high hydrophobicity of the surface would make the divinylbenzene particle the ideal packing for reversed-phase chromatography. However, the slow mass-transfer characteristic of polymeric packings is especially pronounced for divinylbenzene packings and has prevented it from displacing silica-based bonded phases as the preferred packings of reversed-phase chromatography. Although some underivatized divinylbenzene packings are used for reversed-phase chromatography, the most common application of the underivatized packing is size-exclusion chromatography. A divinylbenzene packing derivatized with a C_{18} alkyl chain is also available for reversed-phase HPLC.

The benzene ring is accessible to many reactions. The most common derivatives of styrenic packings are cation exchangers prepared by chlorosulfonation and anion exchangers prepared by chloromethylation and subsequent amination. More about this in the section covering ion exchangers.

Different derivatization schemes have been applied as well. In a proprietary multistep process, large-pore divinylbenzene particles are first hydrophilized and then derivatized to yield ion exchangers suitable for the chromatography of proteins (e.g., Mono-Q, Mono-S packings from Pharmacia, Sweden).

6.3.2   Methacrylate-Based Packings

Methacrylate-based packings are prepared from a methacrylate and an appropriate crosslinking agent like ethyleneglycol dimethacrylate. The commonly used primary monomer is glycidoxymethacrylate. The epoxy group in this monomer can be further derivatized to yield a polymer with functional groups, or it can be hydrolyzed to the diol to yield a highly hydrophilic polymer.

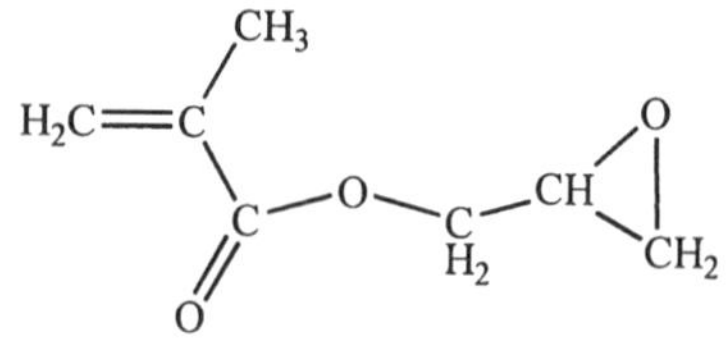

Glycidoxymethacrylate

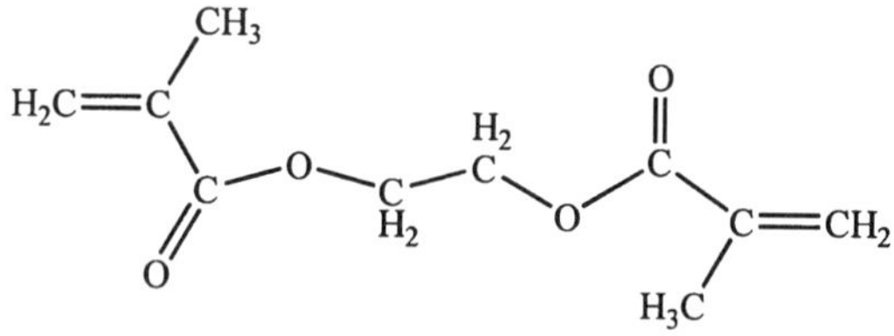

Ethyleneglycol dimethacrylate

As is the case for styrene–divinylbenzene polymers, both microreticulate packings and packings with a permanent pore structure can be prepared. The size of the permanent pores can be manipulated in the same fashion as for the styrenic polymers. Therefore the same range of pore sizes is available for methacrylate packings as for styrene–divinylbenzene packings.

Methacrylate-based packings are generally hydrolytically stable between pH 2 and 12; the weak link is the ester linkage. Depending on the surface chemistry, they have found use in a broad range of application areas. The primary use of the methacrylate-based packings with a polyol surface is aqueous size-exclusion chromatography. But they are also good packings for hydrophilic interaction chromatography. In addition, the alcohol groups can be further derivatized. For example, the reaction with propanesultone results in a strong cation exchanger (SP) with a hydrophilic skeleton that shows very little nonspecific interaction with proteins. The reaction with 1-chloro-2-diethylaminoethane results in a weak cation exchanger for proteins (DEAE). A derivatization of the diol groups with long-chain acid chlorides, on the other hand, results in a hydrophobic packing useful for reversed-phase chromatography. Alternatively, alkylethers can be formed, resulting in packings with similar chromatographic properties, but with improved chemical stability. Low levels of substitution with shorter-chain acid chlorides or benzoyl chloride yield packings for hydrophobic interaction chromatography.

The epoxy ring is also a good starting point for further surface reactions. It can be opened with amines to yield anion exchangers for both protein separation and ion chromatography. Furthermore, different methacrylate monomers can be used in the preparation of the base resin. Methacrylate-based packings are very versatile.

6.3.3 Vinylalcohol-Based Packings

The skeleton of the third type of polymeric packing is formed by polyvinyl-alcohol. The underivatized polymer is very hydrophilic. Since it is based on a hydrocarbon chain, it is stable to both high and low pH values.

Structure of polyvinylalcohol

The alcohol groups can be derivatized with the chlorides of long-chain fatty acids to yield hydrophobic packings suitable for reversed-phase chromatography (19). The key advantage of such a packing over silica-based packings is the lack of acidic sites on the surface, which are the cause of tailing peaks in the chromatography of basic analytes on silica-based packings.

6.4 OTHER SURFACE MODIFICATION TECHNIQUES FOR INORGANIC PACKINGS

Because of the weaknesses of both polymeric packings and silica-based packings, there continues to be an interest in the design of stationary phases based on other inorganic oxides, especially alumina, zirconia, and titania. Porous particles based on all three oxides can be prepared with mechanical and physical properties similar to silica. However, there is no surface derivatization available for either oxide that compares to the silanization technique used in the preparation of the silica-based bonded phases. Simple organic derivatives of these oxides are hydrolytically unstable. While silanization of the oxide surfaces has been reported to result in packings with a good pH stability, the same report also states that the hydrolytic stability of the zirconium and titanium siloxane bond is inferior to the pure siloxane bond.

However, other derivatization techniques have been developed that are suitable for application to any surface and therefore can be used for the modification of the surfaces of these three oxides. In the following, we will first discuss the properties of the surfaces of alumina, titania, and zirconia, and then proceed to examine some surface modification techniques that have been used to date.

6.4.1   Properties of Other Inorganic Packings

The surface of silica is composed of oxygen atoms, which are either neutral (around pH 3) or negatively charged (at pH > 3). A positively charged surface is obtained only in a very acidic environment (pH < 2). The surface of alumina is similar to the surface of silica, but alumina is amphoteric. An uncharged surface is obtained around a neutral pH. At acidic pH values a positively charged surface is obtained, and at basic pH values the surface is negatively charged. Therefore alumina can act as a cation exchanger as well as an anion exchanger. Alumina is hydrolytically stable at pH $\leqslant 13$.

While the silicon in silica is always surrounded by four oxygen atoms in a tetrahedral configuration, the aluminum in alumina can be located in the center of either a tetrahedron or an octahedron. The commonly used form of alumina is γ-alumina, whose surface can be formulated as follows (20):

Whether the different sites on the surface are dissociated or not, is a function of the pH. The five structures shown have been correlated with the five distinct bands that have been observed in the infrared between 3000 and 4000 cm^{-1}.

The surfaces of titania and zirconia are significantly different from silica or alumina, but appear to be similar to each other (21). Zirconia has been studied extensively by Carr and co-workers (22,23). Zirconia is hydrolytically stable from pH 0 to 14. Like alumina, the surface of zirconia can be either positively charged, neutral, or negatively charged. Therefore, at low pH, zirconia acts as an anion exchanger, and at high pH it is a cation exchanger. In between, there is a balance of positively charged, negatively charged and neutral sites that varies as a function of the pH. However, zirconia exhibits an additional interaction mechanism not found in silica: zirconium(IV) sites on its surface, which are hard Lewis acids, strongly interact with analytes through a ligand-exchange mechanism. This significantly complicates the surface interactions compared to silica. Furthermore, the zirconium(IV) sites on the surface may have more than one coordination site each. Up to three fluoride ions have been found to bind per surface zirconium (23). Also, the kinetics of the interaction with these sites can be slow, resulting in low plate counts. These studies indicate that for the preparation of a well-behaved HPLC packing from zirconia it may be even more necessary to hide the surface than it is with silica.

6.4.2   Other Surface Modification Techniques

For alumina, titania, and zirconia, there exists as yet no covalent bonding chemistry that is equivalent to the silanization technique used for silica. Although attempts have been made to silanize these other oxides, the hydrolytic stability of these phases does not match up to the hydrolytic stability of the support itself. Therefore alternative surface modification techniques have been developed that do not rely on the attachment of the modifier to the surface. The coating can be simply insoluble in the intended mobile phases, or a crosslinked coating can be formed that stretches like a net around the skeleton of the particle. Both techniques are, in principle, independent of the nature of the substrate and can be applied to all inorganic or polymeric packings.

The most widely used example of the second technique is based on a process developed by Schomburg (24). He coated a prepolymerized polybutadiene of controlled molecular weight onto the surface of a substrate and then completed the polymerization using radical initiators at elevated temperature. The result is a hydrophobic surface suitable for reversed-phase chromatography that can be applied to silica, alumina, titania, and zirconia. Reversed-phase packings with significantly improved pH stability have been obtained by this method, but the thickness of the coating results in hindered mass transfer and consequently lower column efficiency. A related packing is prepared from a polybutadiene–maleic anhydride copolymer. It is used for the separation of mono- and divalent inorganic cations in a single run.

A wide variety of polymers have been immobilized on silica. A comprehensive table with references can be found in a review article by Petro and Berek (25). They range from very polar to nonpolar coatings. Many of these coatings can be applied to other inorganic oxides as well. The majority of the organic–inorganic composite materials have been designed for the chromatography of biopolymers via hydrophobic interaction, ion exchange, or reversed-phase chromatography. In the following we will only describe packings that are of interest to the user, that is, those that are commercially available.

As an example, Alpert and Regnier (26) coated silica with polyethylenimine 6 and crosslinked the product subsequently with a suitable crosslinker. Their preferred crosslinker was pentaerythrol tetraglycidyl ether. The coating was also applied to alumina and titania, and lately to zirconia (27).

As another example, a prepolymer containing N-(1-pyridinio) amidates can be coated onto the surface of the inorganic substrate (28). Exposure of the polymer film to ultraviolet light results in the formation of nitrene- and isocyanate-type functional groups, which serve to crosslink and thus immobilize the prepolymer on the substrate. Ion exchangers and mildly hydrophobic packings for the chromatography of biomolecules have been synthesized via this route on alumina (Bioprotocol columns from Cohesive, USA). The combination of the base-stable substrate with the polymeric coating makes it possible to clean and depyrogenate the packing at high pH with sodium hydroxide.

The technology of the immobilization of chromatographically desirable surfaces onto mechanically and chemically stable substrates is still under active development, and further improvements can be expected in the future.

REFERENCES

1. O.-E. Brust, I. Sebestian, and I. Halász, *J. Chromatogr.* **83**, 15–24 (1973).
2. U. D. Neue, Diplomarbeit, Universität des Saarlandes, Saarbrücken (1973).
3. K. Karch, I. Sebestian, and I. Halász, *J. Chromatogr.* **122**, 3–16 (1976).
4. K. Karch, I. Sebestian, I. Halász, and H. Engelhardt, *J. Chromatogr.* **122** 171–184 (1976).
5. G. E. Berendsen and L. de Galan, *J. Liq. Chromatogr.* **1**, 403–426 (1978).
6. L. C. Sander and S. A. Wise, *J. Chromatogr.* **656**, 335–352 (1993).
7. J. J. Kirkland, J. L. Glajch, and R. D. Farlee, *Anal. Chem.* **61**, 2–11 (1989).
8. H. Engelhardt, H. Löw, W. Eberhardt, and M. Mauss, *Chromatographia* **27**, 535 (1989).
9. T. H. Walter and B. A. Alden, private communication (1996).
10. H. Engelhardt and M. A. Cuñat-Walter, *Chromatographia* **40**, 657 (1995).
11. A. J. Alpert, *J. Chromatogr.* **266**, 23–37 (1983).
12. A. J. Alpert, *J. Chromatogr.* **359**, 85–97 (1986).
13. A. J. Alpert and P. C. Andrews, *J. Chromatogr.* **443**, 85–96 (1988).

14. Y. Ohtsu, H Fukui, T. Kanda, K. Nakamura, M. Nakano, O. Nakata, and Y. Fujiyama, *Chromatographia* **24**, 380–384 (1987).

15. Y. Ohtsu, Y. Shiojima, T. Okumura, J. I. Koyama, K. Nakamura, O. Nakata, K. Kimata, and N. Tanaka, *J. Chromatogr.* **481**, 147–157 (1989).

16. J. C. Moore, *J. Polym. Sci.* Part A, **2**, 835 (1964).

17. N. Tanaka, K. Hashidzume, M. Araki, H. Tsuchiya, A. Okuno, K. Iwaguchi, S. Ohnishi, and N. Takai, *J. Chromatogr.* **448**, 95 (1988).

18. L. L. Lloyd, *J. Chromatogr.* **544**, 201–217 (1991).

19. T. Ohtani, Y. Tamura, M. Kasai, T. Uchida, Y. Yanagihara, and K. Noguchi, *J. Chromatogr.* **515**, 175–182 (1990).

20. K. K. Unger, "Adsorbents in Column Liquid Chromatography," in *Packings and Stationary Phases in Chromatographic Techniques*, K. K. Unger, ed., Chromatographic Science Series, Vol. 47, Marcel Dekker, New York (1990).

21. U. Trüdinger, G. Müller and K. K. Unger, *J. Chromatogr.* **535**, 111–125 (1990).

22. J. A. Blackwell and P. W. Carr, *Anal. Chem.* **64**, (1992); 853-862; J. A. Blackwell and P. W. Carr, *Anal. Chem.* **64**, 863–873 (1992).

23. J. Nawrocki, M. P. Rigney, A. McCormick, and P. W. Carr, *J. Chromatogr.* **A657**, 229–282 (1993).

24. U. Bien-Vogelsang, A. Deege, H. Figge, J. Köhler, and G. Schomburg, *Chromatog raphia* **19**, 170 (1984).

25. M. Petro and D. Berek, *Chromatographia* **37**, 549–561 (1993).

26. A. J. Alpert and F. E. Regnier, *J. Chromatogr.* **185**, 375–392 (1979).

27. C. McNeff and P. W. Carr, *Anal. Chem.* **67**, 3886–3892 (1995).

28. H. S. Kolesinski, *Am. Lab.* 18C–18H (July 1996).

ADDITIONAL READING

Bergna, H. E. "The Colloid Chemistry of Silica: An Overview," in *The Colloid Chemistry of Silica*, H. E. Bergna, ed., Advances in Chemistry Series Vol. 234, American Chemical Society, Washington DC, 1994.

Berthod, A. "Silica: Backbone Material of Liquid Chromatographic Column Packings," *J. Chromatogr.* **549**, 1–28 (1991) (a review with 62 references).

7 Column Selection

> Aus wie vielen Elementen
> Soll ein echtes Lied sich nähren,
> Daß es Laien gern empfinden,
> Meister es mit Freuden hören?
>
> —J. W. von Goethe, *Elemente*

7.1 INTRODUCTION

"What column do I select for my analysis?" This is clearly the most frequently posed question by novices to the field. It is also a very difficult question. However, the answer to a difficult question does not have to be difficult itself.

In this section, we will ask ourselves several questions that will then, step by step, get us to the point where we have an understanding what our column options are. In the beginning, we will pose a few questions that may appear trivial to the experienced chromatographer, but that get us very close to the goal. Then we will explore the remaining options in more detail.

Before we start the development of a new method, we should ask ourselves whether we really need to develop a new method. In many cases established methods already exist. In some cases, the use of a particular method may even be prescribed by regulatory bodies. Examples are the analysis of pharmaceutical formulations, which can be found in the *U.S. Pharmacopeia* or methods for environmental pollutants established by the Environmental Protection Agency (EPA). If we suspect that an established method exists, it pays to search in the relevant literature. For example, EPA Method 8310 describes the analysis of polynuclear aromatics in groundwater and wastewater. If we are interested in the determination of polynuclear aromatics in tobacco smoke or in charcoal-broiled steaks, we can use the same HPLC method as described for the water analysis, and our analytical problem will be the sample preparation method rather than the HPLC method itself. We may end up modifying the method because of specific interferences encountered in the different matrix, but at least we start off with a method that is known to work for the analytes of interest.

Specific methods have been established for quite a large number of compounds or analytic problems. Among these are amino acids, sugars in food, common organic acids in food, vitamins, and additives (e.g., antioxidants) in polymers. Column manufacturers may have an applications database from which they can recommend a column and a method. However, it is recommended to rely on literature methods only, if the methods are well established and have been proved out in many laboratories. EPA methods or pharmacopoeia methods fall into this category, but a single reference on a method for an uncommon analyte should be viewed with caution. It is not uncommon that such a method does not work or does not work well when duplicated. In such a case, we are actually better off to develop a new method ourselves.

If the prospects for finding a method in the literature are low, we need to proceed to the next step: we should categorize the analyte(s) by their chemical nature. Table 7.1 lists these categories.

**TABLE 7.1   Categories of
Sample Compounds**

Polymers
Proteins
Peptides
Nucleic acids
Carbohydrates
Inorganic ions
Others

Each of these categories has its own distinct characteristics, and therefore specific columns and/or techniques are used for the analysis of these compound classes. The catchall category at the end comprises small organic molecules, the most common analytes in HPLC. In the sections that follow we will briefly discuss the column selection for the various analyte classes, while a detailed discussion of a particular technique is reserved to the chapter dedicated to this technique.

7.2   POLYMER ANALYSIS

If our analyte is a polymer, we are most likely interested in a determination of the molecular-weight distribution of the polymer. The technique used for this is size-exclusion chromatography. To proceed further, we need to know what solvents the polymer is soluble in. Then we need to select a column that does not exhibit any interaction with the polymer in the given solvent and has the correct pore size for a fractionation of the molecular-weight range of the polymer. The details will be discussed in Chapter 8.

If the polymer is soluble in water, the correct column is one that is designed for aqueous size-exclusion chromatography. These columns are based on crosslinked polar polymers such as glycidyl methacrylate, hydroxyethyl methacrylate, or polyvinylalcohol. The columns based on glycidyl methacrylate are especially designed for the high-pressure demands of high-performance size-exclusion chromatography. They are available in a broad range of pore sizes to fit to the need of the analysis. The high crosslinking results in more rigid packings and minimizes swelling and shrinking.

Also, silica-based packings are available for aqueous size-exclusion chromatography. The commercially available ones are coated with a glycidoxypropylsilane bonded phase. While they do not swell or shrink, they are not compatible with basic pH values and contain residual silanols that may interact with the polymer. Also, only a limited pore-size range is available compared to the aqueous size-exclusion chromatography columns based on glycidyl methacrylate.

If the polymer is soluble in an organic solvent, the commonly used packings for organic size-exclusion chromatography are based on styrene–divinylbenzene. They are also highly crosslinked and swell or shrink little. Here also, silica-based size-exclusion packings are available that are modified with a coating of trimethyl silane. While they do not swell and shrink at all, they also contain residual silanols that might interact with the sample polymer. Also, the range of pore sizes is smaller than the range available in styrene–divinylbenzene. Consequently the latter are the by far more commonly used packings for organic size-exclusion chromatography.

7.3 PROTEIN ANALYSIS

Proteins can be separated by any of a number of different HPLC techniques: size-exclusion chromatography, ion-exchange chromatography, hydrophobic interaction chromatography, or reversed-phase chromatography. Affinity chromatography is not usually considered to be an HPLC technique, and is not covered in this book.

All packings for protein chromatography have a large pore size, typically around 100 nm, but at least 30 nm. In smaller pores, mass transfer is restricted by hindered diffusion in the pores, which results in broad peaks. In very large pores, mass transfer is augmented by flow in the pores.

In most protein separations, the goal is a purification rather than an analysis. To obtain a pure protein from a biological matrix, a combination of the separation techniques shown above are usually used. Thus, one would typically first fractionate the raw material by size, then subject the fraction which now has a reasonably uniform molecular weight to a separation by either ion exchange, hydrophobic interaction, or both. Alternatively, one can use more than one ion-exchange technique at different pH values. However, the combination of size separation, ion exchange, and hydrophobic interaction is the most effective, since the separation mechanisms are reasonably orthogonal to each other. (Orthogonal separation techniques are characterized by a total lack of correlation of retention.) Reversed-phase chromatography is less suitable for the purification of proteins, since typical reversed phase mobile phase conditions tend to denature proteins.

Separations of high-molecular-weight compounds such as proteins are — with the exception of size-exclusion separations — always gradient separations. It is very difficult to obtain constant retention values for macromolecules under isocratic conditions. Therefore isocratic separations are very impractical. Also, gradient chromatography has a higher peak capacity than does isocratic chromatography, which is an advantage in the separation of complex mixtures.

If the purpose of the separation is analysis, we can use any of the given techniques that allows us to accomplish the goal of the analysis. Size-exclusion chromatography is the simplest technique, and it should be considered first. It is, for example, most appropriate for establishing content uniformity of a dosage form, where the protein is the only high-molecular-weight ingredient in the formulation. It can also be used to establish the presence of dimers or higher associations of the parent protein. If size-exclusion chromatography is inadequate for the task, our next choice is ion exchange. It is the most flexible technique; we can use weak or strong anion exchangers or cation exchangers and vary pH and ionic strength of the gradient. Strong anion exchangers are based on quaternary amines and are usually designated Q or QMA (quaternary methyl amine). Weak anion exchangers are based on tertiary amines. A common type is designated DEAE (diethylaminoethyl). A weak anion exchanger has carboxylic acid functional groups: CM (carboxy methyl). Strong cation exchangers contain aliphatic sulfonic acids. The designation is S or SP (sulfopropyl). For a detailed discussion of retention mechanisms and techniques, see Chapter 12.

Hydrophobic interaction chromatography is less versatile than ion exchange. In this technique, one relies on the interaction of hydrophobic patches on the surface of the protein with the stationary phase. Packings with low populations of hydrophobic groups, for example, phenyl groups, are used. We need to distinguish these packings from reversed-phase packings, where the entire surface is covered with hydrophobic groups.

The last analytical technique for protein separations is reversed-phase chromatography. It tends to give the sharpest peaks, but is plagued with some problems. Some proteins may be recovered only partially, sometimes resulting in ghost peaks in blank gradients following the analysis. If your sample does not exhibit this problem, reversed-phase is a good high-resolution technique that is useful, for example, for the monitoring of impurities in a protein preparation. There is no a priori advantage of one bonded phase over the other. C_1, C_4, C_8, and C_{18} phases have all been used successfully. The only requirements are that the phases should be endcapped, and, as mentioned above, have a pore size of 30 nm or larger. C_8 and C_{18} phases tend to be hydrolytically more stable than the shorter chains, and may be preferred from that standpoint. On the other hand, it has been reported occasionally that long-chain bonded phases give reduced recovery, but it has not been demonstrated unequivocally that the cause is the longer chain length.

7.4 PEPTIDE ANALYSIS

The primary tool for peptide analysis is reversed-phase chromatography. A standard procedure employs a gradient from high water content to high acetonitrile content with 0.1% trifluoroacetic acid (TFA) in both water and acetonitrile. An alternative is the same gradient with 0.1% HCl instead of TFA. This technique has the advantage that detection can be accomplished in the low UV range (210 nm), but it requires an HPLC instrument that does not contain stainless steel in the fluid path. High resolution is achieved, and the technique can be used for peptide mapping. Both packings with 10-nm pore size and with 30-nm pore size have been used. At least up to decapeptides, there is no sign of restricted diffusion in 10-nm pores. Endcapped C_8 and C_{18} packings are preferred over shorter chains because of their improved hydrolytic stability.

For very hydrophilic peptides or glycopeptides, hydrophilic interaction chromatography can be used as well. For a detailed discussion, see Chapter 11.

7.5 NUCLEIC ACID ANALYSIS

For the analysis of nucleic acids in general, nonporous particles are preferred. While the separation of oligonucleotides up to a length of about 200 basepairs (depending on the pore size of the support) can be carried out on both porous and nonporous packings, there is a clear advantage for nonporous resins for nucleic acids above this length. The primary separation mechanism is ion exchange; the stationary phases are nonporous resins (particle size 2–4 μm) with diethylaminoethyl (DEAE) groups or quaternary amino groups. The separations are effected by a gradient with increasing ionic strength.

An alternative technique is the separation by ion-pair chromatography on nonporous hydrophobic resins, for example, alkylated polystyrene–divinylbenzene particles. The separations of oligonucleotides, DNA fragments and polymerase chain reaction products have been demonstrated. Previous implementations of this technique suffered from reproducibility problems, and whether the newer techniques are an improvement over the past remains to be seen at the time of this writing.

Synthetic oligonucleotides are small enough for chromatography on fully porous packings. Protected oligonucleotides can easily be separated from unprotected ones by reversed-phase chromatography. Synthetic oligonucleotides can be separated from by-products of the synthesis (failure sequences) by reversed-phase or ion-exchange chromatography.

7.6 CARBOHYDRATE ANALYSIS

Several different analytical techniques are used for carbohydrate analysis, some with specialized techniques exclusively developed for carbohydrates and closely related compounds: hydrophilic interaction chromatography, size-exclusion chromatography, reversed-phase chromatography, and high-pH anion-exchange chromatography. Since carbohydrates are of interest in many fields, some highly specific analytical techniques have been developed for some special problems. For example, the analysis of the oligosaccharide composition of corn syrup is accomplished by size-exclusion chromatography on sulfonated styrene–divinylbenzene columns specially designed for this purpose.

The highest resolution with the highest sensitivity is produced by high-pH anion-exchange chromatography, which uses a specially designed superficially porous anion exchanger with quaternary amine functions in conjunction with a pulsed amperometric detector.

If high sensitivity is not required, such as, for example, for carbohydrate analysis in food, hydrophilic interaction chromatography is the best choice. In hydrophilic interaction chromatography, a polar stationary phase such as a propylamino bonded phase is used with acetonitrile–water mobile phases of high (70–80%) acetonitrile content. An increase in water content reduces retention. The most popular phase for the hydrophilic interaction chromatography of carbohydrates is a silica-based propylamino bonded phase. The basic environment promotes a fast mutarotation of the α- and β-anomers, which otherwise would result in two peaks for every sugar. However, this packing slowly bleeds ligand and silica. Polymer-based amino columns, which have become available in the last few years, do not suffer from this problem and are therefore preferred.

The retentivity of normal silica-based reversed-phase packings is barely enough to achieve retention of sugars. Also, not all C_{18} bonded phases are useful for carbohydrate analysis. A typical mobile phase is 100% water. In this mobile phase, well-endcapped packings often undergo hydrophobic collapse, namely, a sudden loss in retention. Nonendcapped C_{18} bonded phases are therefore more suitable for carbohydrate separation by reversed-phase chromatography.

Recently, graphitized-carbon stationary phases have become available for reversed-phase carbohydrate analysis. They exhibit a stronger hydrophobic interaction than C_{18} bonded phases. They are also compatible with a high-pH mobile phase, which speeds up mutarotation and prevents peak splitting due to anomer separation.

further study

7.7   ANALYSIS OF INORGANIC IONS

The natural choice of a separation technique for both inorganic cations and anions is ion-exchange chromatography. The standard detection technique for these analytes is conductivity detection. This requires a low background conductance of the mobile phase. Special techniques have been developed under the name of ion chromatography that combine special low-capacity ion exchangers with special mobile phases. For many analytical problems, "cookbook" methods have been worked out by the suppliers of ion-chromatography equipment. Therefore there is usually little need to develop new methods.

Transition-metal ions can be separated by ion chromatography or as complexes by reversed-phase chromatography. Standard methods are available.

7.8   ANALYSIS OF LOW-MOLECULAR-WEIGHT ORGANIC COMPOUNDS

This group of compounds represents the most common analytes in HPLC. Generally, a separation can be accomplished by a number of suitable techniques. Candidates are listed in Table 7.2.

Normal-phase chromatography is carried out using nonpolar mobile phases and polar stationary phases. Stationary phases include silica, alumina, and polar bonded phases such as aminopropyl, cyanopropyl, and diol. Retention times are strongly affected by small amounts of polar additives in the mobile phase, including water. Equilibration with the mobile phase can be very slow. In general, it is more difficult to control and reproduce retention times in normal-phase chromatography than in any of the alternative techniques. Therefore, one would resort to normal-phase chromatography only, if something speaks against using one of the other techniques. Since most of the other techniques use aqueous mobile phases, one reason to use normal-phase chromatography might be the incompatibility of the analyte with water. In this case, we also must carefully consider the compatibility of the analyte with the

TABLE 7.2   Techniques for the Analysis of Low-Molecular-Weight Organic Compounds

Size-exclusion chromatography using packings with very small pore size
Reversed-phase chromatography
Hydrophilic interaction chromatography
Ion-exchange chromatography
Paired-ion chromatography
Normal-phase chromatography

stationary phases. Silica contains acidic silanol groups; alumina contains both acidic and basic sites. A polar fairly unreactive stationary phase is the cyanopropyl bonded phase, which may be the best choice in such a case. It must be pointed out, though, that bonded phases are not free of silanol groups, but that the silanols may be sufficiently sterically hindered to prevent reaction or interaction with a sensitive analyte. Another option for the analysis of analytes that are incompatible with aqueous mobile phases is size-exclusion chromatography using small-pore styrene–divinylbenzene packings, provided the resolving power is sufficient for the analytical task. An example of such a problem is the determination of dioctadecylsiloxane in octadecylchlorosilane, which is easily accomplished using nonaqueous size-exclusion chromatography.

Size-exclusion chromatography is in general a very simple technique. Because of its simplicity, it tends to be highly reproducible and reliable. However, the separation power is very limited. Nevertheless, we might consider it as a dependable solution to simple analytic problems.

If nothing forces us to use normal-phase chromatography, we should preferentially select one of the other techniques. We can divide the analytes into three categories: very polar, neutral analytes; ionic or ionizable analytes; and everything else.

If the analytes are very polar and neutral, the best analytical technique is likely to be hydrophilic interaction chromatography. It has proved its value for the analysis of carbohydrates. Hydrophilic interaction chromatography is carried out using very polar stationary phases in conjunction with aqueous mobile phases with a high organic content. Useful stationary phases are silica, alumina, polar bonded phases such as aminopropyl or diol, and even ion exchangers. The stability of the aminopropyl bonded phase in aqueous eluents is limited, but the other stationary phases can be recommended without reservation. For a detailed discussion, see Chapter 11.

If the analytes are ionic or ionizable, they can be separated by ion-exchange, ion-pairing, or reversed-phase chromatography. Both polymeric and silica-based ion exchangers are available. Silica-based ion exchangers exhibit better efficiency than polymer-based ion exchangers, but can not be used at basic pH values. Silica itself is a cation exchanger of excellent efficiency and can be used for the analysis of basic sample compounds. For a detailed discussion, see Chapter 12.

Ion-exchange chromatography is used when all analytes of interest are ionic. If we are dealing with a mixture of analytes, some ionic and some nonionic, ion-pair chromatography using reversed-phase packings or simple reversed-phase chromatography may be more appropriate. Ion-pair chromatography is a very powerful technique, since hydrophobic and ionic interactions can be adjusted nearly independently from each other. Any reversed-phase column is suitable for ion-pair chromatography, but fully endcapped bonded phases with a longer chain length such as C_8 or C_{18} are preferred. We will discuss ion-pair chromatography together with reversed-phase chromatography.

Reversed-phase chromatography can be used to separate both neutral and ionic organic analytes. It is therefore a very versatile technique. Its versatility is confirmed by its popularity: some 80% of all HPLC separations are carried out using reversed-phase packings. Both polymeric reversed-phase packings and silica-based bonded phases are available. Polymeric packings exhibit inferior efficiency, which is why silica-based bonded phases are preferred. Bonded phases comprise C_{18}, C_8, C_6, C_4, C_3, C_2, C_1, phenyl, cyclohexyl, and cyanopropyl phases. There is no a priori superiority of one chain length or bonded phase over another that would guide us to the selection of one packing over another based on properties of our analytes. One can nevertheless make some generic recommendations. For a given silane type and bonding level, the phases with the longer alkyl chains are hydrolytically more stable than the shorter chains and the phenyl and cyanopropyl phases; thus longer alkyl chains like C_8 or C_{18} are preferred. All silica-based packings contain silanol groups that may interact with polar functional groups on the analytes, especially amino groups, and may cause peak tailing. Manufacturers of HPLC packings have learned to mitigate this interaction, and newer reversed-phase packings based on high-purity silicas exhibit much less tailing with basic analytes than older packings. Therefore these packings are the generally most preferred reversed-phase packings today.

The selectivity differences between bonded phases of different chain length are small and cannot easily be predicted. However, significant selectivity differences can be encountered when there is additional interaction between the analytes and polar functional groups in the packing. Thus packings with a high silanol activity can exhibit notable selectivity differences compared with the identical packing with low silanol activity. Cyanopropyl packings have always been used as an alternative to the more standard C_8 and C_{18} packings. They have exhibited improved peak shape for basic compounds and often a different selectivity as well. However, columns are notoriously unstable, both hydrolytically and mechanically. Today, phases have become available with a polar functional group embedded in the long-chain hydrocarbon that combine the advantageous alternative selectivity with an improved stability. These packings are the best choice if the standard C_8 or C_{18} phases do not deliver the crucial selectivity necessary for a separation.

Fully endcapped C_8 and C_{18} phases can exhibit hydrophobic collapse in eluents that contain close to 100% water. Polar functional groups on the packing prevent this phenomenon. Thus nonendcapped packings or packings with embedded polar functional groups are preferred for very polar analytes that require mobile phases with no or very little organic modifiers.

A detailed discussion of these issues is found in Chapter 10.

7.9 SUMMARY

In this chapter I intended to give guidance to the novice chromatographer as to the methods that are available for a particular analysis. I provided information on the advantages of some techniques over others to guide the user to the best methodology. But contrary to gas chromatography, and with the exception of size-exclusion chromatography (SEC), selectivity in HPLC is determined not by the column alone but also by the mobile phase. There is therefore no one-for-one assignment between an analytical problem and the "best" column for this problem. If you expected this kind of an advice, you surely would be disappointed. Such a solution does not exist objectively. So, while you may get advice from an experienced chromatographer to "buy column XYZ from supplier ABC," you will get that individual's personal preference, but not an objectively optimal choice.

My approach in this chapter has been to present you with the options and then discuss the alternatives. I placed emphasis on reliability and reproducibility, if the techniques look otherwise similar. Therefore I recommended, for example, size-exclusion techniques over retention chromatography, if the analytical problem can be solved by both techniques. Also, for reasons of reproducibility, I recommended reversed-phase methods over normal-phase methods. Among the options for reversed-phase packings, I recommended C_{18} bonded phases over CN bonded phases. Most of my recommendations are paralleled by the popularity of a particular technique. We can see this as a confirmation of the recommendations; thousands of chromatographers cannot be wrong.

8 Size-Exclusion Chromatography

> In another moment down went Alice after it, never once considering how in the
> world she was to get out again.
>
> — Lewis Carrol, *Alice in Wonderland*

Size-exclusion chromatography (SEC), previously also called *gel-permeation chromatography* and *gel-filtration chromatography*, is the conceptually and mechanistically simplest form of chromatography. The analytes do not interact with the surface of the packing, but are separated by their ability to penetrate the pores of the packing. Smaller analytes penetrate into smaller pores than larger analytes, thus exploring a larger fraction of the pore volume. Therefore the elution volume of smaller analytes is larger than the elution volume of larger analytes. Large molecules elute first.

8.1 SEPARATION MECHANISM

The actual separation mechanism is slightly more complicated. It is depicted schematically in Figure 8.1. We assume rigid spherical molecules of radius r_m and cylindrical bottomless pores of radius R_{po}. Then the fraction of the pore space that can be explored by the molecule is (1)

$$K_0 = \left(1 - \frac{r_m}{R_{po}}\right)^2 \tag{8.1}$$

This is the partition coefficient of size-exclusion chromatography. From this equation we can see that one can actually obtain a size separation already on

140

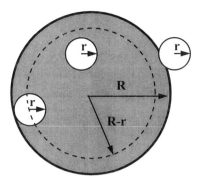

Figure 8.1   Principle of size-exclusion chromatography.

a packing with a single uniform pore size, and a pore-size distribution is not necessary for size-exclusion chromatography. The distribution coefficient is always smaller than 1. A distribution coefficient approaching 1 is obtained for a very small molecule, and it becomes 0 for a molecule that is the same size as the pore. For molecules larger than the pore size, the equation is undefined. This simple model is most useful for a basic understanding of size-exclusion chromatography.

However, most macromolecules are random coils under the conditions used for size-exclusion chromatography, and the packings for SEC have a pore-size distribution. Thus the penetration of a soft, randomly shaped macromolecule into the pores of a packing is better described by (2)

$$K_0 = \int_0^\infty \int_0^\infty P_r P_p P_R \, dR_{\text{po}} \, dr_m \tag{8.2}$$

where P_r denotes the probability that a macromolecule forms a random coil of radius r, P_p is the probability that this random coil will reside in a pore with radius R, and P_R is the distribution of pore radii. [*Note:* This description contains actually a correction over the approach used by Carmichael (2), who assumed that the probability of a molecule with the radius r to reside in a pore with the radius R is either 1 or 0, depending whether $r < R$ or $r > R$. As we have seen in the preceding discussion and Equation (8.1), this assumption is incorrect.]

Since there is no interaction between the analyte and the packing, the solute distribution between mobile and stationary phase is controlled by entropy alone (3):

$$K_0 = e^{\Delta S^\circ / R} \tag{8.3}$$

The immediate consequence of this is the fact that the retention in size-exclusion chromatography is, in principle, temperature-independent. However, a temperature dependence arises from several secondary effects; for instance, the swelling of polymeric packings is temperature-dependent, and the hydrodynamic volume of the random coil of a sample polymer increases with temperature.

The largest elution volume in size-exclusion chromatography is the total mobile-phase volume in the column: the sum of the pore volume and the interstitial volume; thus the separation space is constrained. Therefore, how good a separation can be obtained depends strongly on the width of the individual band, namely, on the plate count. System plate count is very important in SEC.

The smallest elution volume should in principle be the interstitial volume. However, the interstitial space can be viewed as a second pore space with very large pores, of the order of the size of the particles, in which the fluid velocity changes gradually from 0 at the particle wall to a finite velocity in the center of the stream. Therefore, very large molecules exhibit additional "exclusion" in the interstitial space, and the elution volume decreases with molecular size even outside the pores of the particle. This phenomenon is called *hydrodynamic chromatography.*

The retention factor in size-exclusion chromatography is defined relative to the elution volume of an excluded peak:

$$k_e = \frac{V_R - V_e}{V_e} \qquad (8.4)$$

This can also be expressed in terms of the partition coefficient and the interstitial and pore fraction in the column:

$$k_e = \frac{\varepsilon_p K_0}{\varepsilon_i} \qquad (8.5)$$

The ratio of the pore fraction to the interstitial fraction has been used as a measure of quality of a size-exclusion column. It is the ratio of the useful volume to the unused volume in a column. However, a better measure is the pore fraction itself.

The purpose of most size-exclusion separations is the determination of the molecular-weight distribution of a polymer. Since the elution volume of a molecule is determined by its size, which is a function of its molecular weight, one can, in turn, use the elution volume to determine the molecular weight of an unknown. In order to do so, one must know the relationship between the molecular weight and the elution volume. This relationship is obtained empirically by injecting standards of known molecular weight and measuring their elution volume. One then plots the logarithm of the molecular weight of the standard versus the elution volume. This is called the *SEC calibration curve*; a typical calibration curve is shown in Figure 8.2. The largest elution

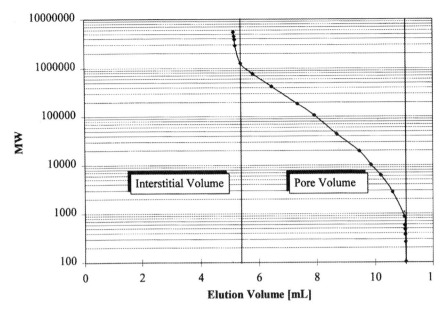

Figure 8.2   Typical SEC calibration curve.

volume is the total permeation volume, which is the sum of the pore volume and the interstitial volume. The points to the left of the total permeation volume represent the elution volumes of standards that are partially excluded from the pores. At roughly half the total permeation volume the calibration curve takes a sharp turn upward; that is, molecules of different size elute at nearly the same elution volume. This is the border between the pore volume and the interstitial volume. The fact that the elution volume continues to drop with increasing molecular weight is due to the size-exclusion effect in the interstitial space.

Now that we have established the relationship between the molecular weight of our example polymer and the elution volume, we can inject a sample of the same polymer with an unknown molecular-weight distribution and determine its molecular-weight distribution, provided the detector response is proportional to the concentration of the polymer in solution. The relationship between the chromatographic response and the molecular weight of the sample is shown schematically in Figure 8.3.

If a different polymer needs to be analyzed, a different calibration curve needs to be established. However, if the relationship between the molecular weight and the size, or more precisely, the hydrodynamic volume, of a molecule is known for both the standard polymer and the unknown polymer, one can simply transform the molecular-weight axis. The procedure for this is called the *universal calibration*, and its principles are discussed below.

The Flory–Fox equation describes the relationship between the size of the

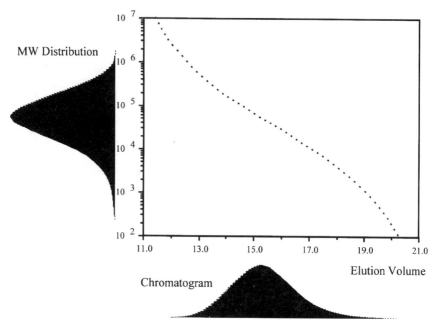

Figure 8.3 Relationship between the chromatogram and the molecular-weight distribution of a polymer.

random coil of a linear polymer in solution and its molecular weight:

$$\sqrt{\langle s^2 \rangle^3} = \frac{1}{\Phi} M[\eta] \tag{8.6}$$

The left-hand side of the equation is the third power of the root-mean-square radius of gyration of the molecule (proportional to the volume or "size" of the molecule). The right-hand side contains the product of the molecular weight M of the polymer and the intrinsic viscosity $[\eta]$. Φ is a proportionality constant. Benoit et al. (4) found that he could obtain a universal curve for all polymers when he plotted the product of the intrinsic viscosity and the molecular weight against the elution volume.

So, how does this help us? The Mark–Houwink equation relates the intrinsic viscosity to the molecular weight:

$$[\eta] = K_M M^a \tag{8.7}$$

The coefficients K_M and a are characteristic for each combination of polymer, solvent and temperature and are known as the *Mark–Houwink constants*. If we have these constants for the polymers under the conditions of interest, we can simply convert one molecular-weight axis into another one [by combining Eqs. (8.6) and (8.7)]:

$$\log(M_2) = \frac{1}{a_2 + 1} \left[\log\left(\frac{K_{M1}}{K_{M2}}\right) + (a_1 + 1) \log(M_1) \right] \tag{8.8}$$

This relationship works well for large polymers, but breaks down in the molecular-weight range of oligomers. A compilation of Mark–Houwink constants is found in Reference 5.

The efficiency of a column is of utmost importance in size-exclusion chromatography. Under standard operating conditions it is dominated by the mass-transfer term of the van Deemter equation and varies significantly over the chromatogram. The phenomenon has been studied in detail by Groh and Halász (6) using polystyrene standards with a narrow molecular-weight distribution as samples. The smallest molecules with the highest diffusion coefficients elute at and near the total permeation volume. The C term is relatively small because of the relatively fast diffusion in and out of the pores. As the molecular size increases, the diffusion coefficient decreases. Furthermore, as the size of the molecule reaches the same magnitude as the size of the pore, the diffusion in the pores becomes hindered. Both effects result in an increase in the C term and a decrease in efficiency. Counterbalancing this is the concomitant exclusion from the pore volume, which results in a decrease in the mass-transfer term until ultimately, at complete exclusion, the mass-transfer coefficient completely disappears and column efficiency is high again. Groh (7) describes the phenomenon by the following empirical equation:

$$C \propto \frac{k_e}{(1 + k_e)^2} d_p^2 M^{b_s} \tag{8.9}$$

The coefficient which includes the molecular weight is a descriptor for the influence of the diffusion coefficient in the pores. The exponent b_s contains the correction factor for restricted diffusion in the pores and is 0.54 for the material that Groh investigated, while the coefficient for the diffusion coefficient in free solution was 0.60.

Since the mass-transfer term decreases with the square of the particle size, it is desirable to use a packing material with a small particle size to maximize efficiency. However, there are some practical constraints, which we will discuss later which force the use of larger-than-optimal particle sizes.

Resolution in size-exclusion chromatography is a function of both the calibration curve and column performance. To normalize resolution to a molecular-weight difference of a decade, the specific resolution has been defined as

$$R_{sp} = \frac{\Delta V}{w[\log(M_2) - \log(M_1)]} \tag{8.10}$$

We can rewrite this using the partition coefficient K and the fraction ε_p of the column occupied by pores

$$R_{sp} = \frac{V_c}{w} \frac{\varepsilon_p \Delta K_0}{\log(M_2) - \log(M_1)} \tag{8.11}$$

The rightmost term on the right side of the equation can be seen as the selectivity in size-exclusion chromatography. It is the inverse of the slope of the dimensionless form of the calibration curve multiplied by the fraction of the column occupied by pores. It should be maximized by selecting the best pore size and pore size distribution for the molecular-weight range of the analyte.

$$S_{\text{SEC}} = \frac{\varepsilon_p \Delta K_0}{\log(M_2) - \log(M_1)} \tag{8.12}$$

8.2   SELECTION OF MOBILE PHASE AND COLUMN TYPE

The first step in the selection of the size-exclusion separation system is the choice of the mobile phase. We need to select a mobile phase in which the analytes, usually polymers, are soluble. This, in turn, determines the selection of the stationary phase, specifically, whether we select a packing designed for organic or aqueous size-exclusion chromatography (the term "aqueous" may include polar solvents). If the goal of the separation is a molecular-weight determination, the requirements for the mobile phase–column combination are quite stringent:

1. The mobile phase must be a good solvent for the polymer, yielding a solution of noninteracting random coils (exceptions: biopolymers such as proteins).
2. There must be no interaction between the polymer and the stationary phase (neither by retention through adsorption or ion exchange nor by exclusion by mechanisms other than size exclusion).

Furthermore, the most preferred conditions use solvents with a low viscosity, which is one of the reasons for the frequent use of elevated temperature in size-exclusion chromatography. The lowered viscosity of the mobile phase maximizes resolution at acceptable run times.

In general, one can divide size exclusion chromatography into two categories: SEC in organic solvents and SEC in aqueous solvents. For polymers that are soluble in organic solvents, styrene–divinylbenzene stationary phases are used for the most part. For packings that are soluble in aqueous eluents, hydrophilic methacrylates are most commonly used.

For most polymers, appropriate mobile phases have been worked out already. An example of this is shown in Table 8.1, which lists a range of

TABLE 8.1   Suggestions for Mobile Phases and Columns for the Size-Exclusion Chromatography of Many Polymers

Polymer	Solvent/Temperature
A. Mobile Phases Used with Styrene–Divinylbenzene Packings	
Polystyrene	Toluene/RT
Polyisobutylene	Toluene/RT or THF/RT
Polybutylene	Toluene/70°C
Chlorinated rubber	Toluene/70°C
Polybutadiene	Toluene/70°C
Polyisoprene	Toluene/70°C
Polydimethylsiloxane	Toluene/70°C
Chlorinated polyethylene	Trichlorobenzene/140°C
Polyethylene–ethylacrylate	Trichlorobenzene/140°C
Polyethylene–vinylacetate	Trichlorobenzene/140°C
Polyethylene–methacrylic acid	Trichlorobenzene/140°C
Polyphenyleneoxide	Trichlorobenzene/140°C
Poly-4-methylpentene(1)	Trichlorobenzene/140°C
Polyethylene	Trichlorobenzene/140°C
Polyethylene, ultra-high-molecular weight	Trichlorobenzene/145°C
Polypropylene	Trichlorobenzene/145°C
Polyetheretherketone	Phenol/TCB 1 : 1/145°C
Polyetherketone	Phenol/TCB 1 : 1/145°C
Polyglycolic acid	γ-Butyrolactone
Acrylonitrile–methylmethacrylate	THF/40°C
Cellulose Acetate	THF/40°C
Cellulose acetate–butyrate	THF/40°C
Cellulose acetate–propionate	THF/40°C
Cellulose nitrate	THF/40°C
Cellulose propionate	THF/40°C
Cellulose triacetate	THF/40°C
Cellulose tricarbanilate	THF/40°C
Ethyl cellulose	THF/40°C
Diallyl phthalate	THF/40°C
Epoxy	THF/40°C
Polyester alkyd	THF/40°C
Polybutene(1)	THF/40°C
Polybutadiene–styrene (SBR)	THF/40°C or toluene/70°C for high-butadiene SBR
Phenol–formaldehyde	THF/40°C
Phenol–furfural	THF/40°C
Polymethylmethacrylate	THF/40°C
Polypropyleneglycol	THF/40°C
Polystyrene	THF/40°C
Polysulfone	THF/40°C

(Continued)

TABLE 8.1   (Continued)

Polymer	Solvent/Temperature
A. Mobile Phases Used with Styrene–Divinylbenzene Packings	
Polyvinylacetate	THF/40°C
Polyvinylbutyral	THF/40°C
Polyvinylchloride	THF/40°C
Polyvinylchloride–acetate	THF/40°C
Polyvinylidenechloride	THF/40°C
Polyvinylformal	THF/40°C
Polystyrene–acrylonitrile (SAN)	THF/40°C or DMF/85°C for high-acrylonitrile SAN
Polystyrene–alphamethylstyrene	THF/40°C
Polyester thermoset	THF/40°C
Phenolics	THF/40°C
Rosin acids	THF/40°C
Polyglycolic acid	THF/40°C
Melamine–formaldehyde	Hexafluoroisopropanol + 0.01 M
Nylon (all types)	Sodium trifluoroacetate/40°C
Polybutylene–terephthalate	or
Polyethylene–terephthalate	*m*-Cresol + 0.05 M LiBr/100°C
Polyacrylonitrile	DMF + 0.05 M LiBr/85°C
ABS (acrylonitrile–butadiene–styrene)	DMF + 0.05 M LiBr/85°C
ASA (acrylic–styrene–acrylonitrile)	DMF + 0.05 M LiBr/85°C
ABA (acrylonitrile–butadiene–acrylate)	DMF + 0.05 M LiBr/85°C
Carboxymethyl cellulose	DMF + 0.05 M LiBr/85°C
ABS/polycarbonate	DMF + 0.05 M LiBr/85°C
Polybutadiene–acrylonitrile	DMF + 0.05 M LiBr/85°C
Polyurethane	DMF + 0.05 M LiBr/85°C
Polyacetal	DMF + 0.05 M LiBr/145°C
Polyoxymethylene	DMF + 0.05 M LiBr/145°C
Polyimide	*N*-Methyl pyrrolidone + 0.05 M LiBr/ 100°C
Polyamide–imide	*N*-Methyl pyrrolidone + 0.05 M LiBr/ 100°C
Polyetherimide	*N*-Methyl pyrrolidone + 0.05 M LiBr/ 100°C
Polyethersulfone	*N*-Methyl pyrrolidone + 0.05 M LiBr/ 100°C
Polyvinylidenefluoride	*N*-Methyl pyrrolidone + 0.05 M LiBr/ 100°C
Polyfuran–formaldehyde	Dimethylacetamide/60°C
Cellulose	Dimethylacetamide + 0.5% LiCl/80°C
B. Mobile Phases Used with Glycidylmethacrylate Packings	
Polyethylene oxide	Water (+0.1 M LiNO$_3$)
Cornstarch, dextrane, pullulane	Water (+0.1 M LiNO$_3$)

TABLE 8.1   (Continued)

Polymer	Solvent/Temperature
B. Mobile Phases Used with Glycidylmethacrylate Packings	
Hydroxyethyl cellulose	Water ($+0.1$ M LiNO$_3$)
Polyacrylamide	Water $+ 0.5$ M LiNO$_3$
Polyacrylamide, partially hydrolyzed	Water $+ 0.3$ M NaCl $+ 0.1$ M KH$_2$PO$_4$, pH 7
Acrylamide/acrylic acid copolymer	Water $+ 0.3$ M NaCl $+ 0.1$ M KH$_2$PO$_4$, pH 7
Acrylamide/dimethylallylammonium chloride copolymer	Water $+ 0.24$ M Na formate
Polyvinylalcohol	Water/acetonitrile 80/20
Polyvinylpyrrolidone	Water/methanol 50/50 $+ 0.1$ M LiNO$_3$
Polystyrene sulfonate	Water/acetonitrile 80/20 $+ 0.1$ M NaNO$_3$
Lignin, sulfonated	Water/acetonitrile 80/20 $+ 0.1$ M NaNO$_3$
Polyacrylic acid	Water $+ 0.1$ M NaNO$_3$
Polyalginic acid	Water $+ 0.1$ M NaNO$_3$
Hyaluronic acid	Water $+ 0.1$ M NaNO$_3$
Carrageenan	Water $+ 0.1$ M NaNO$_3$
Polyvinylpyrrolidone/acrylic acid	Water $+ 0.2$ M LiNO$_3$ $+ 0.1$ M Tris, pH 9
Polyvinylpyrrolidone/acrylic acid	Water $+ 0.5$ M LiNO$_3$ $+ 0.1$ M Tris, pH 7
DEAE dextrane	Water $+ 0.8$ M Na NO$_3$
Polyvinylamine	Water $+ 0.8$ M Na NO$_3$
Polyepiamine	Water $+ 0.1$ M TEA
N-Acetylglucosamine	Water $+ 0.1$ M TEA $+ 1\%$ acetic acid
Polyethyleneimine	Water $+ 0.5$ M Na Ac $+ 0.5$ M HAc
Poly(*N*-methyl-2-vinyl-pyridine)	Water $+ 0.5$ M Na Ac $+ 0.5$ M HAc
Poly(2-vinyl-pyridine)	Water $+ 0.5$ M Na Ac $+ 0.5$ M HAc
Chitosan	Water $+ 0.5$ M Na Ac $+ 0.5$ M HAc
RNA	Water $+ 0.1$ M NaCl $+ 0.1$ M phosphate, pH 7 $+ 1$ mM EDTA
DNA fragments, plasmids	Water $+ 0.3$ M NaCl $+ 0.1$ M Tris-HCl, pH 7.5 $+ 1$ mM EDTA
Oligonucleotides	Water $+ 0.3$ M NaCl $+ 0.1$ M Tris-HCl, pH 7.5 $+ 1$ mM EDTA

polymers and the mobile phases used for these polymers. It is divided into two sections; section *A* lists polymers that are analyzed using organic SEC, and section *B* lists those that are separated using aqueous SEC. The information in Table 8.1 is compiled from a variety of sources, especially References 8 and 9. It is not inclusive; other mobile phase–column combinations may be suitable as well. For a more detailed discussion of the chromatographic conditions, the reasoning for the choice of certain mobile phases, or for alternative options, more specialized literature should be consulted (e.g., Ref. 9).

Instead of styrene–divinylbenzene packings, silica derivatized with trimethylsilane groups can be used as well for organic SEC. However, this packing still contains residual silanols, which complicate the interaction of the polymer with the stationary phase and cause unwanted retention. Because of this effect, styrene–divinylbenzene packings are preferred for organic SEC. In aqueous SEC, silica derivatized with a hydrolyzed glycidoxypropylsilane (diol silica) is a good substitute for the hydrophilic methacrylate-based packings. Once again, silanol-group interactions complicate the picture, but they can be dealt with easily in the same way as the residual carboxylic acid groups in the methacrylate are dealt with.

A few general observations can be made about the selection of mobile phases for aqueous size-exclusion chromatography. Neutral hydrophilic polymers can be chromatographed in water or in water with a small amount of salt added. If the polymer is ionic, a salt should definitely be added to the mobile phase to suppress ionic interactions, both of the polymer with itself and of the polymer with the stationary phase. Ionic interactions can both result in retention via ion exchange and in exclusion due to ionic repulsion of polymers that carry the same charge as the charge of the packing. Methacrylate-based packings are not completely free of charge. For the most part they carry a small amount of negative charge, probably due to small amounts ($< 20\,mmol/mL$) of methacrylic acid. Silica-based packings carry a negative charge at neutral pH due to the existence of residual silanols on the surface of the packing.

In the case of ionizable polymers, it is also advisable to control the pH of the mobile phase to regulate the charge and the degree of ionization. This is especially true for biopolymers.

If the polymer is partially hydrophobic, as is sulfonated polystyrene, hydrophobic interaction must be suppressed as well. This is usually done with the addition of methanol or acetonitrile to the mobile phase; 20% of the organic modifier is usually sufficient.

One can identify the presence of adsorption or exclusion effects through a systematic variation of a mobile-phase parameter and observation of the chromatographic response. The addition of salt can result in a reduction of retention, if ion-exchange interaction is present. It can result in an increase in retention by either breaking ion exclusion or by promoting hydrophobic interaction. The addition of an organic solvent will result in a reduction in the elution volume, if hydrophobic interaction is present. For example, one can eliminate both ion exclusion and hydrophobic interaction by adding both a salt and an organic solvent. Table 8.1, section *B* lists examples for these strategies.

An example of the influence of ionic strength is shown in Figure 8.4. There, the elution volume (normalized by the column volume) of several proteins on a silica-based diol column is plotted in dependence of the salt concentration in the mobile phase at constant pH. The silanol groups of this type of packing are negatively charged at neutral pH; therefore, the positively charged lysozyme and cytochrome *c* are retained slightly by ion exchange at low ionic concentration. This interaction is eliminated at higher ionic strength. Bovine

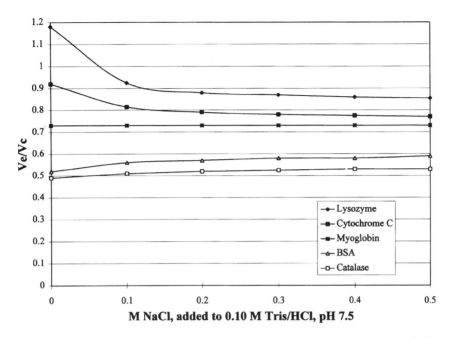

Figure 8.4 Dependence of the retention of proteins on the ionic strength. Column: Protein-Pak 300 SW, 7.5 mm × 300 mm, Waters Corporation. The stationary phase is a diol-derivatized silica.

serum albumin and catalase are negatively charged. Their decreased retention at low ionic strength is attributed to ion exclusion. The slight increase in retention of BSA at high ionic strength may be due to the onset of hydrophobic interaction. The retention of myoglobin with an isoelectric point of 7.0 is unaffected by the changes in ionic strength. Thus a careful examination of the influence of mobile-phase parameters on the retention behavior of selected analytes can be very useful in the elimination of non-size effects in SEC.

8.3   SELECTION OF THE PORE SIZE

After the correct combination of mobile phase and stationary phase has been chosen, the investigator needs to select the best pore size or combination of pore sizes for the analysis. All packings for size-exclusion chromatography are available in different pore sizes to accommodate the needs of a particular analysis. Especially polymer-based packings such as styrene–divinylbenzene-based packings for organic size-exclusion chromatography are available in pore sizes ranging from about 1 nm to about 1 μm. An example of the calibration curves of a family of styrene–divinylbenzene packings is shown in Figure 8.5. Traditionally, a bank of columns has been used in SEC, both for

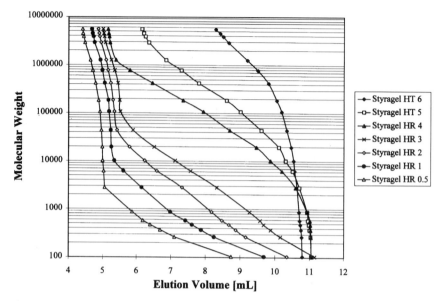

Figure 8.5 Calibration curves for a family of styrene–divinylbenzene packings for size-exclusion chromatography.

increasing the plate count and for adjusting the separation power of the bank to the molecular-weight distribution of the polymer(s) to be analyzed.

First, we have to select a column that has pores large enough for the analyte to penetrate. Manufacturers typically give information about the molecular-weight range over which a particular packing material will separate. This is usually expressed in molecular weight of a standard such as polystyrene. To translate this into the molecular-weight range of the polymer that you want to analyze, you need to know what molecular weight of your polymer gives the same size random coil as the standard. As shown above, this can be done if the Mark–Houwink constants of the standard polymer and the sample polymer are known. But there is a simpler approach that is accurate enough for the current purpose. What we need to do is to calculate which molecular weight of the standard polymer and the sample polymer results in the same (extended) chain length. Styrene, with a molecular weight of 104, contributes a molecular-weight increment of 52 for every methylene group in the polystyrene chain. The molecular-weight increment per methylene group in polyethylene is 14. Thus the polyethylene molecular weight that gives the same chain length and therefore the same size random coil as a 200,000-dalton polystyrene is $200,000 \times \frac{14}{52} = 53,800$.

Often, more than one column is suitable for the molecular-weight range of interest. Then we need to select the column with the highest resolution. Resolution depends on both the plate count and the selectivity. Let us consider the selectivity next.

The selectivity of a column [Eq. (8.11)] depends on the fraction of the column volume occupied by pores ε_p and the inverse slope of the normalized calibration curve:

$$S_{SEC} = \frac{\varepsilon_p \Delta K_0}{\log(M_2) - \log(M_1)} \qquad (8.12)$$

Therefore, we desire to use columns with a large pore fraction and a flat calibration curve in the range of interest. There is seldom much play in the porosity; the porosity of SEC packings is generally maximized to the point that any additional pore volume would weaken the packing too much. However, the slope of the calibration curve may vary significantly, depending on the breadth of the pore size distribution of the packing or whether the packing is a blend of different pore sizes (10). In Figure 8.6 the SEC selectivity of a single-pore-size column is compared to that of a mixed-bed column. The mixed-bed column has a uniform selectivity from low to high molecular weight, but its selectivity is inferior to the single-pore-size column over all the middle

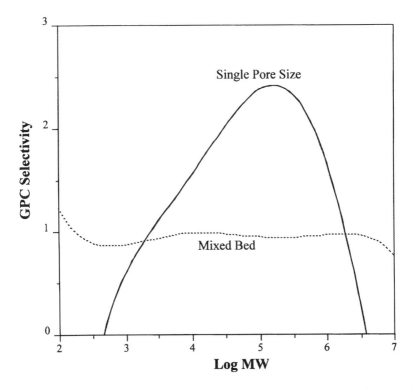

Figure 8.6 Comparison of the SEC selectivity of a single-pore-size column and a mixed-bed column.

range. Therefore, if the molecular weight of the polymer to be analyzed would spread from 2000 to 2,000,000, you would achieve a significantly higher resolution with the single-pore-size column. Unfortunately, manufacturers of SEC columns do not provide the selectivity information anymore, and you have to derive it yourself from the calibration curves.

In order to maximize resolution, you should generally select columns and column combinations that provide selectivity only in the range in which it is needed. This is especially important for column combinations, which are commonly used in size-exclusion chromatography. If a column bank provides a separation capability outside the needed range, it is not only a waste of analysis time but is also likely to destroy resolution in the desired range, since the unnecessary column in the bank contributes to band spreading, but not to the separation. Typical single-pore-size columns have a separation range of 2.5–3 molecular-weight decades. Thus most separation problems can be covered by just two pore sizes, and often just one pore size is sufficient. To increase resolution in a column bank, only columns packed with these optimal pore-size packings should be added to the bank, while columns with a separation capability outside the desired range should be removed. However, there may be some tradeoff in practice. While it is fundamentally desirable to dedicate a well designed column bank to a particular analytical problem, this may be impractical in a lab that is faced with a broad range of problems covering the gamut of molecular weights. In such a case, a column or column bank with a fairly uniform SEC selectivity like the mixed-bed column in Figure 8.6 may be more appropriate.

8.4   SELECTION OF THE PARTICLE SIZE

The selection of the particle size of a column involves several compromises. One is the compromise between analysis time, resolution, and backpressure discussed in general in Chapter 3. The second is the selection of the range in the chromatogram in which we desire maximum resolution and the conditions associated with this. The third is the compromise between the desire to maximize resolution and the prevention of breakdown of the sample due to shear degradation. The second and third are specific to size-exclusion chromatography.

The peak width varies with the linear velocity, and consequently the analysis time, the particle size, and the diffusion coefficient of the analyte. This can be expressed by the following equation (6,11):

$$h = \frac{2D_M(1 + k_e)}{u_i} + 2d_p + 0.6 \frac{k_e}{(1 + k_e)^2} \frac{d_p^2}{D_M} u_i \qquad (8.13)$$

[*Note:* This equation contains the simplifying assumption that the restriction to diffusion is independent of the partition coefficient, which certainly is not true. On the other hand, the dependence of the C term on the retention factor was derived empirically in Ref. 6 and therefore contains implicitly the factor for restricted diffusion in the pores.]

Since the diffusion coefficient varies significantly over the size-exclusion chromatogram, due to the variation in molecular weight, the HETP varies significantly over the chromatogram as well. Furthermore, the dependence of the HETP on the linear velocity and therefore the optimal linear velocity varies significantly over the chromatogram as well. This is shown in Figure 8.7, where the plate count is plotted as a function of the SEC retention factor and the analysis time. The data were calculated from Reference 11 for polystyrene standards in methylene chloride at room temperature using a packing with an exclusion limit of about 150,000 and a particle size of 10 μm packed in a 25-cm column. A reasonably uniform plate count throughout the chromatogram is obtained only at an analysis time of about 45 min for this column. Figure 8.8 shows the same plot for 3-μm particles in the same column. The plate count is higher, and the condition with the most uniform plate count throughout the chromatogram has shifted toward shorter analysis times. It is important to realize that the optimum linear velocity for the largest part of the chromatogram does not correspond to the optimum linear velocity for the totally permeating peak, which is commonly used as an indicator of plate count. Rather, the optimal velocity for most of the chromatogram occurs at a factor of 4–10 lower than the optimum linear velocity for the totally permeating peak.

The range of best overall resolution is best illustrated by plotting the number of peaks that can be resolved against the analysis time. This can easily

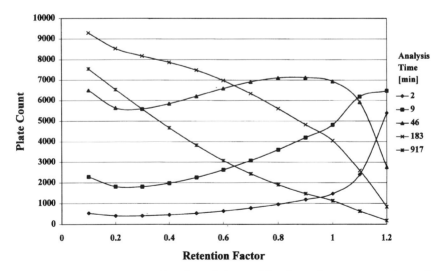

Figure 8.7 Plate count as a function of the position in a chromatogram (retention factor) and the analysis time: 10-μm particles.

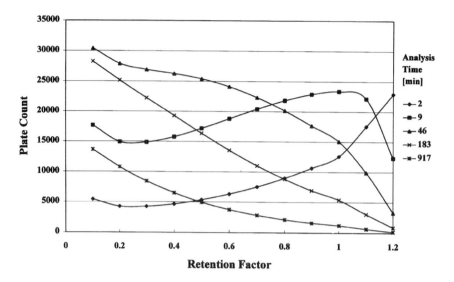

Figure 8.8   Plate count as a function of the position in a chromatogram (retention factor) and the analysis time: 3-μm particles.

be done for SEC, since the volume over which the separation takes place is limited. Once again we use the data generated by Ahr (11). Figure 8.9 shows a plot of the peak capacity for polystyrene standards in methylene chloride at room temperature of the packing described above packed into a 25-cm column. The particle size was varied from 3 μm to 5–10 μm. The maximum peak capacity was obtained at an analysis time of about 100 min for the 10-μm packing, at about 40 min for the 5-μm packing, and at about 15 min for the 3-μm packing.

A typical SEC column has a diameter of 7.8 mm and is used at a flow rate of 1 mL/min. This results in an analysis time of 10 min per column. From Figure 8.9 we can see that such a short analysis time is too fast for a 10-μm packing and that a lot of resolution is sacrificed for the high-molecular-weight part of the chromatogram. The analysis also demonstrates that a significantly improved chromatogram could be obtained by reducing the flow rate to 0.25 mL/min for a 10-μm column. On the other hand, a column packed with 3-μm particles and the same internal diameter would do very well at a flow rate of 1 mL/min. From the standpoint of resolution a particle size of $\sim 3 \mu$m is probably optimum for most SEC applications (12).

Contradicting this requirement for small particle size is the fact that polymers, especially high-molecular-weight polymers, are sensitive to shear and can degrade during transport through the column. This can best be observed with polystyrene standards of narrow molecular-weight distribution. Poly-styrene standards of a molecular weight of above 3 million can show late elution and distorted peaks at a flow rate of 1 mL/min in toluene in standard

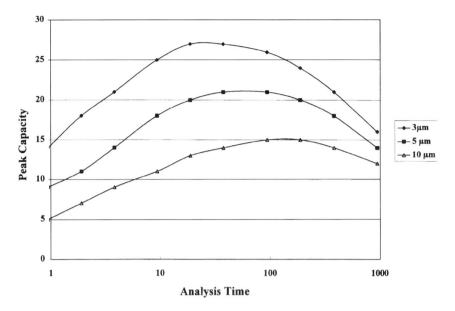

Figure 8.9   Plot of the peak capacity for three different particle sizes versus the analysis time. Column length 25 cm.

SEC columns (7.8 mm i.d.) packed with 5-μm particles. For this reason manu-facturers offer columns for high-molecular-weight applications packed with 10- or even 20-μm particles (e.g., for ultra-high-molecular-weight polyethylene). In this case the practical need to obtain unadulterated results overrides the theoretical need for high resolution.

In conclusion, for maximum resolution you should generally select the smallest particle size available, typically 5- or 3-μm particles. On the other hand, if you are analyzing polymers with a high molecular weight, you should pick larger particle sizes to avoid shear degradation of your sample in the column.

8.5   POLYMERIC PACKINGS FOR ORGANIC SEC

All available polymeric packings for size-exclusion chromatography in organic solvents are based on styrene–divinylbenzene copolymers. They are prepared by a suspension polymerization technique. Droplets containing the monomers and a pore-forming solvent (porogen) are suspended in an aqueous medium and polymerized using radical initiators. Macroporous packings are formed as follows. After initiation, the polymer in the droplet grows to a size determined by the porogen and then precipitates. The precipitated spherules in the droplet agglomerate and are bound together by further crosslinking. They form the skeleton of the particle. This way particles of a fairly high rigidity are generated. In microporous, also called microreticulate, packings the pores are formed by the crosslinking network itself. These packings do not have a permanent pore structure. Instead, the pores are created through the swelling of the particles in solvent. Microporous packings tend to be mechanically weaker than macroporous packings. Packings with a polystyrene exclusion limit of < 10,000 daltons are microporous packings, those with an exclusion limit of above 40,000 dalton are macroporous. Packings with an exclusion limit between these two values can be prepared either way.

Since the microporous packings depend on swelling for the formation of the pores, they also can be subject to shrinking when exposed to nonswelling solvents. You should consult the manufacturers' literature for solvent compatibility of these packings. Macroporous packings swell and shrink as well, but the phenomenon is limited and accordingly these packings tolerate solvent changes much easier than the microporous packings.

The skeleton of the styrene–divinylbenzene packings is very hydrophobic. This is favorable for the size-exclusion chromatography in organic solvents, since hydrophobic interaction can easily be prevented in organic solvents. However, every packing contains small amounts of polar functional groups, partly from the initiator and partly from unknown origin. Although the actual recipes are the proprietary knowledge of the manufacturers, one can expect cyano groups to be present in the packings stemming from the commonly used initiator azo-bis-isobutyronitrile. Small amounts of alcohol and aldehyde groups have also been found. When polymers that contain reactive groups that can bind to these residual functional groups are analyzed, they can become permanently attached to the packing.

Manufacturers of styrene–divinylbenzene packings for organic size-exclusion chromatography are Jordi (Jordigel, USA), Polymer Labs (Plgel, UK), Showa Denko (Shodexgel, Japan), Tosoh (TSK gel, Japan), and Waters (Styragel, USA).

8.6   POLYMERIC PACKINGS FOR AQUEOUS SEC

Most polymeric packings for size-exclusion chromatography in aqueous mobile phases are based on crosslinked glycidoxymethacrylate that is further hydrophilized and contains ($-CH_2-CHOH-CH_2O-$) as the surface moiety. The exact composition, the nature of the crosslinker, the initiator, and the method of rendering the packings hydrophilic are all proprietary information that is not commonly disclosed by the manufacturers. But on the basis of studies available in the academic literature, one can surmise that the basic particle formation process is similar to the one described for styrene–divinylbenzene-based packings.

Although these packings are highly crosslinked, they exhibit some swelling and shrinking in the commonly used solvents, such as water with some electrolyte. Usually, they swell most in water. One should consult the manu-facturer's recommendations before attempting to use these packings in other solvents.

The interacting surface of the packings for aqueous SEC carry alcohol or polyol functions. These functional groups are very hydrophilic and prevent the interaction with polar polymers in an aqueous medium. However, hydrophobic sample polymers might still interact weakly with the packing. Also, most packings carry a small amount of residual charge. The majority carry a negative charge, probably due to methacrylic acid stemming from hydrolysis of the packing. But some carry a positive charge of unknown origin. Consult the manufacturer's literature.

An alternative to the methacrylate-based packings are packings based on crosslinked polyvinylalcohol (Asahipak GPC columns). These packings are available in several pore sizes with a permanent pore structure, but their exact chemical composition is not disclosed by the manufacturer. They can be used both with aqueous solvents and with organic solvents.

Suppliers of polymeric packings for aqueous size-exclusion chromatography are Polymer Labs (Plgel, UK), Showa Denko (Shodex OHpak, Japan), Tosoh (TSKgel PW, Japan), and Waters (Ultrahydrogel, USA).

Application examples for polymeric packings for aqueous SEC are found in Table 8.1. One of the more difficult applications examples is the SEC of nucleic acids. The parameters of the separation using methacrylate-based packings were studied in detail by Kato et al. (13). The problem is that the elution volume increases with ionic strength, and there is no retention plateau as encountered with other polymers. Nevertheless, useful chromatograms can be obtained under standardized conditions.

There is one special application of size-exclusion chromatography that is different from the techniques discussed so far: it is the SEC of oligosaccharides, which is carried out on sulfonated styrene–divinylbenzene packings in various ionic forms, most frequently the Ca form. The mobile phase is water at elevated temperature. An example of this is the determination of the oligomeric composition of corn syrup shown in Figure 8.10.

8.7 SILICA-BASED PACKINGS

Silica-based packings have the advantage that they neither swell nor shrink when exposed to different solvents. Silica itself can be used in some simple cases for size-exclusion chromatography, both in organic solvents and in water, but for many polymers of practical interest the silica surface is too adsorptive. Therefore, the surface of silica is deactivated by either derivatization with trimethylsilane for organic SEC or with glycidylpropylsilane for aqueous SEC. As with any bonding technique to silica, we have to keep in mind that there is a large number of silanols left on the surface after bonding. They cannot be removed, due to steric hindrance, but they usually are still accessible enough

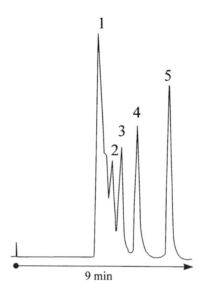

Figure 8.10   SEC analysis of corn syrup on a sulfonated styrene–divinylbenzene resin in the Ca form. Peak identification: (1) high-MW oligomers; (2) DP-4; (3) maltotriose; (4) maltose; (5) glucose. (*Chromatographic conditions*: Column: Sugar-Pak 1, 6.5 mm × 300 mm. Mobile phase: water. Flow rate: 1 mL/min.) (Chromatogram courtesy of Waters Corp.)

to interact with sample molecules. In aqueous SEC they form a negative charge on the surface at neutral pH, which can interact with charged analytes. Therefore it is always advisable to add a salt to the mobile phase. In this aspect this is similar to the situation with the methacrylate gels.

Manufacturers or suppliers of silica-based packings for SEC are Beckman (UltraSpherogel), Bio-Rad (Bio-Sil SEC), Mac-Mod (Zorbax PSM and GF), Merck (Lichrospher), Synchrom (SynChropak GPC), Showa Denko (Protein KW), Tosoh (TSKgel SW), and Waters (Protein-Pak).

A common application of the diol-type packings, as the glycidylpropyl-silane-derivatized packings are commonly called, is the size-exclusion chromatography of proteins. Classic, non-HPLC SEC of proteins has commonly and successfully been carried out on packings based on polydextranes such as Sephadex. Since these packings are too weak to withstand the pressures employed in HPLC, the diol-derivatized silica has become a substitute with reasonably similar properties as the traditional packings. However, one needs to take into account the presence of the acidic silanols on the surface. The most effective way to suppress this interaction is to increase the ionic strength of the elution buffer. Usually, less than 0.5 mol/L of salt is sufficient. Too high an ionic strength promotes hydrophobic interaction, which is equally undesirable. One can plot the elution volume of the proteins of interest against the ionic strength. If one encounters a plateau of unchanging elution volume in the middle of the ionic strength range, one can safely exclude interaction with the stationary phase in this range. Alternatively, one can change the charge of the protein and/or the silica surface by a change in pH, but one must keep in mind that silica limits one to a nonalkaline pH range. The best pH is typically slightly above the isoelectric point of the protein. What plays a role in the interaction between the surface of the packing and the protein is not the total charge of the protein, but rather the charge on the surface of the protein. For a more detailed discussion, see the section covering the ion-exchange chromatography of proteins.

Size-exclusion chromatography is especially useful for the preparative chromatography of proteins. The loadability is higher than for any technique of retention chromatography, since no surface of the packing is involved. Thus the limiting factor is only the solubility of the analyte in the mobile phase. This was traditionally exploited in the purification of proteins long before the event of HPLC. Because of the high cost of HPLC packings and the need for a costly high-pressure apparatus, truly large-scale SEC of proteins is still carried out with classic packings.

8.8 PRACTICAL ASPECTS OF SEC

Most size-exclusion chromatography is carried out on polymeric columns. Polymeric columns are soft and swell or shrink during solvent changes. Any solvent changeover should be done with utmost care, after consultation of the manufacturer's recommendations. Usually, a changeover at slow flow rates is recommended. When changing from a room-temperature mobile phase to a high-temperature solvent, special care needs to be taken. The high-temperature solvents are more viscous than the solvents used at room temperature. Thus one first increases the temperature, but not higher than about 10°C below the boiling point of the solvent that is in the column. Then one converts to the high-temperature solvent at slow flow rate. Then finally the temperature is increased to the operating temperature. Occasionally, such a conversion may even require an intermediate solvent. In general, solvent changes are to be avoided as much as possible and columns should be dedicated to a particular solvent.

The best injection volume in SEC is a compromise. One desires on one hand to keep the volume small to avoid band broadening and loss of resolution due to volume overload of the column; on the other hand, a more concentrated polymer sample is highly viscous, which can result in shear degradation of the sample or a distortion of the sample band at the top of the first column due to an effect that has been aptly called "viscous fingering." Table 8.2 lists the recommended concentration for samples of different molecular weights (14).

High-molecular-weight samples are prone to degradation for a variety of reasons. Shaking of the samples should not be practiced; the samples should

TABLE 8.2 Recommended Sample Concentrations

MW Range	Concentration (%)
<25,000	<0.25
25,000–200,000	<0.1
200,000–2,000,000	<0.05
>2,000,000	<0.02

be dissolved slowly. Sometimes the exposure to light or oxidizing agents should be avoided as well.

If a large injection volume is used, it is important that the injection volume of all samples and standards be the same for an accurate molecular-weight determination. Otherwise there would be an offset in the elution volume corresponding to half the difference in the injection volumes of the different samples.

When a column bank of different pore sizes is used, the columns should be arranged in the order of descending pore size with the column with the largest pore-size packing closest to the injector. In this arrangement the portion of the sample with the highest molecular weight is separated first; the sample is thus diluted and the viscosity reduced, which lessens the chance of a degradation of the sample due to shear in the column.

When analytes of very high molecular weight are analyzed, one should choose columns with a larger particle size. Although this limits resolution, it also avoids shear degradation of the high-molecular-weight component of the analyte. To avoid excessive loss of resolution at the low-molecular-weight end, one can combine column banks of different particle sizes arranged as outlined in the previous paragraph.

Occasionally, the analyte contains components that are still reactive. These compounds may react with the packing material and bind to it, changing the character of the packing. They also might make the packing reactive to other samples. In such a case one is advised to dedicate the column to the reactive analytes.

REFERENCES

1. E. F. Casassa, *J. Phys. Chem.* **75**, 3929 (1971).

2. J. B. Carmichael, *J. Polym. Sci.* **6**, Part A-2 517 (1968).

3. E. G. Malawer, "Introduction to SEC," in *Handbook of Size-Exclusion Chromatography*, C.-s. Wu, ed., Chromatographic Science Series, Marcel Dekker, New York, 1995.

4. H. Benoit, Z. Grubisic, P. Rempp, D. Decker, and J. G. Zilliox, *J. Chim. Phys.* **63**, 1507 (1996).

5. *Polymer Handbook*, 3rd ed., J. Bandrup and E. Immergut, eds., Wiley Interscience, New York, 1989.

6. R. Groh and I. Halász, *Anal. Chem.* **53**, 1325–1335 (1981).

7. R. Groh, dissertation, Universität des Saarlandes, Saarbrücken (1979).

8. "Solvent Selection Chart," in *Brochure WB012*, Waters Corporation, Milford (USA). 1995.

9. *Handbook of Size-Exclusion Chromatography*, C.-s. Wu, ed., Chromatographic Science Series, Marcel Dekker, New York, 1995.

10. U. D. Neue, *International GPC Symposium Proceedings*, 1994, pp. 779–785.

11. G. Ahr, dissertation, Universität des Saarlandes, Saarbrücken (1982).

12. H. Engelhardt and G. Ahr, *J. Chromatogr.* **282**, 385 (1983).

13. Y. Kato, M. Sasaki, T. Hashimoto, T. Murotsu, S. Fukushige, and K. Matsubara, *J. Chromatogr.* **266**, 341–349 (1983).

14. *Care and Use Manual for Styragel Columns*, Waters, 1990.

9 Normal-Phase Chromatography

Before the development of reversed-phase bonded phases, normal-phase chromatography was the most popular separation technique. It relies on the interaction of analytes with polar functional groups on the surface of the stationary phase, which is strongest when nonpolar solvents are used as mobile phase. Previously, it was also called *adsorption chromatography*. However, the technique has expanded from the exclusive application of metal oxide adsorbents such as silica and alumina as stationary phases to the use of polar bonded phases. Thus the name adsorption chromatography has become too narrow.

Normal-phase chromatography is a very powerful separations tool because of the wide range of solvents available that can be used to fine-tune the selectivity of a separation. However, it has fallen in disfavor with many chromatographers because of some of the complexities involved. Under some circumstances, lengthy equilibration times or reproducibility problems may be encountered, which are due largely to the sensitivity of the technique to the presence of small concentrations of polar contaminants in the mobile phase. If these problems are controlled, the technique typically gives chromatograms superior to reversed-phase methods, due to the low viscosity of the commonly used solvents.

9.1 RETENTION MECHANISM

The origin of retention in normal-phase chromatography is the interaction of polar functional groups of the analytes with polar functional groups on the surface of the packing. This interaction is mediated by the interaction of the mobile phase with the polar functional groups on the surface of the packing. Characteristically, the interactions are mostly dipole–dipole and hydrogen-

164

bonding. The surface of the stationary phases used in normal phase chromatography is occupied by polar functional groups capable of these interactions. Classical stationary phases are silica and alumina, with a dense population of $\equiv$Si—OH and $\equiv$Al—OH functional groups. There are several bonded phases that exhibit a retentivity that is comparable to the oxides: aminopropyl —$(CH_2)_3$—NH_2, cyanopropyl —$(CH_2)_3$—CN, and diol —$(CH_2)_3$—O—CH_2—CHOH—CH_2—OH. A few more specialized phases are available as well.

One can generalize the strength of the interaction of analytes and solvents with these stationary phases. The strength of the interaction increases in the following sequence: aliphatic hydrocarbons < olefinic hydrocarbons < aromatic hydrocarbons $\approx$ chlorinated   hydrocarbons < sulfides < ethers < ketones $\approx$ aldehydes $\approx$ esters < alcohols < amides (1). Amines and carboxylic acids exhibit a still larger interaction, but the strength depends on the acidity or basicity of the packing. On neutral packings such as the diol packing, the strength of the interaction of amines and acids is roughly equal. Amines are retained more on silica, and acids interact more strongly with the aminopropyl packing. Water is the strongest eluent for all normal-phase packings.

If a sample molecule has multiple functional groups, then the most polar one dominates the interaction. If it contains multiple functional groups of comparable interactive strength, they cooperate in the retention process. The strength of this cooperation depends on the steric position of the functional groups relative to each other. If they are located such that a simultaneous interaction with the stationary phase is possible, the interaction is stronger than if this were not the case. This principle makes normal-phase chromatography especially suitable for the separation of stereoisomers. On the other hand, normal-phase chromatography is not very effective for the separation of members of a homologous series, like methyl, ethyl, propyl, butyl derivatives of a parent compound. In contrast, this is the strong point of reversed-phase chromatography, as we will see later.

The retention mechanism is viewed as the competition of analyte molecules with mobile-phase molecules that cover the surface. The interaction of solvent molecules with the surface obeys the same rules as the interaction of analyte molecules. Hydrocarbons interact only weakly and non-specifically with the surface. Thus their elution strength is weak, which means that only analytes that interact weakly with the stationary phase can be eluted within a reasonable retention factor. Solvents with polar functional groups interact more strongly with the surface and thus have a higher elution strength. The elution strength of a mobile phase can be adjusted by mixing weakly eluting solvents with more strongly eluting solvents. This makes it possible to continuously vary the elution strength of the mobile phase to bring the analytes of interest into the desired retention window.

One can easily see that mixtures of different solvents can be prepared that have the same elution strength. But since the exact nature of the competition between analyte molecules and mobile-phase molecules depends on their respective functional groups, one can expect that solvent mixtures of similar elution strength but with different polar functional groups exhibit different selectivities; that is, the relative position of peaks in the chromatogram can vary. This can be used to manipulate or even enable the separation between sample constituents while maintaining the overall retention time. On the other hand, the nonpolar component of the mobile phase plays no role in the selectivity and functions largely as a diluent of the polar component of the mobile phase. Similarly, the hydrocarbon side chains of the polar solvent in the mobile phase play only a subordinate role in determining selectivity. Thus the addition of alcohols to a mobile phase results in similar selectivities, while aldehydes or ketones give a selectivity different from that of alcohols.

The elution strength of various solvents and their selectivity has been studied extensively in the literature (Refs. 2–7, and references cited therein). A quantitative model for the elution strength of solvent mixtures has been developed on the basis of fundamental molecular parameters of the solvents (2–4), and solvents have been classified according to their selectivity (5). The knowledge derived in these studies helps in the understanding of the retention behavior of analytes and enables us to achieve the desired selectivity of a separation.

The relationship between the retention factor and the elution strength of the mobile phase is given by the following equation (4):

$$\log(k) = \log(\beta) + S^\circ - \varepsilon^\circ A_A + \Delta_{eas} \tag{9.1}$$

where k is the retention factor, and β is the phase ratio, which in this case is defined as the volume of the adsorbed layer of mobile phase to the nonadsorbed mobile-phase volume. The parameter S° is the dimensionless interaction energy of the analyte with the sorbent surface, which is obtained when the elution strength of the mobile phase is 0. A_A is the dimensionless area of the adsorbed analyte, measured in units of the area of the adsorbed mobile phase molecule. ε° describes the elution strength of the mobile phase. The parameter ε°, called the *solvent strength*, will be discussed in detail in the following section. Δ_{eas} is a second-order term that corrects for effects that are not covered by the assumptions underlying the linear relationship between the strength of the mobile phase and the logarithm of the retention factor. An example of such an effect is the strong interaction of analyte molecules with the mobile phase in solution. (Note: In this discussion we omit the influence of the adsorbent activity term, in accordance with the treatment in Ref. 4.)

If secondary effects can be excluded, the retention factor of an analyte is simply a function of the solvent strength of the mobile phase:

$$\log\left(\frac{k_1}{k_2}\right) = A_A(\varepsilon_2^\circ - \varepsilon_1^\circ) \tag{9.2}$$

where the subscripts refer to mobile phases 1 and 2. Solvent strength parameters have been compiled for many different solvents (see Table 9.1). Of particular interest is the solvent strength of mobile phases obtained by mixing two or more solvents. In the simplest case, the solvent strength of a mixture obeys the following equation:

$$\varepsilon_{AB} = \varepsilon_A + \frac{\log(x_B^{n_B(\varepsilon_B - \varepsilon_A)} + 1 - x_B)}{n_B} \quad (9.3)$$

where the ε values designate the solvent strengths of solvents A, B and their mixture AB; the value n_B is the molecular area of solvent B in the adsorbed state relative to the molecular area of solvent A under the same conditions; and x_B is the mole fraction of more polar solvent B in the mobile phase. From this equation we can simply calculate the composition of a mixed mobile phase that has the desired solvent strength ε_{AB}:

$$x_B = \frac{10^{n_B(\varepsilon_{AB} - \varepsilon_A)} - 1}{10^{n_B(\varepsilon_B - \varepsilon_A)} - 1} \quad (9.4)$$

Table 9.1   Solvent Strength of Some Solvents Commonly Used in Normal-Phase Chromatography (Obtained on Alumina)

Solvent	Solvent Strength ε°
n-Pentane	0
n-Hexane	0
n-Heptane	0
Isooctane	0.1
Xylene	0.26
Diisopropyl ether	0.28
Toluene	0.29
Diethyl ether	0.38
Methylene chloride	0.39
Methyl ethyl ketone	0.51
Acetone	0.56
Dioxane	0.56
Methyl acetate	0.60
Tetrahydrofuran	0.62
tert-Butylmethyl ether	0.62
Ethyl acetate	0.62
Dimethyl sulfoxide	0.62
Diethylamine	0.63
Nitromethane	0.64
Acetonitrile	0.65
Isopropanol	0.78
Ethanol	0.88
Methanol	0.95

This concept works well for solvents that do not exhibit a strong interaction with the stationary phase. Examples for this are mixtures of hydrocarbons such as hexane with aromatic hydrocarbons (e.g., toluene) or chlorinated hydrocarbons (e.g., methylene chloride). These solvents tend to cover the surface of the adsorbent uniformly without interacting with specific adsorption centers on the surface of the packing. More polar solvents tend to interact specifically with the adsorption centers on the surface of the packing. Thus they are not evenly distributed over the surface but are localized by these adsorption centers. This violates the primary assumption of the simple model that the surface is covered uniformly with the solvent molecules. In order to accommodate this effect, one can either assume unrealistically high and variable molecular surface areas for the polar solvent, or modify the basic model to take account of the localization effect. The effect can be treated rigorously, but the calculation of the solvent strength of a mixed mobile phase becomes nontrivial. However, a method has been published and computer programs have been described that can accomplish this task (8). Also, several references contain nomograms that describe the solvent strength of binary mixtures (1,8,9). The solvent strength of some important solvent combinations are shown in Figure 9.1.

The determination of the solvent strength of mixtures of three or more solvents has been described in Reference 6 and can be accomplished with the computer program described in Reference 8. Since the stronger solvents all displace the weaker solvent in the adsorbed layer on the surface according to their concentration and the strength of their interaction with the surface, the solvent strength of a mixture of multiple solvents can simply be expressed through the concentration of the weaker solvent that remains adsorbed on the surface:

$$\varepsilon_{ABC \cdots M} = \varepsilon_A + \frac{\log(x_A/\theta_A)}{n_B} \tag{9.5}$$

where θ_A is the molar fraction of the solvent A that remains adsorbed on the surface. In this case, n_B is the average molecular area of all stronger solvents. The basic procedure for calculating the solvent strength of a multicomponent mixture involves the calculation of the surface concentration of all species involved. However, in practice, one would start with binary solvent mixtures of equal elution strength and then mix those to obtain ternary or quaternary mixtures. Although the elution strength of the higher mixtures is not the same as the elution strength of the binary component mixtures, the departure is sufficiently small that it can be compensated empirically without too much difficulty.

Multicomponent solvent mixtures are used for fine-tuning the selectivity of a separation. The selectivity, that is, the relative retention of peaks in the chromatogram, depends on the choice of the polar solvent(s) used to adjust retention. The primary cause of selectivity differences between different mobile

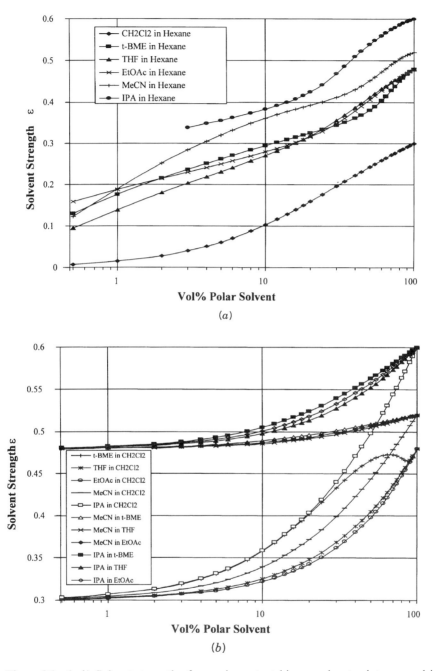

Figure 9.1 (a, b) Solvent strength of some important binary solvent mixtures used in normal-phase liquid chromatography. (Data courtesy of V. Meyer and M. D. Palamareva, based on a program written by H. E. Palamarev and previously published in part in Ref. 9.)

phases of equal elution strength lies in the interaction between the localization effect of the solute and the polar component of the mobile phase (7). The localization effect is a function of the configuration of the adsorbed molecules, which, in turn, should depend on the type of polar functional groups of the adsorbed molecules and the polar functional groups on the surface of the packing. Further effects can arise from the interaction of the sample with the polar solvent in solution, such as via hydrogen bonding.

The properties of solvents has been studied extensively by Snyder (5), who created a classification of the solvent properties of common solvents. It has been found (7) that (excluding proton donors such as alcohols) the maximum difference in mobile-phase selectivity is obtained if the polar solvents have a large difference in basicity. Thus, for maximum selectivity differences, one solvent should have a low basicity. Solvents of this type are acetonitrile, ethyl acetate or other esters, and acetone or other ketones. The other solvent should have a high basicity; examples are ethers such as *tert*-butyl methyl ether, diethyl ether or tetrahydrofuran, or amines such as triethylamine. Between these groups and alcohols, large differences in chromatographic selectivity can be obtained in normal-phase chromatography (10).

In Figures 9.1a, b, the solvent strength of several binary mixtures of normal-phase solvents is plotted against the volume percent of the polar modifier. The data were previously published in the form of nomograms in Reference 9. Additional data were supplied by M. D. Palamareva, who used the program described in Reference 8. Figure 9.1 makes it possible to select binary solvent compositions of the same solvent strength but significantly different selectivity. Also, within the same selectivity group, several options are given. Many more solvents can be used in normal-phase chromatography, but the combinations shown in Figure 9.1 are designed to cover a maximal range of selectivity options and should be sufficient for many separation problems. Also, the selected solvents have a low viscosity and can be used at 254 nm (even ethyl acetate, with a UV cutoff of 260 nm, can be used at low concentration).

The retention mechanism and solvent selectivity have been studied most carefully with alumina or silica as stationary phases. The knowledge of both for bonded phases used in normal-phase chromatography is much more limited. Nevertheless, it is safe to assume that similar selectivity rules for solvent strength and selectivity can be applied, especially since the results obtained for alumina and silica correlate well with each other.

Water has a very high solvent strength in normal-phase chromatography. Thus, even small amounts of water in the mobile phase can strongly affect retention, especially when very non-polar mobile phases are used with silica or alumina as adsorbents. Since the natural water content of hydrocarbon solvents is very low, an equilibration with a dry hydrocarbon-based mobile phase may take a very long time. It is not unusual to observe shifting retention times over a period of several days. This nasty problem is probably the main reason for the high popularity of reversed-phase methods compared to normal-phase methods. However, these difficulties diminish significantly with the polarity of the mobile phase and are not a problem any more with a solvent strength above $\varepsilon° = 0.35$. Also, it is much lessened with polar bonded phases (11). Moreover, an effective means is available that mitigates the problem even further: the use of mobile phases that are half-saturated with water. In order to prepare a "half-saturated" mobile phase, the nonpolar constituent of the mobile phase is split in half and to one half water is added. The portion containing the water is stirred for a while, approximately an hour, then the excess water is removed. Both halves are recombined to form a solvent that is "half-saturated"with water. This procedure results in a significantly accelerated equilibration of the column and in an increased reproducibility of the retention times.

Another advantage of using mobile phases with a controlled water content arises from the fact that it will first adsorb to the most active centers on the surface. Consequently, the surface becomes less heterogeneous, and more symmetric peaks are obtained. Water can be removed from silica or alumina by washing with polar solvents, especially methanol.

9.2 SILICA

The most commonly used normal-phase packing is silica. Silica columns with a broad range of properties are commercially available. Standard particle sizes are 3, 5, and 10 μm, and the most commonly used pore sizes are around 10 nm (6–15 nm). Specific pore volumes and specific surface areas of the silicas can vary over a wide range, but as pointed out in Chapter 4, the actual impact of both parameters on the actual phase ratio in the packed column is much smaller. Thus, a typical HPLC column is packed with a stationary phase of a pore size of 10 nm and contains a surface area of between 100 and 150 m^2 per milliliter of mobile-phase volume. Therefore, the retentivity of different silica columns is, in principle, comparable.

However, the similarities end when we look at the silicas more closely. The active groups on the surface of silica are the silanols, and there are several different types of silanols:

Single versus geminal   Bridged versus lone

Silanols

The distribution of the silanols depends on the treatment of the silica. High-temperature treatment removes geminal and bridged silanols, and more single, lone silanols are formed. They are more acidic than either the geminal or bridged silanols. Thus, silicas subjected to a high-temperature treatment contain less silanols, but the silanols are more active. The actual ratio of lone silanols to bridged or geminal silanols depends on the treatment temperature. The preparation procedure of some silicas includes, by necessity, a high-temperature step to burn off fillers or other contaminants. These silicas then need to be rehydroxylated by exposure to a suitable aqueous medium.

The surface silanols are acidic and can function as ion exchangers:

$$\equiv\!Si\!-\!OH + Na^+OH^- \leftrightarrow \equiv\!Si\!-\!O^-Na^+ + H_2O$$

In some preparation procedures, the surface of the silica can be covered with sodium or ammonium ions. They can be removed by treatment with an acid. Similarly, any surface contamination by other metal ions can be removed by acid treatment.

A further complication is the contamination of the silica matrix itself by metal impurities. Common impurities include iron and aluminum, which are present in the raw materials used if the silica synthesis is not based on organic silanes. These metal impurities can increase the acidity of surface silanols. The impact of these impurities on the properties of bonded phases has been studied carefully, but an equally extensive study for normal-phase chromatography is not yet available.

There are no standardized pretreatment procedures for commercially available HPLC silicas. As a consequence, one can observe significant differences in the retention behavior of a set of standards on different silicas (12). Therefore, one cannot a priori expect that a chromatographic method developed on one silica can be transferred to another silica without difficulties. However, since the mobile-phase composition plays such an important role in normal-phase chromatography, the conversion of a method from one silica to another is easier in normal-phase chromatography than in reversed-phase chromatography.

Nevertheless, we have to be aware that large differences exist between different commercial silicas. Müller (13, 14) measured the pH of suspensions of 5% silica in water. This is a measure of the surface acidity or basicity. The results are shown in Table 9.2. Subsequent measurement of the sodium content of some of the silicas demonstrated that at least for some of the silicas a high sodium content was responsible for the alkaline pH of the suspension.

Müller (13) studied the retention characteristics of the different silicas for different analytes with different functional groups. For benzene derivatives with polar functional groups, a significant influence of the surface pH of the silica was found only for acidic and basic analytes, such as phenol and anilines. Compared to methyl phenyl ketone, phenol was less retained on the acidic Zorbax silica and more retained on the basic Spherisorb and Hypersil silicas. It also exhibited strong tailing on the basic silicas. Conversely, aniline,

TABLE 9.2   pH Values of Suspensions of 5% Silica in Water[a]

Silica	pH Value
Zorbax BP-Sil	3.9
Lichrospher Si 100	5.3
Lichrospher Si 300	5.5
Nucleosil 100 V	5.7
Lichrosorb Si 100	7.0
Porasil	7.2
Partisil 10	7.5
Polygosil 60-5	8.0
Lichrosorb Si 60	8.1
Spherosil XOA	8.1
Hypersil	8.1, 9.0
Lichrospher Si 500	8.8, 9.9
Lichrospher Si 1000	9.2
Spherisorb S10W	9.5

[a]Duplicate values reflect different batches of the packing.

methylaniline, and dimethylaniline eluted with good peak shape and short retention on the Hypersil batch whose pH was measured to be 9.0, while they coeluted with increased retention on the batch with the slightly lower pH. On the acidic Zorbax BP-Sil, aniline eluted with a longer retention time and strong tailing, while the substituted anilines failed to elute at all. This clearly demonstrates the importance of the surface pH at least for acidic and basic analytes. On the basis of these findings, Müller recommends the use of aniline and phenol as test samples to characterize the acidity of a silica.

Schwarzenbach (15) took advantage of the differences between acidic and basic silicas by pretreating the silica with solutions of buffer salts. This can be done either with the packing itself or in situ in a prepacked column. The most important parameter is the pH of the buffer solution, the nature of the buffer is of only secondary importance. Figure 9.2 shows a comparison of a separation of cinnamic acid from its esters on untreated and buffer-treated silica. The improvement in peak shape for the acid is significant.

As we will see in Chapter 10, many studies have demonstrated that the use of a fully hydroxylated high-purity silica as the basis for reversed-phase bonded phases results in superior chromatographic performance of these phases. In normal-phase chromatography, the situation is not as clearcut. Good results can be obtained on several different silicas, and the dominant factor for the differences is probably the acidity/basicity as described in the previous paragraphs.

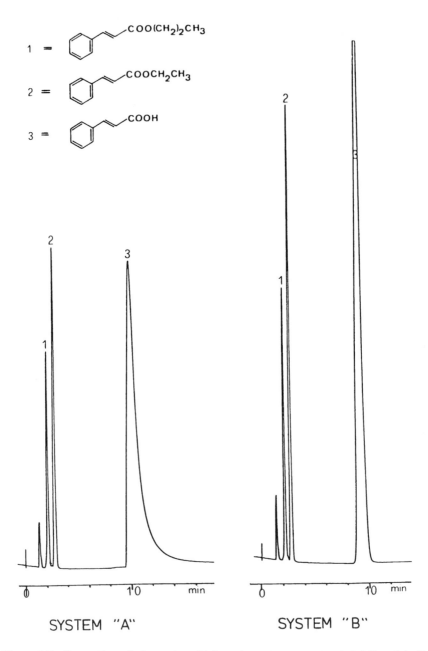

Figure 9.2 Separation of cinnamic acid from its esters an untreated (*A*) and buffer-treated (*B*) silica. (*Chromatographic conditions:* Columns: Lichrospher Si 100, 5 μm, 250 mm × 3 mm. Mobile phase: hexane/ethyl ether 9:1, 1 mL/min.) (Reprinted from Ref. 15, p. 211, by courtesy of Marcel Dekker, Inc.)

9.3 SILICA-BASED BONDED PHASES

The most popular bonded phases for normal-phase chromatography are aminopropyl, cyanopropyl, and diol. Other polar bonded phases, including bonded ion exchangers, can be used as well. Another, less commonly used polar bonded phase is the nitrophenyl bonded phase. The retentivity of the aminopropyl and the diol phase is comparable to that of the oxides. Cyanopropyl and nitrophenyl phases have a smaller retentivity. Although differences in selectivity can be found for the different bonded phases, they are typically small. Thus the elution order of most members in a family of related compounds is largely preserved from packing to packing. For example, in a study of the elution behavior of steroids on silica, and aminopropyl and cyanopropyl bonded phases, an inversion in elution order was observed only for two to four pairs out of 15 analytes (16).

Commercially available aminopropyl phases are prepared from trimethoxy or triethoxy aminopropylsilanes. The structure of the phase is shown below:

Aminopropyl bonded phase

Because of its high retentivity, this is a good substitute for silica or alumina. It replaces alumina in one important application: the class separation of hydrocarbons (Fig. 9.3). Using hexane or heptane as solvent, a mixture of hydrocarbons is separated into the group of saturated hydrocarbons that elute unretained; the group of olefins, eluting with very little retention; and then the aromatics, which, in turn, elute predominantly in groups containing the same number of rings. This separation is also possible on alumina, but not on silica.

As expected, the aminopropyl bonded phase is more retentive toward acidic samples than either silica or cyanopropyl and diol bonded phases. For example, steroids with phenolic functional groups are more retained on the aminopropyl bonded phase (16).

A few precautions are necessary when using the propylamino bonded phase: the NH_2 group is quite reactive and can react with aldehydes and ketones to form imines, or it can be oxidized, for example, by peroxides. Also, it retains organic acids quite strongly. In water, a partial hydrolysis occurs as a result of the strongly alkaline environment in the pores; the commonly encountered ligand density of the propylamino bonded phase translates to a roughly 1 molar concentration of base in the pores. The pH in the pores is then about 10 or higher, and under these conditions the silica surface is attacked. Bonded phase is lost, acidic silanols are formed on the surface, and the pH shifts to lower values. At some point, the system reaches equilibrium, but the resulting phase is not the same any more as the original phase. Thus the washing of an

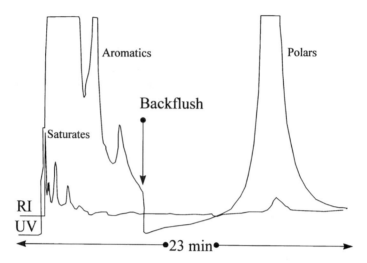

Figure 9.3   Group separation of hydrocarbons on an aminopropyl column. The sample
is a crude oil. The ranges of the elution of the different fractions are indicated. After
elution of the aromatic fraction, the column is backflushed to elute the polar fraction.
(*Chromatographic conditions:* Column: Energy Analysis Column, 3.9 mm × 300 mm.
Mobile phase: hexane. Flow rate: 2 mL/min.) (Chromatogram courtesy of W. Dark,
Waters Corp.)

aminopropyl column with water, for example, for cleaning, should be avoided.
    The diol phase is very retentive as well. Its structure is shown below:

Diol bonded phase

Commercial diol phases are prepared from trimethoxy or triethoxy glycidoxy-
propyl silane. The epoxy group is then hydrolyzed to form the diol function.
    Diol phases are about as retentive as silica or aminopropyl bonded phases
(17). Compared to silica and alumina, diol phases are much less sensitive to
the water content of the mobile phase. They are also less reactive than
aminopropyl phases. Furthermore, they can be washed with water without
difficulty. Thus they are the best choice among the stationary phases for
normal-phase chromatography. Although the stationary phase is readily available from many sources, it is most commonly found as a packing for aqueous
size-exclusion chromatography.
    The cyanopropyl packing is used in both normal-phase and reversed-phase

chromatography. It is less retentive in normal-phase chromatography than silica, alumina, aminopropyl, or diol phases. Thus one can view it as a deactivated normal phase. Retention times are lower by a factor of ~ 2 compared to the parent silica and an aminopropyl bonded phase based on the same silica (16). The structure of the cyanopropyl bonded phase is shown below:

Cyanopropyl bonded phase

Several different versions are commercially available. They may be based on a mono-, di-, or trifunctional silane. From the standpoint of normal-phase chromatography, there is no principal advantage of one type over the other, except that the reproducibility of a bonded phase based on a monofunctional silane is easier to control.

It is unclear how much of the polar interaction of this packing is due to the cyano functional group or due to residual silanols. Most likely, both partici-pate in the retention of analytes. Detailed studies of the retention mechanism of this packing are not available.

Cyanopropyl columns have the nasty habit of collapsing in solvents of intermediate polarity. The instability is mechanical in nature, not chemical; the packed bed collapses into a more dense state. The phenomenon is akin to the events that occur when a bucket full of wet beach sand is tapped from the side; the sand settles, and water rises to the surface. In both nonpolar and polar solvents, this catastrophic collapse is prevented by the adhesion of particles to each other. In solvents of intermediate polarity, this adhesion vanishes and bed collapse can occur. Therefore, many manufacturers offer both normal-phase and reversed-phase versions of the cyanopropyl packing.

If a cyanopropyl column needs to be exposed to solvents of intermediate polarity, for example during cleaning procedures, the flow should not be interrupted. The pressure actually holds the particles in place, and a collapse of the bed is less likely to occur. The preferred storage solvent for normal-phase cyanopropyl columns is a hydrocarbon like hexane.

Nitrophenyl packings are designed to exhibit π-donor/π-acceptor interac-tions with suitable analytes. The structure of such a packing is shown below:

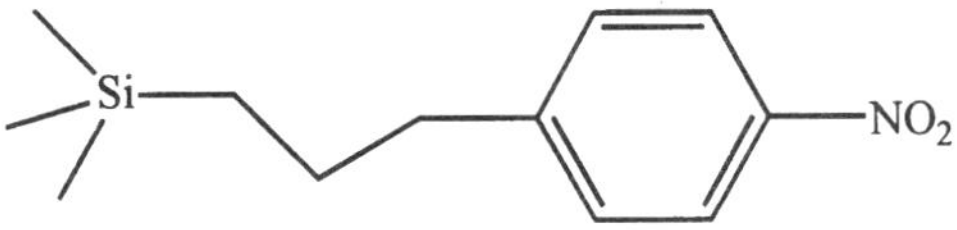

Nitrophenyl bonded phase

Therefore, one use of such a packing is the separation of polynuclear aromatic hydrocarbons (PNAs or PAHs) (18). Besides the desired interaction, it exhibits less specific polar interactions due to the presence of residual silanols. The combination of both can give rise to unique selectivities.

9.4 ALUMINA

Alumina is the second inorganic oxide in use in HPLC. However, it is by far not as widespread as silica, and there are fewer types commercially available. The reason is primarily that a simple, high-quality surface derivatization technique akin to the silanization of silica is not available, which restricts applications largely to the use of underivatized alumina.

The commercially available HPLC-grade aluminas consist largely of γ-alumina, a form prepared by heat treatment between 500 and 800°C. The structure of γ-alumina is spinel, where the unit cell contains 32 oxygen atoms arranged in 64 tetrahedrons and 32 octahedrons (19). The aluminum atoms are heterogeneously distributed over the tetrahedrons and octahedrons of the unit cell. This leads to a formation of Lewis base sites for the vacant positions and Lewis acid sites for the occupied positions. Like the surface of silica, the surface of alumina is covered by hydroxy groups. However, the number of hydroxy groups and their coordination is different (11):

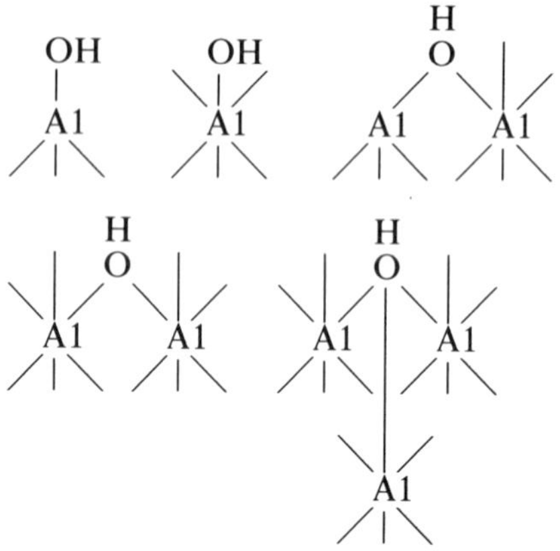

Unlike silica, the surface is occupied not only by these Brönsted acid sites, but also by Lewis acid and base sites. The distribution of the active sites is a function of the pretreatment of the alumina. The primary factors are the level of hydroxylation and the pH of the surface. Alumina is available in three grades — basic, neutral, and acidic alumina — depending of the pretreatment of the alumina. Thus the surface properties of an alumina packing can vary a lot, just as it was the case for silica.

Overall, alumina is even more polar, that is, more retentive, than silica. This is especially true for analytes with electron-rich functional groups. The primary use of alumina in normal-phase chromatography has been the separation of olefinic hydrocarbons from saturated hydrocarbons, and the group separation of polynuclear aromatic hydrocarbons, which is due to the interaction of the aromatic π-electrons with Lewis acid sites.

With the multitude of possible interactions with sample molecules, it is not uncommon to encounter strong interactions, especially with polar analytes, for example, organic acids. For polar analytes, one may encounter strongly tailing peaks or even difficulties to elute the analyte. Occasionally, a decomposition of analytes has been observed as well.

An interesting form of alumina is available from Biotage, Inc., under the trade name Unisphere. Its spherical particles are made from platelets. The external surface of the particles has the appearance of a sand rose. It is claimed that this unique morphology results in an improved permeability compared to spherical and irregular particles.

9.5 SPECIAL TECHNIQUES

There are a few techniques related to normal-phase chromatography that are worth mentioning. Silica can be coated with the salts of heavy metals. This results in unique selectivities for analytes that form complexes with the metal ions. An example of this is "argentation chromatography" (20). The silica is coated with silver nitrate, which gives it a special selectivity for compounds with aliphatic double bonds. The technique has acquired a broad range of applications. However, while Ag^+-coated TLC plates are commercially available, HPLC columns have to be prepared by the user. Other metal salts can be used in a similar manner for different applications.

In another special technique, the pores of the silica are filled with a polar liquid. This technique was popular before the development of bonded phases under the names of partitioning chromatography or "heavily loaded" columns. Recently, a version of it was resurrected for use in sample preparation. The retention mechanism is largely dominated by the choice of the liquids, and the underlying silica plays only a secondary role. The preferred silica has a large porosity and a small pore size. Small pore sizes hold the stationary phase better than do large pore sizes.

Columns are prepared by equilibrating a silica column with a mobile phase that is saturated with the stationary phase. Since this may take a long time, the process can be accelerated by injecting the stationary phase onto the column. It is highly recommended to keep the column and the solvent reservoir at a controlled temperature to prevent the column from bleeding. The capacity of these heavily loaded columns is by up to 2 orders of magnitude higher than the same column used in regular normal-phase chromatography. This makes this technique attractive for preparative chromatography.

Silica can be coated with many polar compounds. If the adsorption of these compounds onto the silica in the mobile phase of use is high enough, a new stationary phase is formed. This is an easy way to create stationary phases with special properties, which are difficult to create in other ways.

9.6 PRACTICAL ASPECTS OF NORMAL-PHASE CHROMATOGRAPHY

Normal-phase chromatography can be very powerful in the hands of the experienced chromatographer. Because of the complications associated with the lengthy equilibration times or sample decomposition when silica or alumina are used as stationary phases, and owing to the column stability problems associated with cyano columns, many chromatographers shy away from this technique. But today even the novice should no longer ignore this technique. Diol columns do not exhibit any of the problems just described, and the use of this stationary phase can make normal-phase chromatography about as easy as reversed-phase chromatography. I still would recommend, though, to pay close attention to the water content of the mobile phase and to practice the use of "half-saturated" mobile phases, as outlined above. Table 9.3 lists the solubility of water in various commonly used normal-phase solvents, largely collected from Reference 21.

If the sample contains very polar constituents, it is advisable to remove them through sample preparation techniques like solid-phase extraction. Samples dissolved in water should not be injected directly onto a normal-phase column. When an aminopropyl column is used, additional precautions are necessary, which were outlined above when we discussed this stationary phase.

When the HPLC instrument is converted from a reversed-phase mobile phase to a normal-phase mobile phase, care should be taken to flush all

TABLE 9.3 Solubility of Water in Solvents Used in Normal-Phase Chromatography (21, 22)

Solvent	% wt/wt ($^\circ$C)
Hexane	0.0111 (20)
Heptane	0.0091 (25)
Isooctane	0.0055 (20)
Toluene	0.0334 (25)
Dichloromethane	0.198 (25)
Chloroform	0.072 (23)
Carbon tetrachloride	0.010 (24)
o-Dichlorobenzene	0.309 (25)
Ethyl acetate	2.94 (25)
Butyl acetate	1.86 (20)
Diethyl ether	1.468 (25)
Methyl-t-butyl ether	1.5 (RT)

aqueous solvent from all components of the system. If this is not done, water or polar solvents can bleed slowly from unpurged parts of the instrument onto the column, changing the activity of the packing and causing shifting retention times or even distorted peaks.

Normal-phase solvents are especially volatile. This is an advantage for preparative chromatography, but also can cause trouble, if this is not taken into consideration in analytical chromatography. Retention times can shift as a result of selective evaporation of some components of the mobile phase. Also, safety should be considered. Some normal-phase solvents, especially hydrocarbons, are highly flammable; others have adverse health effects. Consult the applicable material safety data sheets. All containers including waste containers should be closed. When not in use, the mobile-phase container should be sealed.

REFERENCES

1. L. R. Snyder and J. J. Kirkland, *Introduction to Modern Liquid Chromatography*, 2nd ed., Wiley, New York, 1979.

2. L. R. Snyder, *Principles of Adsorption Chromatography*, Marcel Dekker, New York, 1968.

3. L. R. Snyder, *Anal. Chem.* **46**, 1384 (1974).

4. L. R. Snyder and H. Poppe, *J. Chromatogr.* **184**, 363–413 (1980).

5. L. R. Snyder, *J. Chromatogr. Sci.* **16**, 223–234 (1978).

6. L. R. Snyder and J. L. Glajch, *J. Chromatogr.* **214**, 1–19 (1981); L. R. Snyder and J. L. Glajch, *J. Chromatogr.* **214**, 21–34 (1981).

7. L. R. Snyder, J. L. Glajch, and J. J. Kirkland, *J. Chromatogr.* **218**, 299–326 (1981).

8. M. D. Palamareva and H. E. Palamarev, *J. Chromatogr.* **477**, 235–248 (1989).

9. M. D. Palamareva and V. R. Meyer, *J. Chromatogr.* **641**, 391 (1993).

10. J. L. Glajch and J. J. Kirkland, *J. Chromatogr.* **495**, 51–63 (1989).

11. K. K. Unger, "Adsorbents in Column Liquid Chromatography," in *Packings and Stationary Phases in Chromatographic Techniques*, K. K. Unger, ed., Chromatographic Science Series, Vol. 47, Marcel Dekker, New York, 1990.

12. J. J. Kirkland, C. H. Dilks, Jr., and J. J. DeStefano, *J. Chromatogr.* **635**, 19–30 (1993).

13. H. Müller, dissertation, Universität des Saarlandes, Saarbrücken (1983).

14. H. Engelhardt and H. Müller, *J. Chromatogr.* **218**, 395 (1981).

15. R. Schwarzenbach, *J. Liq. Chromatogr.* **2**(2), 205–216 (1979).

16. S. Hara and S. Ohnishi, *J. Liq. Chromatogr.* **7**(1), 59–68 1984; S. Hara and S. Ohnishi, *J. Liq. Chromatogr.* **7**(1), 69–82 1984.

17. M. Verzele, F. Van Damme, C. Dewaele, and M. Ghijs, *Chromatographia* **24**, 302–308 (1987).

18. E. P. Lankmayr and K. Müller, *J. Chromatogr.* **170**, 139–146 (1979).

19. C. J. C. M. Laurent, Ph.D. thesis, Technische Hogeschool, Delft (1983).

20. D. Cagniant, "Argentation Chromatography: Application to the Determination of Olefins, Lipids, and Heteroatomic Compounds," in *Complexation Chromatography*, D. Cagniant, ed., Chromatographic Science Series, Vol. 57, Marcel Dekker, New York, 1992.

21. *Techniques of Chemistry*, Vol. II, *Organic Solvents, Physical Properties and Methods of Purification*, 3rd ed., J. A. Riddick and W. B. Bunger, eds., Wiley-Interscience, New York, 1970.

22. *Merck Index*, 11th ed., Merck & Co., Inc., Rahway, NJ, 1989.

ADDITIONAL READING

H. Engelhardt and H. Elgass, "Liquid Chromatography on Silica and Alumina as Stationary Phases," in *High-Performance Liquid Chromatography, Advances and Perspectives*, Vol. 2, Cs. Horváth, ed., Academic Press, New York, 1980.

10 Reversed-Phase Chromatography

In normal-phase chromatography, the stationary phase is polar and the mobile phase is nonpolar. In reversed-phase chromatography, the situation is reversed — the stationary phase is nonpolar, and the mobile phase is polar. A typical stationary phase consists of a long-chain hydrocarbon attached to a support, while a typical mobile phase comprises mixtures of water or buffer with polar solvents such as methanol, acetonitrile, or tetrahydrofuran.

Reversed-phase chromatography is today by far the most popular HPLC technique. About 90% of all analytical separations of low-molecular-weight samples is carried out using a reversed-phase method. In the analysis of intermediate-molecular-weight compounds such as peptides and oligonucleotides, reversed-phase chromatography still plays an important role. In the analysis of high-molecular-weight compounds such as proteins and industrial polymers, it plays only a small role, but its application especially for industrial polymers is slowly increasing.

Reversed-phase chromatography owes its popularity to two facts:

1. The separation principle is readily understood.
2. Good results are obtained for many compounds with few technical complications.

Although many details of the separation mechanism are still under discussion, it is clear that retention is a function of the hydrophobicity of the analyte, which, in turn, is largely a function of the size of the hydrophobic area of a molecule. Therefore, reversed-phase chromatography is ideal for the separation of members of a homologous series. For example, analytes differing from each other by the number of methylene groups in an aliphatic side chain are easily separated by reversed-phase chromatography. In many cases, it is also possible to assign the elution order of the analytes based on their structures. Retention increases with an increase in the water content of the mobile phase in a predictable manner. With a little experience, a chromatographer can look at the structure of an analyte and make some good guesses as to the composition of a mobile phase that will elute the analyte within a reasonable time.

The equilibration of the columns with the aqueous mobile phases is rapid; in most cases only a few column volumes are needed. This is in stark contrast to normal-phase chromatography, where the equilibration of the column with the ubiquitous water is a steady challenge. For the majority of analytes, good peak shapes and reproducible retention times are obtained without difficulty. For some compounds, there are some difficulties due to secondary interactions, but these secondary effects are well understood and can be dealt with using proven procedures. Furthermore, reversed-phase packings have steadily improved, and some of these difficulties have been reduced substantially.

10.1 RETENTION MECHANISM

The surfaces of reversed-phase packings are hydrophobic. The most popular types of reversed-phase packing are silica-based bonded phases, where the ligand is a long-chain hydrocarbon, typically 8 or 18 carbons long. This hydrophobic surface interacts with the hydrophobic part of the analyte. One will find that the larger the hydrophobic part of the analyte, the longer is the retention. More specifically, if one plots the logarithm of the retention factor of a homologous series of analytes (e.g., alcohols) in a given mobile phase against the number of methylene groups in the side chain, one obtains a linear relationship. Furthermore, if one plots this relationship for different homologous series using the same mobile phase, one will find that the lines are parallel to each other (Fig. 10.1). In other words, every methylene group contributes the same increment to the logarithm of the retention factor independent of the nature of the analyte. Other hydrophobic groups contribute as well; a CH group attached to a double bond or an aromatic CH group

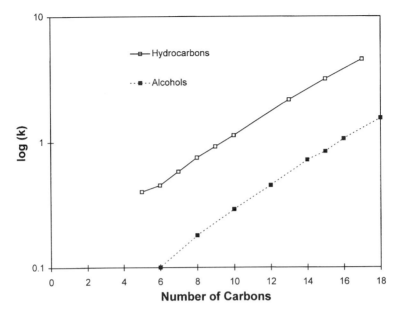

Figure 10.1 Plot of the logarithm of the retention factor versus the number of methylene groups for several homologous series. (Data from Ref. 1.)

contributes a little bit less than a methylene group, a chlorine group attached to a hydrocarbon contributes about as much as a methyl group, and a CF_2 or CF_3 group contributes more than a methylene group. The methylene-group increment changes somewhat with the packing and the mobile phase, increasing with increasing water content of the mobile phase.

Polar functional groups reduce retention also in a rather predictable way, but their incremental contribution depends on their position on the molecule and on the solvent. Aldehyde, keto, and ester groups reduce retention weakly; alcohol or amide groups, more strongly. Carboxylic acid groups reduce retention about as much as amides when they are not ionized. When ionized, however, retention is reduced drastically. This immediately points to the important role that the mobile phase pH plays in controlling retention in reversed-phase chromatography.

Amino functions play a special role in reversed-phase chromatography. It is important to realize that silica-based reversed-phase packings contain a large amount of acidic silanols, which interact with amines via an ion-exchange mechanism. A fully hydroxylated silica contains about $8\ \mu mol/m^2$ of silanols on the surface. As a result of steric hindrance, even the best derivatization reactions can remove only just under $4\ \mu mol/m^2$. Thus about half of the initially present silanols are still present after bonding. See also Sections 6.2 and 10.2.1 for a more detailed discussion. The number of negatively charged silanols on the surface increases with increasing pH. Thus the interaction of amines with the surface increases with pH, while they are positively charged. If the pH is sufficiently high such that the positive charge is removed, the interaction decreases, but for many basic analytes that happens outside the pH range that is compatible with a silica-based packing. This silanophilic interaction was neglected in the early studies of reversed-phase chromatography, but is now recognized. As a matter of fact, silanophilic interactions seem to play a much broader role in reversed-phase retention even for other polar functional groups than has been realized until recently.

If one plots the logarithm of the retention factor against solvent composition expressed in volume fraction (v/v%), one obtains a good linear relationship for methanol–water mixtures. For other solvents (i.e., aqueous acetonitrile or THF), a slight curvature is observed. There is no good fundamental reason why there should be a linear relationship between "solvent composition" and the logarithm of the retention factor; thus one can regard the linearity obtained with aqueous methanol to be an accident. However, the departure from linearity is sufficiently small that the assumption of linearity can be used with advantage in the treatment of gradient elution and in the estimation of a solvent composition with acceptable retention times from data obtained from gradients.

There are three competing explanations of retention in reversed-phase chromatography: the hydrophobic theory, partitioning, and adsorption. All three are still actively discussed at the time of this writing [see *J. Chromatogr.* **A656** (1993)]. The major observable effects can be explained by either theory; thus a distinction between them is of little consequence for the practitioner. The oldest theory is the hydrophobic theory, which was introduced to HPLC by Horváth et al. (2). The retention mechanism can be viewed as the chromatography of cavities in the solvent, which are created by the hydrophobic portion of the analyte. It is dominated by surface tension and the dipole–dipole interaction of the polar functions of the analyte with the solvent. In this theory, the surface of the packing plays only a passive role (3).

The partitioning of the analyte between the aqueous mobile phase and the hydrocarbonaceous bonded phase is a competing theory. It draws support from the fact that there is for the most part a good correlation between retention in reversed-phase HPLC and the octanol–water partition coefficient. However, special effects arise that are due to the fact that the ligand is attached on one end to the surface of the packing. For example, a pure partitioning mechanism cannot explain the shape selectivity observed for aromatic hydrocarbons on phases with different ligand densities.

The third theory treats the mechanism of reversed-phase chromatography as another form of adsorption. However, a pure adsorption mechanism fails to explain the discontinuities observed in plots of the logarithm of the retention factor versus the chain length of the bonded phase.

One can conclude that a complete theory of the retention mechanism of reversed-phase chromatography must include components of both partitioning and adsorption. Although the discussions are far from over, the theory of

Jaroniec (4) represents the currently most comprehensive explanation. It treats the retention mechanism as a mixture of partitioning and displacement. According to this treatment, one should distinguish between two stages of the process of reversed-phase chromatography: (1) the formation of a combined solvent–surface stationary phase and (2) the partitioning of the solute between the mobile phase and this stationary phase. In the first stage, the hydrophobic ligand attached to the silica surface incorporates solvent molecules to form the stationary phase. The solvents representing the organic component of the mobile phase are enriched on the surface of the packing. The enrichment is a function of the thermodynamic equilibrium with the mobile phase. It should depend on the mobile-phase composition and the nature of the stationary phase, including chain length, ligand density, and silanol population. Generally, the surface excess concentration of the organic modifier goes through a maximum at high water concentrations and drops to 0 at 100% water. For packings with a high silanol activity, an excess of water can be observed at high concentration of organic modifier. It has been found that the excess surface concentrations are similar for packings with similar physical and chemical properties and rather insensitive to temperature (over a limited temperature range).

Different solvents are enriched on the surface of the bonded phase to different degrees. Among the commonly used reversed-phase cosolvents, methanol is adsorbed less than acetonitrile, which, in turn, is adsorbed less than tetrahydrofuran. The excess surface concentration has been measured for these solvents on different packings. Both the ligand density and the ligand type play a role in the enrichment.

In the second stage, the solute partitions between the mobile phase and the stationary phase. In this process, the solute displaces the solvents from the stationary phase depending on the molecular sizes r of the solute s and the solvents i. On the basis of this model, one can write the equation for the equilibrium constant $K_{s,i}$ between the solute and the solvent:

$$K_{s,i} = \left(\frac{a_{s,S}}{a_{s,M}}\right)^{r_i}\left(\frac{a_{i,M}}{a_{i,S}}\right)^{r_s} \tag{10.1}$$

where a denotes the activity. From this, one can write a general equation for the distribution coefficient K_s of a solute s between stationary phase and mobile phase, defined as the ratio of the volume fraction of the analyte in the stationary phase to its volume fraction in the mobile phase:

$$\ln(K_s) = \ln(K_s^*) + \ln\left(\frac{\gamma_{s,M}}{\gamma_{s,S}}\right) + n\ln\left(\frac{\varphi_{i,S}}{\varphi_{i,M}}\right) + n\ln\left(\frac{\gamma_{i,S}}{\gamma_{i,M}}\right) \tag{10.2}$$

where γ is the activity coefficient of the solute s or the solvent i in the respective phase; n is the ratio of the molecular size of the solute r_s to the size of the solvent r_i:

$$n = \frac{r_s}{r_i} \tag{10.3}$$

K_s^* is defined as:

$$K_s^* = K_{s,i}^{1/r_i} \tag{10.4}$$

The relationship between the retention factor and the distribution coefficient is simply given by the phase ratio β:

$$\ln(k) = \ln(\beta) + \ln(K_s) \tag{10.5}$$

If we neglect all solute–solvent and solvent–solvent interactions, Equation (10.2) simplifies to Equation (10.6):

$$\ln(k) = \ln(\beta) + \ln(K_s^*) + n \ln\left(\frac{\varphi_{i,S}}{\varphi_{i,M}}\right) \tag{10.6}$$

One can see that this equation contains the composition of the stationary phase as well as the composition of the mobile phase together with a size-dependent factor, which describes the displacement of solvent molecules in the stationary phase by the solute. The latter explains the linear relationship observed experimentally between the logarithm of the retention factor and the hydrophobic surface area of analytes (1). Another experimental verification of this is the linear dependence of the logarithm of the retention factor on the number of methylene groups n_c in a homologous series, which has been reported by many investigators:

$$\ln(k) = O + Pn_c \tag{10.7}$$

where O and P are constants. Whereas P is only a function of the mobile-phase composition and the stationary phase, O is the natural logarithm of the retention factor of the head group of the homologous series on a particular stationary phase in a particular mobile phase. P is the logarithm of the methylene-group selectivity.

Equation (10.6) explains why the methylene-group selectivity is a function of only the mobile phase and the stationary phase, but not the nature of the analyte; the last term in Equation (10.6) contains only the composition of the mobile phase and the stationary phase. Furthermore, Equation (10.6) gives a reason for the commonly encountered observation that plots of the logarithm of the retention factor versus the number of methylene groups in a homologous series exhibit a common intersection point, when the mobile-phase composition or the temperature are varied (Fig. 10.2).

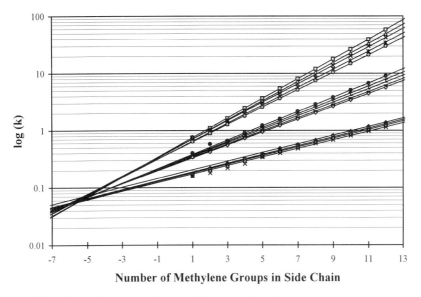

Figure 10.2 Plot of the logarithm of the retention factor versus the number of methylene groups in the side chain of *n*-alkylbenzenes. The data were obtained at three different solvent compositions (0%, 10%, and 20% water in methanol) and at various temperatures around room temperature. The plot shows the linear relationship of the logarithm of the retention factor with the number of methylene groups and the common focal point of the lines frequently found in RPLC. (Data from Ref. 5.)

If the stationary-phase composition of different columns is similar, one should also expect, on the basis of Equation (10.6), that the methylene-group selectivity of different columns is similar. This has been found indeed (6). Also, the same assumption implies a collinear relationship of retention factors between similar columns. In some cases, this has been observed. But also exceptions have been found, which can be explained on the basis of differences in the stationary phase composition, due to the influence of silanols. Thus the partitioning-displacement model presents an intuitive description of the reten-tion process that results in an accurate description of the experimental observations.

Another regular pattern can be observed when the dependence of the retention on temperature is studied. We can derive the expected temperature dependence from the application of van't Hoff's equation to chromatography:

$$\ln k = \ln \beta - \frac{\Delta G^\circ}{RT} = \ln \beta - \frac{\Delta H^\circ}{RT} + \frac{\Delta S^\circ}{R} \qquad (10.8)$$

If the adsorption enthalpy itself does not vary with temperature, one expects a linear relationship between the logarithm of the retention factor and the inverse of the absolute temperature (van't Hoff plot). This is indeed observed in many cases. Conversely, one can use the departure of linearity as an indication of a change in the retention mechanism, for example, due to a change of the conformation of the bonded-phase ligands.

If such effects are absent, the van't Hoff plots of related analytes, such as homologous series, exhibit a common focal point at a specific temperature. This point of "enthalpy–entropy compensation" can be characterized by a specific temperature, the compensation temperature T_c and the free energy of adsorption at this temperature $\Delta G^\circ_{T_c}$. With the concept of the compensation temperature, Equation (10.8) can be written as

$$\ln(k) = \ln(\beta) + \frac{\Delta H^\circ}{R}\left(\frac{1}{T} - \frac{1}{T_c}\right) - \frac{\Delta G^\circ_{T_c}}{RT} \tag{10.9}$$

Compensation temperatures found in the literature tend to be around 600 K.

While the driving force of reversed-phase retention is the interaction of the hydrophobic part of the analyte with the hydrophobic stationary phase, the interaction of the polar functional groups of the analyte with the mobile phase and with the residual silanols on a silica-based packing are responsible for the selectivity of a separation. This has two important consequences:

1. The selectivity of a separation can be influenced significantly by the type of solvent that is used as the organic modifier of the mobile phase.
2. The amount and activity of surface silanols on silica-based packings or their suppression by modification of the bonded phase or the mobile phase are significant.

Similarly, other polar functional groups in the stationary phase can influence the selectivity of the separation as well. Thus one can find significant differences in selectivity between reversed-phase packings based on silica and those based on polymers. Also, reversed-phase bonded phases with polar functional groups incorporated in the ligand can exhibit significant differences in selectivity compared to their purely hydrophobic counterparts.

One can determine the free energy of transfer from the mobile phase to the stationary phase from chromatographic retention data, if the phase ratio is known. The free energy of transfer from the same mobile phase to hexadecane as a solvent can be obtained from measurements of the partition coefficient. The ratio of the free energy of partition to the free energy of the reversed-phase retention has been compared by Carr et al. (7). The ratio is unity for methylene groups or chlorine groups, indicating that the energetics of both processes are similar. Mechanistically, this can be interpreted by assuming that the methylene groups freely penetrate the bulk stationary phase in reversed-phase chromatography. However, the same ratio for polar functional groups results in quite different and usually fairly large values. The values also vary with the solvent composition, while they were fairly constant for methylene groups. The large values indicate that for polar functional groups the partitioning model is inadequate, and that the retention mechanism is more akin to an adsorption mechanism; specifically, the polar functional group does not penetrate the stationary phase. A complete penetration of the hydrophobic portion of the analyte into the stationary phase and an exclusion of the polar portion from the stationary phase can also be deduced from plots of retention factors as a function of the chain length of the stationary phase (8).

10.2   SILICA-BASED BONDED PHASES

The most popular reversed-phase columns today are manufactured from silica-based bonded phases, and among those the C_{18}-type bonded phase is most frequently used. As we have seen in Section 6.2, these bonded phases are prepared by reacting silica with a reactive silane that carries the ligand after which the phase is named. Commonly, chlorosilanes are used as the reactive silane in the presence of a scavenger of the hydrochloric acid, which is a byproduct of this reaction:

$$
\begin{array}{ccc}
& R_1 & R_1 \\
\diagdown & | & \diagdown \quad\quad | \\
-Si-OH \; + \; Cl-Si-R_2 & \xrightarrow[-HCl]{} & -Si-O-Si-R_2 \\
\diagup & | & \diagup \quad\quad | \\
& R_3 & R_3
\end{array}
$$

where R_2 is the ligand; R_1 and R_3 can either be additional functional groups that can bind to the surface or a neighboring silane, or they can be short hydrocarbon groups. For the most common bonded phases, the side chains R_1 and R_3 are methyl groups. More bulky chains, such as isopropyl or *tert.*-butyl groups, impart an improved hydrolytic stability in acidic mobile phases to the packing, but also cause a lower coverage of the surface due to additional steric hindrance.

The amount of ligands that can be bonded to the surface is limited by the steric hindrance of the head group that attaches the ligand to the surface. With di- or trifunctional silanes, higher surface coverages can be achieved, but the reaction is sensitive to the water content in the reaction medium. Water causes the hydrolysis of additional functional groups, which, in turn, results in the attachment of additional silane to the surface. This process is more difficult to control than the process involving monofunctional silanes. This is the reason why the majority of bonded phases available today are based on monofunctional silanes rather than polyfunctional silanes.

The higher surface coverage available with polyfunctional silanes may have an advantage with respect to the hydrolytic stability of the packing, but the results are not clearcut. At acidic pH, a reduced rate of hydrolysis was observed for C_{18}-type bonded phases prepared from di- and trifunctional silanes (9).

However, the dominating factor at alkaline pH is the dissolution rate of the silica itself rather than the removal of the ligand, and at equal surface coverage, monofunctional dimethylsilyl ligands protect the silica better than do their di- and trifunctional equivalents. Variation of the surface coverage results in different shape selectivity of the phase, which is especially important for the separation of polyaromatic hydrocarbons (see below). On the other hand, a high surface coverage can also result in hindered mass transfer, especially for silicas with a small pore size.

The most popular types of bonded phases for reversed-phase chromatography are octadecyl (ODS, C_{18}), octyl (C_8), phenyl, and cyanopropyl (CN, Cyano, Nitrile) bonded phases, in that order. The ligands designated by octadecyl, octyl, and cyanopropyl are unambiguous. The phenyl type comprises several different ligands: one would think that the normal type consists of a phenyl group directly attached to the silane, but the more common ligands are phenylpropyl (e.g., Nova-Pak), phenylethyl (e.g., Zorbax), or α-methylphenylethyl (e.g., μBondapak). This adds to the uniqueness of each stationary phase.

Any imaginable hydrophobic silane has been attached to silica. Usually the selectivity shifts are small and are within the range of differences commonly observed between packings based on the traditional ligands. Therefore, only a few alternative hydrophobic ligands are commercially available. These include butyl (C_4), hexyl (C_6), cyclohexyl, propyl (C_3), duododecyl (C_{22}) and methyl (C_1). The C_3 designation is ambiguous, since it has also been used to describe the trimethylsilyl ligand.

In general, longer chains result in higher retention, with all other parameters held constant. However, after a critical chain length, the retention increase levels off (8). The point where this occurs depends on the nature and the size of the analyte. This point can be interpreted as the instant, where an increase in chain length does not result in an increase in the penetration of the analyte into the bonded phase. The cause of this could be steric or due to a preferred solvation environment of a functional group.

In general, selectivity shifts from one chain length to another are small to nonexistent, especially among the longer alkyl chains. However, when a cyanopropyl bonded phase or other shorter-chain bonded phases are used, the water content of the mobile phase needs to be increased in order to achieve comparable retention to C_8 or C_{18} phases. The shifts in selectivity observed under these circumstances have been attributed to the change in mobile-phase composition rather than to the stationary phase (10). This means in practice that if a selectivity change is desired, there is little merit to change from a C_{18} to a C_8 phase, but that a change to a shorter chain or a cyano column is preferred.

This does, however, not apply to a new generation of bonded phases, which have been designed to reduce the influence of surface silanols on retention (11–14). In these bonded phases, a hydrophilic layer is incorporated into the structure of the bonded phase between the silica surface and the reversed-phase layer. On one hand, this sublayer mitigates the interaction of analytes with surface silanols, which commonly results in tailing peaks, especially for basic compounds. On the other hand, the selectivities obtained are significantly different from those obtained for classic reversed-phase packings, especially for polar analytes. This is due to the combination of the effect of the attenuated influence of silanols and the change in the composition of the stationary phase, which is likely to show an increased adsorption of water. An example of such a selectivity shift is shown in Figure 10.3.

The commercially available bonded phases of this type are the SymmetryShield RP$_8$ column (Waters Corporation, USA), the Supelcosil ABZ column (Supelco, USA), and the Prism column (Keystone Scientific, USA). The Supelcosil ABZ column is prepared from an amino bonded phase by derivatization with a long-chain acid chloride, thus forming an amide group close to the silica surface. The disadvantage of this derivatization technique is the presence of residual amino groups on the surface, which result in increased retention of acidic analytes. The SymmetryShield RP$_8$ column is synthesized in a single step from a silane that has a carbamate group incorporated into the

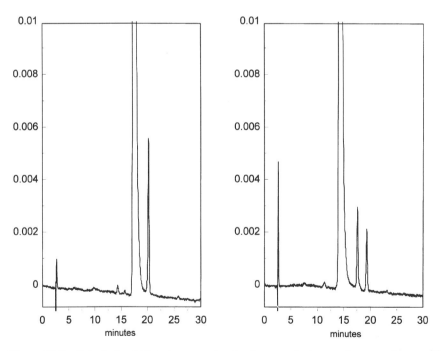

Figure 10.3 Selectivity shift between a standard reversed-phase column and one with an incorporated polar functional group. Sample: impurities of furazolidone. Only one impurity peak is obtained on the Symmetry C$_8$ column, but two peaks are obtained with the SymmetryShield RP$_8$ column using the same mobile phase. (Chromatograms courtesy of M. Z. El Fallah, Waters Corp.)

ligand chain. This avoids the problem of the presence of unwanted functional groups. The composition of the Prism column has not been published at the time of this writing.

Silicas can be prepared with different pore sizes, specific pore volumes, and specific surface areas, and all these different silicas can be used to prepare reversed-phase bonded phases. A general discussion of the interplay of pore size, pore volume, and surface area can be found in Chapter 4. The primary parameter of interest is the pore size. The commonly used packings for the reversed-phase chromatography of small analytes are based on silicas with a pore size of $\sim 10\,nm$. Packings with an average pore size of $<5\,nm$ are not used because of the restricted mass transfer in the micropores. As the pore size increases, the specific surface area of the packing decreases. Reversed-phase packings based on a large-pore-size silica are predominantly used for the chromatography of larger analytes, such as proteins. The standard packings for the reversed-phase chromatography of proteins have a pore size of $\geqslant 30\,nm$.

The specific pore volume is of secondary concern. Packings with a low specific pore volume are mechanically and hydrolytically stronger than packings with a large specific pore volume. As pointed out in Chapter 4, the specific surface area of a packing can be used only in conjunction with the specific pore volume as an indicator of the retentivity or loadability of a packing. For both retentivity and loadability, the surface area per unit bed volume is the relevant parameter and for a packing with a given pore size, this parameter is nearly invariant for packings with different specific surface areas.

10.2.1 Silanol-Group Activity

As discussed also in Section 6.2, the surface of a fully hydroxylated silica contains about $8\,\mu mol/m^2$ of silanols. Even with the best bonding conditions, only $<4\,\mu mol/m^2$ of these silanols can be removed. The ligand concentration of typical bonded phases is somewhere around $\leqslant 3.5\,\mu mol/m^2$. Therefore, a typical silica-based bonded phase contains on a molar basis more "residual" silanols than ligand. Although these residual silanols are not available for further derivatization reactions due to steric hindrance, they are still sufficiently accessible for interaction with analytes and the mobile phase. They contribute to an increased retention and tailing of basic analytes.

A common practice in the preparation of reversed-phase packing is the "endcapping" of the bonded phase with a smaller, less sterically hindered silane. Although some additional silanols are removed, especially on bonded phases where the first step was not very efficient, a complete removal of the silanols is impossible. This is not a surprise, considering the size of a trimethylsilyl ligand compared to the spacing of silanol groups. Nevertheless, endcapping results in many cases in an improvement of the character of a reversed-phase packing. But this is only a partial solution. Also, the endcapping is more prone to hydrolysis than the main ligand and is the weak point of a packing. A much more powerful attack on the silanol problem is offered through the correct preparation of the parent silica prior to bonding.

The activity of silanols depends largely on the pretreatment and the purity of the silica itself. Contrary to intuition, the best silica for the preparation of a reversed-phase bonded phase is a fully hydroxylated silica. The silanols on a fully hydroxylated silica exhibit hydrogen bonds to neighboring silanols. Techniques designed to remove silanols, including treatment at elevated temperature, leave many silanols without the ability to form these hydrogen bonds. This renders the surface energetically less homogeneous. Also, these unbridged silanols are more acidic than the bridged silanols, and therefore exhibit a stronger interaction with basic analytes. This, in turn, results in an increased tailing for this kind of compound. Unfortunately, the preparation procedure of some silicas requires a treatment at elevated temperature to remove porogens. These types of silicas have to be rehydroxylated thoroughly to eliminate highly active silanols.

Metal contamination of the matrix of the silica, especially by aluminum and iron, also increases the acidity of surface silanols and the heterogeneity of the surface. This is a problem that largely plagues older stationary phases, which are based on silicas derived from inorganic raw materials. Many modern silicas are manufactured from organic silanes. In such processes, a high purity of the silica can be maintained with the appropriate precautions. In turn, this results in reversed-phase bonded phases with superior behavior toward basic compounds, that is, without excessive retention or tailing.

The interaction of analytes with the "residual" silanols can be a nuisance to the user of reversed-phase packings. They make the retention behavior of basic analytes more difficult to interpret and less predictable. In some extreme cases, especially for basic peptides, plots of the retention factor versus the percent organic modifier exhibit a minimum; while for pure hydrophobic interaction, a monotonous decline with increasing organic content of the mobile phase is expected. The increased tailing caused by the interaction of silanols with polar, but especially basic, compounds makes peak integration more difficult and reduces column performance.

Several types of interactions are possible between silanols and analytes: hydrogen bonding, dipole interaction, and ion exchange. The latter is the predominant mode of interaction with basic compounds, where most of the undesirable influence of the silanols becomes apparent. Silanols act as cation-exchange groups of intermediate strength. This means that their activity is a function of the mobile phase pH. At around pH 3, all except the most acidic silanols are protonated, and therefore do not undergo the ion-exchange interaction with positively charged analytes. As the pH is increased, more and more silanols become negatively charged and interact with the analyte. This results in an increase in retention and in tailing. This relationship is plotted in Figure 10.4. As one can see from this figure, too, the degree of tailing also depends on the nature of the bonded phase. A well-endcapped bonded phase based on a properly prepared, high-purity silica exhibits less tailing than does a reversed-phase packing based on a conventional silica.

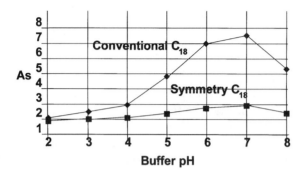

Figure 10.4 Plot of the tailing factor of amitriptyline versus mobile-phase pH for two different reversed-phase bonded phases. At low pH, tailing is suppressed as a result of suppression of the ion-exchange activity of the silanols. As the pH is increased, tailing increases due to the increased ion-exchange activity of the silanols. At high pH, tailing decreases again as one approaches the pK_a of the analyte. The Symmetry reversed-phase packing is based on a deactivated, high-purity silica and exhibits less of this effect than the conventional C_{18} packing. (Data courtesy of T. H. Walter and B. A. Alden, Waters Corp.)

From this discussion one can see that one way to eliminate or at least re- duce peak tailing is the use of acidic mobile phases. This, however, is a severe constraint on the choice of mobile-phase conditions used for the optimization of a separation. Another commonly used approach is therefore the addition of organic amines to the mobile phase. They compete with the analyte for the active sites, and can thus attenuate or even eliminate the undesirable behavior. The concentration of the additive should be in the range of 10–100 mM to be effective. A commonly used mobile-phase additive is triethylamine. However, it is less effective than longer-chain amines, which interact with the bonded phase more strongly. When triethylamine fails to eliminate tailing, one should try octylamine, or if that does not help, either, a cetyl trimethylammonium salt. The latter is most effective, but results in long equilibration times.

Whenever amines are added to the mobile phase, a buffering counterion should be added as well to keep the pH within the stability range of silica. Suitable buffers can be prepared from triethylamine with phosphoric acid (for the pH ranges around 7 and 2), triethylamine with acetic acid ($pK_a = 4.75$), or triethylamine with citric acid (buffering range from pH 2.5 to 6.5). Buffers based on octylamine tend to be a bit more difficult to prepare, since the solubility of octylamine in water is limited, but the same counterions can be used. Quater- nary amines do not shift the pH of a given buffer and should be used just as additives to a pre-prepared mobile phase.

10.2.2 Characterization of Packings

Over the years, many efforts have been made to fully characterize silica-based reversed-phase bonded phases. Of course, primary parameters are the physical characteristics of the packing, including specific pore volume, specific surface area, pore size and pore size distribution, and particle size and particle size distribution, all of which are easily accessible to measurement. The next parameters are the primary chemical characteristics of a packing, such as the type of silane used in the derivatization, the surface coverage, and whether the stationary phase is endcapped. The type of silane is characterized by the chain length of the ligand; whether the silane is a mono-, di-, or trifunctional silane; and what the nature of the side chain is (in the case of di- and monofunctional silanes). The surface coverage is expressed in micromoles per square meter of ligand. It is a significantly more meaningful parameter than the percent carbon commonly found in the manufacturers' literature. It is derived from elemental analysis (% carbon) or from thermogravimetric analysis.

Further information about the nature of the bonded phase can be obtained through solid-state nuclear magnetic resonance and infrared spectrometry. These techniques allow us to obtain semiquantitative information about the distribution of the different types of silanols on the surface on both the parent silica and the bonded phase. Since it has become understood that metal contamination in the matrix of the silica can significantly influence the activity of silanols on the silica surface, the examination of a silica for trace-metal content can be very enlightening. A low content (< 10 ppb) of especially aluminum and iron is desirable.

Nevertheless, the best overall characterization of a reversed-phase bonded phase is done by well-designed chromatographic tests. Such a test should enable us to characterize at least the hydrophobicity (retentivity) of a packing and the activity of the "residual" silanols. As we will see, tests can also be designed to characterize the shape selectivity of a packing or its behavior toward complexing agents.

There have been many attempts to create chromatographic tests that check the quality and reproducibility of chromatographic packings for a given set of analytes and chromatographic conditions. But there are relatively few tests that address the fundamental properties of reversed-phase packings. One of the early ones was a test developed by Dreyer (15) and published more recently by Engelhardt and Jungheim (16). It consists of hydrophobic samples, toluene and ethylbenzene, a mildly acidic sample, phenol, and aniline and the toluidine isomers as basic samples. The mobile phase is a mixture of 49% methanol/51% water. Important are the relative retentions of phenol and the toluidines relative to toluene. Coelution of the toluidine isomers indicates a silica with a low silanol activity. If a silica has a high silanol activity, the toluidines are at least partially separated and exhibit tailing peaks. Also, an increased tailing of aniline and the toluidines is an indicator of increased silanophilic interaction.

A derivative of the Engelhardt test has been developed by Serowik and Neue (17). Here again a mixture of neutral, basic, and acidic compounds is used to test both hydrophobic and silanophilic properties of a packing. The mobile phase is a mixture of methanol and a 100 mM phosphate buffer at pH 7.00. If the test is used for the qualification of a batch of packing, the actual composition of the mobile phase and the actual compounds used for testing vary according to the surface chemistry to be tested. For C_{18} packings, and as a general comparative test, the mobile phase consists of 65.0% methanol and 35.0% buffer at 25.0°C. The samples used are amitriptyline or another member of the family of tricyclic antidepressants and propranolol as basic analytes, a member of the lower parabens as mildly acidic compound, and acenaphthene or naphthalene, toluene, and a phthalate such as dibutylphthalate as hydrophobic compounds. The basic analytes were selected because they exhibit an especially strong interaction with surface silanols, even in the presence of a buffered mobile phase. The relative retentions of all analytes relative to acenaphthene are measured and specified in this test, together with tailing factors of the basic compounds. The retention factor of the hydrophobic analytes is used for the measurement of the hydrophobicity of the packing. Both the tailing of the basic analytes and their relative retention to acenaphthene are measures of the activity of silanols at neutral pH. The test is more rugged than the Engelhardt test and can be used for the precision test of batch-to-batch reproducibility of a packing, but is less suitable for general use, since some of the compounds used are pharmaceuticals with restricted access. A typical chromatogram obtained with this test is shown in Figure 10.5.

Figure 10.6 shows a plot of the silanol activity at pH 7, as measured by the relative retention of amitriptyline to acenaphthene, versus the hydrophobicity of a packing, as measured by the retention factor of acenaphthene, for several commercially available C_{18} and C_8 packings. The test indeed groups together packings on the basis of their chemical properties (e.g., C_{18} packings based on high-purity silicas or unendcapped reversed-phase packings). Also, the three different J'Sphere packings from YMC fall practically on a line that slopes from the upper left corner down to the lower right corner. According to the manufacturer, these packings are based on the same silica using the same ligand and the same bonding technology, but with different ligand densities. Thus one would expect a high silanol activity with a low hydrophobicity for the packing with the low coating and a somewhat lower silanol activity and a higher hydrophobicity for the highly coated packing.

Both the Engelhardt test and the Neue test capture the total silanol activity at neutral pH. They do not discriminate between acidic and regular silanols. Therefore Walter (18) designed a test capable of measuring the amount of acidic silanols alone. The test uses a phosphate buffer at pH 3.0 in a methanol/buffer mobile phase. The basic analytes that probe the acidic silanols are propranolol and chlorpheniramine, and the neutral reference compound is toluamide. Results of this test were correlated with the metal contamination in the matrix of the parent silica, which is believed to control the acidity of the surface silanols.

The contamination of the surface of a packing with metals can be assessed

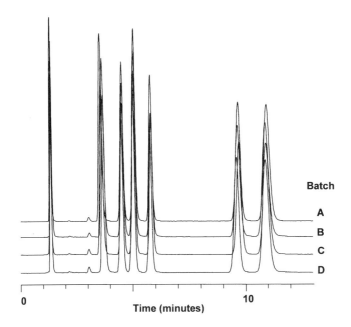

Figure 10.5 Batch-to-batch reproducibility test according to the method developed by Serowik and Neue. The packing is SymmetryShield RP$_8$, and the mobile phase is 65% methanol/35% 0.1 M K$_2$HPO$_4$, pH 7. The peaks are, in the order of elution: uracil, propranolol, butyl paraben, dipropylphthalate, naphthalene, acenaphthene, amitriptyline. The figure shows an overlay of the chromatograms obtained on four different batches of the packing. (Chromatograms courtesy of T. H. Walter and B. A. Alden, Waters Corp.)

using chelating compounds as probes. 1,5-Dihydroxyanthraquinone and a-cetylacetone have been used by Tanaka (19) in a mobile phase consisting of 30% v/v 0.02 M phosphate buffer at pH 7.6 and 70% methanol. Metal contamination is indicated by tailing of the complexing agents. Another popular probe for chelating sites is hinokitiol. The structures of the chelating agents is shown below:

1,5,-Dihydroxyanthraquinone   Acetylacetone   Hinokitiol

Characteristics of some C8 and C18 Reversed-Phase Packings

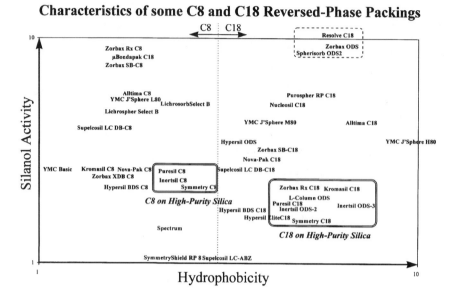

Figure 10.6   Plot of the silanol activity at pH 7 versus the hydrophobicity of a packing. (The graph is based on data provided by B. A. Alden, Waters Corp.)

However, one has to be careful with tests for metal contamination of the silica surface. HPLC systems are usually made of steel, as is the column hardware, including the frits. All steel will slowly bleed metals, especially iron, which accumulates on the surface of the packing. Thus the test for metal contamination is a test of the status of the contamination of a column rather than a test of a permanent property of the packing. The test is meaningful only, when carried out using an HPLC system with a metal-free fluid path and a nonmetallic column.

Sander and Wise (20, 21) developed a test that is sensitive to the shape selectivity of a C_{18}-packing, which, in turn, depends on the spacing of the chains, or as Sander and Wise express it, on the monomeric versus polymeric character of the packing. The test mixture, which is available as a NIST Standard Reference Material (SRM 869), consists of benzo[a]pyrene (BaP), phenanthro[3,4-c]phenanthrene (PhPh), and 1,2–3,4–5,6–7,8-tetrabenzonaphthalene (TBN). The elution order of these compounds on C_{18} packings depends on the chain spacing of the bonded phase. For "monomeric" packings, that is, packings with a surface coverage $< 3.5\,\mu\text{mol/m}^2$, the elution order is BaP $\leqslant$ PhPh $<$ TBN. For "polymeric" phases, the elution order is PhPh $<$ TBN $\leqslant$ BaP. The relative retention between TBN and BaP has been found to vary from about 0.6 to 2.2 for commercial C_{18} columns. One can use this test mixture to assess the properties of a C_{18} material prepared from multifunctional silanes where a simple measurement of the surface coverage is an insufficient characterization of the packing. It is less useful for the quality control (QC) of monofunctionally derivatized packings; the surface coverage based on the measurement of the carbon content and the specific surface area of a packing is a more accurate and more precise indicator of the reproducibility of a packing.

10.2.3   Reproducibility of Reversed-Phase Packings

Reversed-phase packings are used in the majority of HPLC methods in many industries around the world. Consequently, the reproducibility of these packings is critical. From the discussion in the previous paragraphs, we understand that different brands of packings have different properties. But how closely can the properties of a single brand be reproduced?

Many different parameters influence the reproducibility of a packing. First, we have the physical properties of the silica itself: the specific surface area, the specific pore volume, the pore size distribution, and the particle size distribution. Among these, the specific surface area and the specific pore volume are the most critical parameters, since both together will determine the surface area per column volume, which, in turn, determines retention. However, more important than the physical properties of a packing are its chemical properties: the amount and character of the silanols and the amount and nature of the primary ligand and secondary ligands, if more than one surface reaction is carried out (e.g., endcapping). From elemental analysis or thermogravimetry, the amount of ligand per unit surface area (surface coverage) can be calculated. However, the best test of chromatographic reproducibility is a well-designed chromatographic test, as described in the previous chapter.

The reproducibility of the surface modification of a bonded phase depends first on the choice of the silane. Di- and trifunctional silanes can polymerize, and their reaction with the silica surface is less controllable than that of monofunctional silanes. Thus packings based on monofunctional silanes are intrinsically more reproducible.

However, even with bonding technology based on monofunctional silanes, there are wide differences, since the final properties of a bonded phase depend on the pretreatment of the silica as well. Also, the experience of a manufacturer, especially at the time that the technology was developed, plays an important role in the final reproducibility of the product.

Table 10.1 gives some data for the reproducibility of some commercial packings manufactured by Waters Corporation. Listed are the relative standard deviations for the percent carbon, the retention factor for a neutral hydrophobic compound (acenaphthene), and the relative retentions between neutral compounds and the relative retention between a hydrophobic base and a neutral compound, as described in the last section (17). The packings listed were developed in different decades: μBondapak C_{18} in 1973, Nova-Pak C_{18} in 1981, and Symmetry C_{18} in 1994. One can clearly see the improved reproducibility of the packings with increased maturity of the technology. One

TABLE 10.1 Reproducibility of Some Commercial C_{18} Packings

Brand	Relative Standard Deviations for Batch-to-Batch Reproducibility			
	Carbon (%)	Retention Factor of a Neutral Analyte (%)	Relative Retention between Two Neutral Analytes (%)	Relative Retention between a Neutral Analyte and a Basic Analyte (%)
μBondapack C_{18}	2.9	8.8	1.4	6.8
Nova-Pak C_{18}	2.2	3.9	1.1	4.3
Symmetry C_{18}	1.4	2.8	0.5	2.2

can also see that the reproducibility depends on the nature of the analytes — the relative retention between neutral analytes is over 4 times more reproducible than the relative retention between a basic analyte and a neutral analyte, which stresses the importance of good testing procedures discussed in the previous section.

To get an appreciation of the data in Table 10.1, it should be pointed out that the data are obtained under strict control of the mobile-phase composition and the column temperature. A 1% error in the mobile-phase composition results in an error in the retention factor of between 7% and 15%. Under the test conditions, the retention factor of a basic analyte will shift by 10% for an error in mobile-phase pH of 0.1 pH units, while the retention factor of a neutral compound is unaffected. Thus an error of 0.1 pH units also changes the relative retention by 10%. Temperature changes retention factors by between 1–2.5% per 1°C, and relative retentions of a pair of neutral and basic compounds are affected by 1–2%. Considering these other influences, the batch-to-batch reproducibility of a modern packing such as Symmetry C_{18} is quite excellent. An example of the good batch-to-batch reproducibility of modern reversed-phase packings is shown in Figure 10.7, which shows the impurity profile of taxol on three different batches of Symmetry C_8.

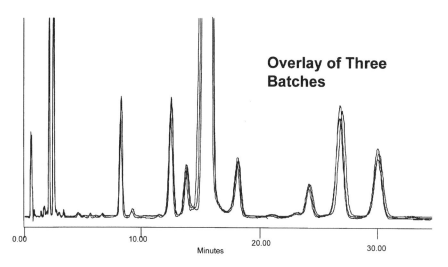

Figure 10.7 Reproducibility of modern reversed-phase packings. Overlay of the impurity profile of taxol obtained on three different batches of 3.5 μm Symmetry C_8 packing. [Chromatographic conditions: Column: 3.5 μm Symmetry C_8 (4.6 mm × 100 mm), Waters Corp. Flow rate: 1.8 mL/min. Mobile phase: methanol/acetonitrile/water 9–34–57. Detection: UV at 230 nm. Sample: 25 μL of 1 mg/mL of paclitaxel (taxol).] (Chromatogram courtesy of D. J. Phillips.)

10.3   POLYMERIC REVERSED-PHASE PACKINGS

The major disadvantage of silica-based reversed-phase packings is their hydrolytic instability, especially at alkaline pH. Many chromatographic procedures would benefit from an alkaline mobile phase. Basic compounds that are positively charged at acidic or neutral pH loose the charge at high pH. This would result in a significantly different selectivity of the separation and potentially improved peak shapes. As another example, saccharides give two peaks per compound at neutral to acidic pH, due to the separation of the anomers. At alkaline pH, the speed of the mutarotation increases significantly, and the two anomer peaks merge into a single narrow peak.

Polymer-based packings are compatible with alkaline mobile phases. Styrene-divinylbenzene packings are stable over the entire pH range. Methacrylate-based packings are limited by the stability of the ester group, and acrylamide-based packings are limited by the amide group. Polyvinylalcohol-based reversed-phase packings are prepared by esterification of polyvinylalcohol, and their pH stability is also determined by the stability of the ester group.

All polymer-based packings available today exhibit inferior efficiencies compared to well-designed silica-based packings for low-molecular-weight analytes. Polymer-based packings have micropores, probably in the size range of $\leqslant 1$ nm, which inhibit mass transfer for small molecules. This is the reason why polymeric packings have not replaced silica-based packings.

The pressure capability of polymeric packings is lower than that of silica. Most silica columns can be used without difficulty up to 40 MPa or even higher. Although the pressure capability of polymeric packings varies widely with pore volume and degree of crosslinking, they are typically constrained to pressures under 20 MPa. However, 10 MPa is quite sufficient for many HPLC applications.

10.3.1   Styrene-Divinylbenzene-Based Packings

Styrene–divinylbenzene-based packings have been used in the size-exclusion chromatography of organic polymers long before HPLC was born. But it took a long time before they were first used for reversed-phase chromatography. The reason for this was the difficulties associated with the swelling and shrinking of the packing. Columns packed for size-exclusion chromatography collapse as a result of shrinking of the packing, when one attempts to convert them to reversed-phase solvents. They have to be packed in an aqueous environment specifically for use in reversed-phase chromatography.

Most unmodified styrene–divinylbenzene packings exhibit generally much larger retention than C_{18} bonded phases. Retention factors can be over one order of magnitude higher than on a C_{18} bonded-phase packing, requiring a higher organic content in the mobile phase for reasonable retention times. This has been attributed to the specific interaction of analytes with the π-electrons of the aromatic ring of the packing. However, this explanation is not likely to be correct, since this effect is not observed for phenyl bonded phases. A better explanation is the increased available surface area due to the microporosity of the packings, which also is the cause for hindered mass transfer and broad peaks. Furthermore, there is generally a good correlation between the retention on silica-based bonded phases and on styrenic packings, which also does not support the concept of a specific interaction due to the presence of the aromatic ring. Exceptions are, of course, analytes with an electron-deficient aromatic ring, such as di- and trinitrobenzenes, which exhibit strong interaction with benzene rings. Several different packings are commercially available that all share the characteristics described above.

A styrene–divinylbenzene packing derivatized with C_{18} is commercially available, too. This packing exhibits lower retention than underivatized packings, but the pore size is larger and the surface area is lower.

Packings with different pore sizes are commercially available as well. The packings with the larger pore sizes have been used for the reversed-phase chromatography of proteins and peptides. For these kinds of analytes, the microporosity of polymeric packings is irrelevant, and chromatographic performance is comparable to silica-based reversed-phase packings with the added advantage of hydrolytic stability. Because the packing consists entirely of C—C bonds, the styrene–divinylbenzene packings are hydrolytically stable from pH 0 to 14.

10.3.2 Methacrylate-Based Packings

A wide range of methacrylate-based packings are available for reversed-phase chromatography, exhibiting a graded range of hydrophobicities. Similar to silica-based reversed-phase packings, the hydrophobicity is obtained through variations in the aliphatic chain of the methacrylate and through the overall content of the aliphatic chains. The commonly used crosslinker is ethylene-glycol dimethacrylate.

The pH range is more limited than the range available to styrenic packings, but extends significantly into the alkaline pH range. The values quoted for related packings by different manufacturers vary somewhat, but the general rule seems to be a good compatibility within the range of pH 2–12, with the capability to short time exposure to pH 13 for cleaning purposes.

Related to the methacrylate-based packings is a C_{18}-type polyacrylamide-based packing. Its pH range has been reported to extend from pH 1 to 13 (22).

10.3.3 Polyvinylalcohol-Based Packings

Reversed-phase packings have been obtained by derivatization of a poly-vinylalcohol packing with octadecanoyl, octanoyl, and butyryl groups. Because of the ester linkage of the functional group, the pH stability is comparable to the methacrylate packings. The commercially available packings are derived from a base particle with a pore size of 250 Å, which makes the phases suitable for the analysis of peptides and proteins. The hydrophobicity of the different types follows the expected pattern.

10.4   OTHER TYPES

Several packings have been developed that represent unique alternatives to the standard silica-based bonded phases. There has been a long-standing effort to design a porous graphitized-carbon packing suitable for HPLC. Inorganic ceramics other than silica have been explored as alternatives. Among those, alumina coated with polybutadiene or silicone polymers are commercially available. A third interesting type of reversed-phase packing has been developed specifically for the analysis of metabolites from biological fluids: internal reversed-phase packings. These types of packings are covered in the following paragraphs.

10.4.1 Porous Graphitized Carbon

Porous graphitized carbon, commercially available since 1988, is one of the newer additions to the arsenal of reversed-phase packings (24). Its behavior is quite different from that of reversed-phase bonded phases. On one hand, this makes it an attractive alternative to these packings; on the other hand it also has been an impediment to its broad acceptance.

Porous graphitized carbon is fairly rigid and compatible with pressures of ≤ 40 MPa. It is stable from pH 1 to 14 and does not swell or shrink in the presence of organic solvents. While the skeleton of silica is formed by the coalescence of microbeads of molecular dimensions, the backbone structure of porous graphitized carbon is formed from intertwined bands of graphite. The detailed nature of the surface is still under continued investigation.

For purely hydrophobic analytes, porous graphitized carbon is about as retentive as a C_{18} packing. When polar functional groups are added, the drop in retention is much more pronounced for a C_{18} packing than for porous graphitized carbon. This indicates that other interactions between the analyte and the packing influence retention (25). More specifically, the retention mechanism on porous graphitized carbon is not just hydrophobic interaction. Apparently, the delocalized π-electrons of the graphite bands are responsible for this additional interaction. It should be noted that retention is also observed with nonpolar mobile phases. This has been attributed to the electron-pair-acceptor properties of the graphitic bands, but more trivial explanations such as the presence of unwanted polar functional groups cannot be entirely rejected. Furthermore, one often observes tailing peaks, which are indicative of a heterogeneity of the surface. On the other hand, the change of retention of basic analytes with a change of the pH of the mobile phase is more predictable with porous graphitized carbon than with a silica-based reversed-phase packing (26), which is complicated because of the presence of silanols. Also, porous graphitized carbon easily resolves structural isomers, which often are not or only barely resolved on reversed-phase bonded phases.

While studies of the retention mechanism on porous graphitized carbon are still ongoing, it has found a stronghold in at least one application area: the reversed-phase chromatography of complex carbohydrates. Reversed-phase chromatography on silica-based packings is unsuitable for these compounds, since at neutral pH the anomers of each carbohydrate are resolved, resulting in two peaks for every compound. Porous graphitized carbon can be used with basic mobile phases, which speeds up the mutarotation such that the anomer peaks merge into one peak. Furthermore, the retention of derivatized and underivatized carbohydrates is increased on porous graphitized carbon.

10.4.2 Polymer-Coated Alumina

Alumina is as mechanically stable as silica, but is hydrolytically more stable, especially in the alkaline pH range. Therefore it is attractive as a carrier for reversed-phase surfaces. However, no surface-derivatization technique comparable to the silanization of silica is available. Coating techniques have been developed by Schomburg et al. (27), which add a stable hydrophobic polymer to the alumina surface. The most popular packing type, which is available from several sources, is alumina coated with a crosslinked polybutadiene. The second commercially available type is obtained by applying a polysiloxane coating to the alumina.

Indeed, an improved pH-stability is obtained with both types of packings, up to a pH of 13. However, the commercially available types of packings exhibit a significantly reduced mass transfer, resulting in low column efficiencies. This constrains the widespread applicability of this type of packing.

10.4.3 Internal-Surface Reversed-Phase Packings

It has been observed that reversed-phase columns used for the analysis of biological fluids exhibit a short column life, sometimes as low as 100 sample injections. Under common reversed-phase conditions, proteins can adsorb to the outside of the packing, which soon leads to a partial clogging of the flow passages, high backpressure, and eventually a destruction of the packed bed. The commonly used solution is sample pretreatment (protein precipitation) together with the use of guard columns. But sample pretreatment adds time-consuming steps to the analytical protocol. The use of a guard column alone without sample pretreatment is unsatisfactory also, since the guard column needs to be replaced too frequently to make routine, automated analysis convenient. An ingenious solution to this dilemma has been developed by Pinkerton and co-workers (28): the external surface of the packing and the surface of large pores, which are both accessible to proteins, are rendered hydrophilic and noninteracting, while the surface of small pores, which is accessible to small molecules, is hydrophobic. This was originally achieved by bonding the hydrophobic ligand to the surface through a peptide linkage, which was subsequently digested by an enzyme. The enzyme can operate on the surface only in the larger pores that are accessible to it. The surface in the smaller pores remains hydrophobic. Molecules of the size of the enzyme or larger will only be exposed to a hydrophilic noninteracting surface and will pass through the column unretained. Small molecules will interact with the entire surface, including the hydrophobic surface in the small pores, and can be separated by a reversed-phase mechanism.

Many different phases can be designed using this technology. Later, other techniques were developed that accomplished the same goal. In one alternative technology, the hydrophobic bonded phase is covered with a crosslinked polyethyleneoxide, which prevents proteins from reaching the hydrophobic surface, but lets small molecules penetrate.

Internal-surface reversed-phase packings, also called *restricted-access packings*, indeed prolong the lifetime of a column, but they are not a complete solution to the problem. Some protein still adsorbs, although the amounts are small. Columns can also clog by simple precipitation of the proteins. Consequently, one still observes a limited column life. Also, many analytical methods for pharmaceuticals and their metabolites have originally been developed on conventional packings. For both reasons, an alternative way to use the internal-surface reversed-phase packings has emerged — they are used for automated, on-line sample cleanup, while the actual separation is performed with classical reversed-phase columns. The serum sample is injected directly onto a small guard column (typical length < 2 cm) packed with an internal-surface reversed-phase packing. The majority of the proteins are eluted from the guard column unretained, while the analytes of interest are retained. The effluent of the guard column is first directed to waste, until the protein has been eluted. Then, the effluent is switched to the analytical column, where the separation of the analytes takes place. This technique is instrumentally more complex, but protects the analytical column more effectively and also uses the more expensive internal-surface reversed-phase packings in a more cost-effective way.

10.4.4   Immobilized Artificial Membranes

The ability of drugs to penetrate cell membranes has been of utmost importance in the design of drugs. The classical technique used to measure this ability is the partitioning of the drug between octanol and water. Many attempts have been made to use reversed-phase chromatography for the same purpose. Although generally a reasonable correlation is obtained between the logarithm of the water–octanol partitioning coefficient and the logarithm of the retention factor in reversed-phase chromatography, deviations do exist, which are most likely attributable to the influence of silanols on retention, especially of basic drugs. To get around this problem and design a better model for the cell membrane, phosphatidylcholine was immobilized on an aminopropyl-d-erivatized silica (29). The resulting phase is reported to be a better model of the cell membrane than both classical reversed-phase packings and the o-ctanol-water partitioning technique. Immobilized-artificial-membrane packings are also reportedly more suitable for the chromatography of membrane proteins. Several versions of the immobilized artificial membrane have been synthesized and are commercially available (Regis, USA).

10.5   SPECIAL TECHNIQUES

There are two special techniques of reversed-phase chromatography that deserve a brief description: (1) paired-ion chromatography, a technique used to enhance the retention of ionic sample constituents; and (2) the fairly standardized technique for the reversed-phase chromatography of peptides.

10.5.1   Paired-Ion Chromatography

In paired-ion chromatography, a hydrophobic cation or anion is added to the mobile phase. This enhances the retention of oppositely charged analytes. Paired-ion chromatography is a very powerful tool for manipulating the selectivity of a separation. The retention of neutral analytes is nearly unaffected. Analytes with opposite charge to the paired-ion reagent are retained longer, and the retention of analytes with the same charge as the pairing reagent is reduced.

Typical anionic pairing reagents are long-chain sulfonic acids such as hexyl-, heptyl-, or octylsulfonic acid. Cationic pairing reagents are, for example, the tetrabutyl ammonium ion or the cetyltrimethylammonium ion. These reagents are typically added to the mobile phase at a concentration around 5–10 mmol/L. At low concentration of the reagent, the retention of the oppositely charged analytes increases directly proportionally to the concentration of the reagent. At higher concentrations, the increase in retention levels off. At equal concentration, more hydrophobic ion-pairing reagents, specifically, those with a longer chain length, increase the retention more than less hydrophobic reagents. At what concentration the increase in retention levels off depends on the mobile-phase composition and the chain length of the pairing reagent. Figure 10.8 shows the dependence of the retention of water-soluble vitamins on the concentration of ion-pairing reagent.

The reagent is adsorbed onto the surface of the packing. Its surface concentration depends on the mobile-phase composition and the concentration of the reagent. It can be shown that the retention change of an analyte depends primarily on the molar surface concentration of the pairing reagent (30). One can superficially interpret the increased retention of the analytes as an ion-exchange mechanism that is superimposed on the normal reversed-phase

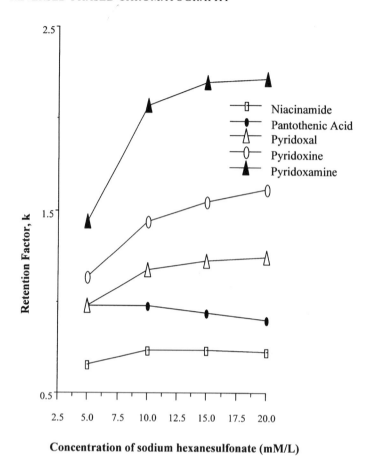

Figure 10.8 Dependence of the retention factor on the concentration of ion-pairing reagent in the mobile phase. (Data courtesy of D. J. Phillips, Waters Corp.)

retention mechanism. However, UV-absorbing ion-pairing reagents such as a cetylpyridinium ion have been used to detect non-UV-absorbing analytes. A positive signal is obtained at the point where the analytes elute, accompanied by a negative signal in a fixed position of the chromatogram. When the pairing reagent is injected itself, a positive signal is obtained at the same retention time as the negative peak obtained when the analytes are injected. This phenomenon is interpreted as stemming from the need to maintain charge balance as the analyte is adsorbed on the stationary phase (31).

Retention can be influenced by the type and the concentration of the pairing reagent, the ionic strength, and the pH of the mobile phase, as well as the concentration and type of the organic modifier to the mobile phase. The technique therefore opens the window for additional parameters that can be used to influence the selectivity of a separation.

A disadvantage of this technique is the long equilibration times required. As mentioned above, typical concentrations of the pairing reagent are around 5 mmol/L. A typical surface concentration of the pairing reagent is around $1 \mu mol/m^2$ or about 0.15 mmol/mL of packed bed. For a typical column volume of 2.5 mL it would take at least 75 mL (or 30 column volumes) to equilibrate the column with the pairing reagent. In practice, it may take even longer to completely equilibrate the column, since the surface concentration of the pairing reagent may be 3–4 times higher than the $1 \mu mol/m^2$ used for this estimate.

Many investigators recommend that columns used for paired-ion chromatography should be dedicated for this application and should not be used for regular reversed-phase chromatography. The reason for this is not clear. Apparently, ion-pairing reagents are difficult to remove from silica-based reversed-phase packings.

10.5.2   RPLC of Peptides

The dependence of the retention factor on the mobile-phase composition is a function of the molecular weight of the analyte. The larger the molecule, the faster the retention changes with a change in the mobile-phase composition. This makes the isocratic elution of large molecules impractical. Therefore, the analysis of peptides and proteins is carried out exclusively using gradients. The standard reversed-phase packings with a pore size of 10 nm are suitable for small peptides at least up to the size of insulin (51 amino acids). For large peptides and proteins, reversed-phase packings with a pore size of 30 nm are typically used. The diffusion of these larger molecules is less hindered in the larger pores of a 30-nm packing.

The shorter-chain reversed-phase packings bonded with C_4 or C_8 ligands were initially used for peptide analysis, but now the standard octadecyl packings are most widely used. From the standpoint of the separation, any chain length can be used successfully, albeit with some differences in selectivity. The longer-chain packings are hydrolytically more stable than their shorter-chain equivalents.

A standard protocol has been developed for proteins and peptides. The organic modifier used is acetonitrile, which has a lower viscosity than other alternatives, resulting in a lower backpressure and sharper peaks. To both the water and the acetonitrile, 0.1% trifluoroacetic acid (TFA) is added. This provides an acidic environment, which suppresses the interaction of the basic groups of the peptide or protein with surface silanols. Unfortunately, TFA absorbs strongly in the low UV, which otherwise would be useful for the sensitive detection of peptides and proteins. Therefore, TFA has been substituted with hydrochloric acid. However, this is only possible with a nonmetallic HPLC instrument, since dilute hydrochloric acid attacks steel.

Figure 10.9 shows an example of peptide separations obtained with reversed-phase liquid chromatography. The sample is the tryptic digest of bovine

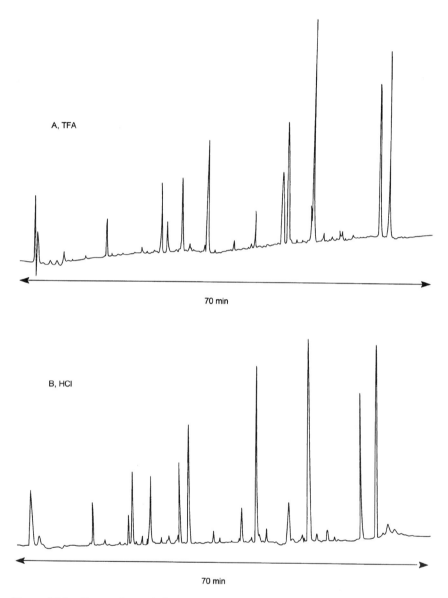

Figure 10.9 Comparison of chromatograms obtained for the tryptic digest of bovine cytochrome *c* using an acetonitrile gradient with either 0.1% TFA (top) or 6 mM HCl (bottom). The column used is the nonmetallic column Delta-Pak HPI C_{18} from Waters. (Chromatograms courtesy of Waters Corp.)

cytochrome *c*. The column used is a nonmetallic column with PEEK tubing and end fittings, and polyethylene frits. Two chromatograms are shown. In one chromatogram 0.1% TFA is used to acidify the mobile phase; in the other case 6 mM HCl is used. A gradient from water with the acidic modifier to 60% acetonitrile with the acidic modifier is applied in both cases. Small selectivity differences are obtained, but the overall chromatograms are very similar.

Isopropanol is occasionally used as the organic modifier for peptide gradients. Generally, isopropanol–water mixtures have a significantly higher viscosity (up to 4-fold) than acetonitrile–water mixtures. This results in a higher backpressure and/or broader peaks. Therefore, acetonitrile-based gradients are generally preferred.

10. 6   PRACTICAL ASPECTS OF REVERSED-PHASE CHROMATOGRAPHY

In this section we will briefly discuss a few aspects of the practice of reversed-phase chromatography. The tips and hints given here are designed to help the practitioner develop solid reliable methods and avoid some common pitfalls. We first discuss column selection, then the mobile phase, and finally some aspects of column care.

10.6.1   Column Selection

A broad range of packings with widely varying properties is available for reversed-phase chromatography. Practitioners must decide which packings to choose for their applications.

Generally, the most important factor is the reproducibility of a packing. For this, you should consider your previous experience, the experience of your colleagues, and the reputation of the manufacturer. Also, check what information the manufacturer supplies to support his claim of reproducibility.

The second criterion depends on the nature of your sample. If your analyte is sensitive to the activity of silanols, you should choose a packing that is designed for this kind of analysis. On the other hand, if this does not matter, for example, when you are trying to separate purely hydrophobic analytes, your choices are much broader.

If your analytes are very polar and require mobile phases that contain nearly 100% water or buffer, you should choose a packing with an embedded polar functional group such as SymmetryShield RP_8 or Supelcosil ABZ. The best hydrophobic packings with the best endcapping exhibit in highly aqueous mobile phases a phenomenon called "hydrophobic collapse", in which the mobile phase is driven out of the pores as the result of an unfavorable wetting angle. Packings with an embedded polar functional group or badly endcapped packings do not exhibit this phenomenon, but the latter may not be the best choice because of excessive silanol activity.

10.6.2 Mobile-Phase Considerations

If your sample contains ionic or ionizable compounds, you should always use a buffered mobile phase to ensure reproducible results. Under unfavorable circumstances, pH changes of as little as 0.1 pH units can have a significant effect on the separation. On the other hand, a properly used buffer allows you to control the pH easily. Remember that a buffer works best at the pK_a of its acid. At this pH, the concentration of the acidic form and the basic form of your buffering species is equal, and the buffering capacity is maximum. Phosphate has two pK_a values in the range of interest for silica-based chromatography: one at pH 2 and the other at pH 7. The pK_a of the acetate buffer is 4.75. Citrate has three pK_a values: 3.08, 4.77, and 6.40. Between citrate and phosphate buffers, the entire pH range useful for silica chromatography can be covered.

In many cases, silanophilic interactions cause tailing, mostly due to ion-exchange interactions. This can usually be reduced or suppressed by the use of amine-based buffers or by using acidic mobile phases, or a combination thereof. Note that some of the newer reversed-phase packings have been designed to minimize this interaction, and good results can be obtained at neutral pH without amine modifiers.

Whenever buffers or other mobile-phase additives are used, you should carefully check the solubility in the mobile phase. This is especially true for gradient applications or when the mobile-phase composition is generated automatically by the HPLC instrument. Buffer precipitation is one of the most common column problems. A precipitated buffer can be redissolved by slowly purging the column with a mobile phase, in which the buffer is soluble.

From the standpoint of pressure, acetonitrile is the preferred organic modifier in reversed-phase chromatography. Acetonitrile-based mobile phases can give an up to 2-fold lower pressure drop than can methanol-based mobile phases at equal flow rate. This also means that column efficiency is higher.

The elution strength increases in the order methanol, acetonitrile, tetrahydrofuran. The retention changes by roughly 10% for every 1% change in the concentration of organic modifier.

Maximum hydrolytic stability of silica-based reversed-phase packings is obtained at pH values around 4. At pH 2 and pH 8 one may encounter a faster aging of the column, visible through decreased hydrophobic interaction and increased silanophilic interaction. However, the actual rate depends on many different parameters, including the choice of buffer, temperature, and the organic modifier. Columns tend to be more stable in acetonitrile-based mobile phases than in methanol-based mobile phases, and recent results indicate that citrate or Tris buffers are preferred over phosphate buffers (32).

10.6.3 Column Care

For long-term storage, columns should be converted to 100% methanol or acetonitrile to minimize hydrolysis. For short-term storage, columns are best left in the mobile phase in which they are used. Cyanopropyl packings have the tendency to collapse in organic solvents of intermediate polarity and are therefore best stored in mobile phase.

The primary column aging process for reversed-phase columns is the hydrolysis of the ligands. The endcapping groups are released more easily than the longer-chain ligands. Thus, an increase in silanol activity is encountered much earlier than a change in hydrophobic retention. Consequently, a column may still be perfectly suitable for the separation of compounds that do not interact with silanols, while the chromatography of compounds that interact strongly with silanols has already changed significantly. It is generally advisable to dedicate a column to a single application, especially in the quality-control laboratory. If a large number of samples of the same type need to be analyzed over the period of several months, it is a false economy to use a single column for more than one application.

Hydrolysis of the stationary phase or other kinds of column aging may even occur during methods development. A finished method that needs to be used for a long time or by other workers should therefore always be verified on a new column, preferentially prepared from the same batch of packing. Column kits for this purpose are available from some manufacturers.

REFERENCES

1. K. Karch, dissertation, Universität des Saarlandes (1974).
2. Cs. Horváth, W. R. Melander, and I. Molnár, *J. Chromatogr.* **125**, 129–156 (1976).
3. W. R. Melander, Cs. Horváth, "Reversed-Phase Chromatography," in *High-Performance Liquid Chromatography*, Advances and Perspectives Series, Vol. 2, Cs. Horváth, ed., Academic Press, New York, 1980.
4. M. Jaroniec, *J. Chromatogr.* **A656**, 37–50 (1993).
5. E. Grushka, H. Colin, and G. Guiochon, *J. Chromatogr.* **248**, 325–339 (1982).
6. W. Melander, J. Stoveken, and Cs. Horváth, *J. Chromatogr.* **199**, 35–56 (1980).
7. P. W. Carr, L. C. Tan, and J. H. Park, *J. Chromatogr.* **A724**, 1–12 (1996).
8. G. E. Berendsen and L. de Galan, *J. Chromatogr.* **196**, 21 (1980).
9. N. Nagae, and D. Ishii, *Am. Lab.* 20–30 (March 1995).
10. P. E. Antle, A. P. Goldberg, and L. R. Snyder, *J. Chromatogr.* **321**, 1–32 (1985).
11. U. D. Neue, C. L. Niederländer, and J. S. Petersen, US Patent 5,374,755 (1994).
12. C. L. Niederländer, dissertation, Universität des Saarlandes (1993).
13. T. L. Ascah, B. Feibush, *J. Chromatogr.* **506**, 357–369 (1990).
14. J. E. O'Gara, B. A. Alden, T. H. Walter, J. S. Petersen, C. L. Niederländer, and U. D. Neue, *Anal. Chem.* **67**, 3809–3813 (1995).
15. B. Dreyer, dissertation, Universität des Saarlandes (1984).
16. H. Engelhardt, M. Jungheim, *Chromatographia* **29**, 59 (1990).
17. E. Serowik, U. D. Neue, unpublished results (1985).
18. T. H. Walter, presentation at HPLC '94 (1994).

19. Y. Ohtsu, Y. Shioshima, T. Okumura, J.-I. Koyama, K. Nakamura, O. Nakata, K. Kimata, and N. Tamaka, *J. Chromatogr.* **481**, 147–157 (1989).

20. L. C. Sander and S. A. Wise, *J. Chromatogr.* **A656**, 335–351 (1993).

21. L. C. Sander and S. A. Wise, *Anal. Chem.* **67**, 3284–3292 (1995).

22. J. V. Kawkins, N. P. Gabbott, L. L. Lloyd, J. A. McConville, and F. P. Warner, *J. Chromatogr.* **452**, 145–156 (1988).

23. T. Ohtani, Y. Tamura, M. Kasai, T. Uchida, Y. Yanagihara, and K. Noguchi, *J. Chromatogr.* **515**, 175–182 (1990).

24. J. H. Knox and M. T. Gilbert, UK Patent 7,939,449 (1979) and US Patent 4,263,268 (1981).

25. M.-C. Hennion, V. Coquart, S. Guenu, and C. Sella, *J. Chromatogr.* **A712**, 287–301 (1995).

26. Q.-H. Wan, M. C. Davies, P. N. Shaw, and D. A. Barrett, *Anal. Chem.* **68**, 437–446 (1996).

27. G. Schomburg, J. Köhler, H. Figge, A. Deege, and U. Bien-Vogelsang, *Chromatographia* **18**, 265 (1984).

28. I. H. Hagestam, and T. C. Pinkerton, *Anal. Chem.* **57**, 1757 (1985).

29. C. Pidgeon, US Patents 4,927,879 (1990) and 4,931,498 (1990).

30. J. H. Knox and R. A. Hartwick, *J. Chromatogr.* **204**, 3–21 (1981).

31. B. A. Bidlingmeyer, S. N. Deming, W. P. Price Jr., B. Sachok, and M. Petrusek, *J. Chromatogr.* **186**, 419–434 (1979).

32. H. A. Claessens, M. A. van Straten, and J. J. Kirkland, *J. Chromatogr.* **A728**, 259–270 (1996).

ADDITIONAL READING

The Journal of Chromatography **A656** (1993) has been dedicated to reversed-phase chromatography.

11 Hydrophilic-Interaction Chromatography

Hydrophilic interaction chromatography (HILIC) has been practiced for a long time, but its name has only recently been coined by Alpert (1). Previously, the techniques of HILIC were lumped together with other unrelated chromatographic techniques, often in ignorance of the actual separation mechanism. This has led to an unfortunate confusion in the literature.

Hydrophilic interaction chromatography can be viewed as an extension of normal-phase chromatography into the realm of aqueous mobile phases. Analytes are very polar compounds such as carbohydrates or polar peptides. The most important application area is the analysis of sugars and oligosaccharides, but the recognition of the broad applicability of HILIC is steadily growing. It has become an indispensable tool in the analysis of glycopeptides and complex carbohydrates derived from glycoproteins.

11.1 RETENTION MECHANISM

In HILIC, retention increases with the polarity of the analyte and decreases with the polarity of the mobile phase. This is similar to normal-phase chromatography and the opposite of reversed-phase chromatography. Hydrophilic interaction chromatography is distinguished from normal-phase chromatography by the fact that the mobile phases are mixtures of water or buffer with organic solvents. The stationary phases are very hydrophilic: polar adsorbents such as silica, polar bonded phases, polar polymeric packings, and ion exchangers. The common factor of all these stationary phases is that they easily adsorb or imbibe water, hence the categorization of "hydrophilic." Because of the strong interaction of these packings with water, the stationary phase contains more water than does the mobile phase. The more hydrophilic the stationary phase is, the larger is the difference in the water composition between the stationary phase and the mobile phase, and the stronger is the retention of an analyte.

Consequently, the retention mechanism of HILIC is the partitioning of the analyte between a water-rich stationary phase and a water-poor mobile phase. This common retention mechanism explains the similarities between the separations obtained on vastly different packings, especially for neutral polar analytes. However, dissimilarities are observed as well. For example, when silica coated with different polyamines is used as a stationary phase for the separation of carbohydrates, subtle selectivity differences are observed between different amines, indicating that the stationary phase is not merely a carrier of water. The retention mechanism of HILIC has not been as thoroughly studied as have the mechanisms of reversed-phase or normal-phase chromatography. In addition, secondary interaction mechanisms can be encountered. A common secondary mechanism is ion exchange, since many of the commonly employed carriers of the water-rich stationary phase have ion-exchange properties. This secondary mechanism can be used with advantage in manipulating the selectivity of the separation.

The addition of salt or buffer to the mobile phase can also be used to manipulate retention. Retention decreases with the addition of salt. The influence of the pH of the mobile phase depends on the nature of the analyte; charged forms are more polar and generally more strongly retained than uncharged forms of the analyte.

11.2   PACKINGS FOR HILIC

The probably earliest application of HILIC is the separation of sugars using ion exchangers as the stationary phase and water/ethanol mixtures as the mobile phase (2). Both strong cation exchangers and anion exchangers can be used. The ion exchangers can be based on an organic matrix, or they can be silica-based bonded phases. As is the case in other types of retention chromatography, the permanent pore structure of inorganic carriers results in improved mass-transfer behavior compared to packings based on organic polymers.

Carbohydrate separations are the most common application of HILIC, and silica bonded with an aminopropyl silane is the most commonly used stationary phase for this separation (3). A separation of several sugars is shown in Figure 11.1. The aminopropyl silica packing has a significant advantage for the separation of carbohydrates. Carbohydrates exist in two anomeric forms, which interconvert. However, the kinetics of this interconversion is rather slow at acidic and neutral pH values. The separation of both anomers by HPLC is very easy. Consequently, one obtains two anomer peaks for every sugar at acidic to neutral pH. This is undesirable, since the chromatogram becomes unnecessarily crowded. In an alkaline environment, the interconversion rate of the anomers is increased significantly, and the two anomer peaks merge into a single peak. Aminopropyl bonded phases provide such an environment, and are therefore commonly used for the separation of carbohydrates.

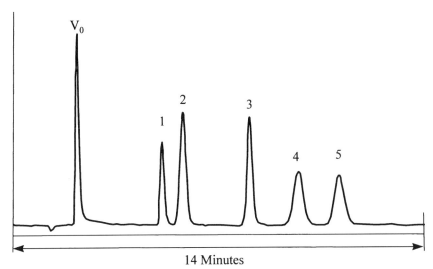

14 Minutes

Figure 11.1 Separation of sugars on an aminopropyl bonded phase. Peak identification: (1) fructose; (2) glucose; (3) sucrose; (4) maltose; (5) lactose. (*Chromatographic conditions:* Column: Waters High-Performance Carbohydrate Column. Mobile phase: acetonitrile/water 75–25 v/v.) Chromatogram courtesy of D. J. Phillips, Waters Corp.)

Aminopropyl derivatized silica has a disadvantage, though; the high concentration of amino groups in the pores of the packing provides a strongly basic environment. The concentration of amino groups in the pores is roughly 1 mol/L. At the resulting alkaline pH, the silica itself is attacked, and the surface dissolves slowly. However, this dissolution process forms acidic silanols, which, in turn, change the local pH. This slows down the hydrolysis, and ultimately a stable surface chemistry results. Aminopropyl columns that have been prestabilized are commercially available specifically for the analysis of carbohydrates.

An alternative for the same kind of analysis are silica columns that have been dynamically coated with a polyamine. Triethylene tetramine or natural amines such as spermidine or putrescine are suitable for this purpose. Small differences in selectivity are observed when different amines are used. Dynamically coated packings give a more stable chromatography than do aminopropyl bonded phases, but the silica is still subject to slow dissolution. Therefore a silica with a strong skeleton (= small specific pore volume) is recommended for this application.

Silica coated with a crosslinked polyimine serves the same purpose. Yet packings based on an organic polymer are hydrolytically more stable. An example is Asahipak NH_2, which is a polyamine-derivatized polyvinylalcohol-based packing. Since there is no concern about any dissolution of the polymer, this packing is currently probably the best choice for carbohydrate separations via HILIC.

Silica itself can be used as a carrier of the water-rich phase. Becaue of the silanols, it is also a cation exchanger of intermediate strength. But it is not as retentive as an aminopropyl bonded phase based on the same silica. Neutral silica-based packings are provided by the glycidylpropyl ("diol" bonded phases) or acetamidopropyl ligands. Alternatives are coated phases such as the TSK-gel Amide 80 or a poly(2-hydroxyethylaspartamide) silica prepared by the technique (1) described in Section 6.2.5. The latter is especially useful for HILIC because it combines the neutral polar alcohol and amide functional groups. Its applicability has been demonstrated for amino acids, peptides, nucleic acids, and other polar compounds. Recently, a related packing with a covalent polysuccinimide coating has been described (4).

For silica-based bonded phases, it has been demonstrated that the retentivity of a packing increases with the number of functional groups. For example, silica derivatized with silanes that carry one, two, or three amino groups per ligand exhibit increased retention with the number of amino groups. The same has been demonstrated for silica derivatized with polyol functions. It also has been shown that the amount of water bound by these bonded phases increases with the number of functional groups (5). One can use this as an argument for the supposed mechanism of HILIC, which postulates a partitioning of the analyte between a water-poor mobile phase and a water-rich stationary phase.

Neutral hydrophilic polymeric packings are also useful for HILIC (6). Examples are the hydrophilic methacrylate phases, which were originally designed for aqueous size-exclusion chromatography. The surface of these packings consists of polyol groups, which are very hydrophilic. These phases tend to be more retentive than silica-based bonded phases.

Surprisingly, it seems that the capabilities of unmodified alumina, zirconia or titania have not been explored for HILIC. Only zirconia bonded with aminopropylsilane has been used for the separation of carbohydrates.

11.3 APPLICATIONS

The classical application of HILIC is the separation of sugars and derivatized and nonderivatized oligosaccharides. The potential for other applications was not generally realized until the recent publication by Alpert (1). In this publication, Alpert demonstrated separations of amino acids, peptides, oligonucleotides, polar organic acids and bases, and even proteins. Consequently, the applicability of HILIC is actually fairly broad.

Sugars are typically separated on propylamino bonded phases using isocratic elution with about 70–80% acetonitrile, depending on the stationary phase. Oligosacharides require a larger concentration of water for elution. A polymeric packing such as Asahipak NH_2, which does not have the hydrolysis problems of silica-based propylamino phases, is an excellent substitute. Commonly, carbohydrate separations are carried out isocratically, since a refractometer is used for detection. But one can also use a UV detector at <210 nm, which makes gradient elution feasible. Derivatized oligosaccharides tagged with a strong chromophore or fluorophore are commonly separated using gradient techniques.

Amino acids have been separated using PolySulfoethyl Aspartamide and PolyHydroxyethyl Aspartamide columns (PolyLC, USA). The mobile phase was 80% acetonitrile/20% 5–25 mM triethylammonium phosphate buffer, pH 2.8. Small differences in elution order were observed between both packings because electrostatic effects combined with the hydrophilic interaction for the PolySulfoethyl Aspartamide column. Retention decreases with increasing buffer concentration. At neutral pH, the PolySulfoethyl Aspartamide retains basic amino acids even in the absence of organic solvent, and acidic amino acids are excluded as a result of ion-exclusion effects. Above 50% acetonitrile, hydrophilic-interaction effects dominate retention.

Dipeptides can be separated also using the PolyHydroxyethyl Aspartamide column. Alpert used a mobile phase of 80% acetonitrile with 10–40 mM triethylammonium phosphate buffer, pH 2.8 or 5.0. The elution pattern can be changed by variation of the pH.

As an example of longer peptides, Substance P fragments with amino acids sequentially removed from the *N*-terminus were separated on both using PolySulfoethyl Aspartamide and PolyHydroxyethyl Aspartamide columns (Fig. 11.2). The separation on both columns was carried out in 80% acetonitrile, 10 mM triethylammonium–methyl phosphonate buffer, pH 3.0 with an increasing salt gradient from 0 to 125 mM sodium perchlorate.

Oligonucleotides were separated on the PolyHydroxyethyl Aspartamide column using a shallow dual gradient of increasing salt concentration with concomitant decrease in acetonitrile. Mobile phase A was 40 mM triethylammonium phosphate buffer, pH 5.0, in 70% acetonitrile; mobile phase B was 75 mM triethylammonium phosphate buffer, pH 5.0, in 80% acetonitrile.

These widely different application examples should be useful for the development of new HILIC applications.

11.4   PRACTICAL ASPECTS

Although many water-soluble organic solvents can be used in HILIC, the preferred solvent is acetonitrile because of its low viscosity, resulting in good column efficiencies at low backpressure. This is similar to the situation in reversed-phase chromatography. Common solvent compositions vary between 50 and 90% acetonitrile. Sugars are typically separated using 70% acetonitrile 30% water.

The most common HILIC application is the separation of carbohydrates on aminopropyl-derivatized silica. As we have discussed in the previous section, this column has only a limited stability under common use conditions. Modern alternatives are available that do not have this weakness; examples are polymer-based amino packings.

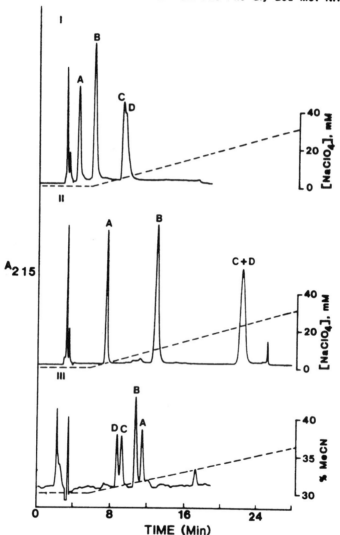

N-TERMINAL VARIANT PEPTIDES

SUBSTANCE P (4-11):   Pro-Gln-Gln-Phe-Phe-Gly-Leu-Met-NH$_2$

Figure 11.2 Separation of peptides via hydrophilic interaction chromatography on two different packings. [*Chromatographic conditions*: Sample: substance P fragments. Columns: (I) PolyHydroxyethyl A; (II) PolySulfoethyl A, 200 mm × 4.6 mm. Flow rate: 0.8 mL/min. Elution: linear gradient 0–100% B over 90 min; eluent A—10 mM TEA methylphosphonate (pH 3.0) with 80% acetonitrile; eluent B—eluent A + 125 mM sodium perchlorate.] (Reprinted from Ref. 1, p. 186, with permission from Elsevier Science-NL.)

All amino columns share with each other the problem that they can be easily contaminated with acidic constituents of the mobile phase, which, in turn, can affect retention. To prevent this, precolumns packed with the same packing can be used if the amount of contamination is small. The columns can be regenerated readily through a wash with a dilute solution of an amine, which removes the adsorbed acids. Other contaminants are removed by washing the column with water.

When samples other than neutral polar compounds are analyzed, one has to take into account ion exchange as a secondary retention mechanism. An example of this is the separation of sialylated oligosaccharides. In this case, a buffered mobile phase is used to control the retention. Another example is the analysis of polar acids such as glycolic acid. Careful consideration should be given to the concentration of the buffer. A high concentration can lead to a reduction in retention, but a low concentration (e.g., 10 mM) leads to excessive equilibration times. See the previous paragraph for additional examples.

Because of solubility constraints, buffers based on an organic amine such as triethylamine are most preferred. The solubility of the buffer should be carefully checked.

For small molecules, the retention in HILIC increases by about 10% for every 1% change in the organic content of the mobile phase. This effect is identical to the effect in reversed-phase chromatography.

REFERENCES

1. A. J. Alpert, *J. Chromatogr.* **499**, 177–196 (1990).
2. H. Rückert, O. Samuelson, *Acta Chem. Scand.* **11**, 315 (1957).
3. F. M. Rabel, A. G. Caputo, and E. T. Butts, *J. Chromatogr.* **126**, 731–740 (1976).
4. A. J. Alpert, M. Shukla, A. K. Shukla, L. R. Zieske, S. W. Yuen, M. A. J. Ferguson, A. Mehlert, M. Pauly, and R. Orlando, *J. Chromatogr.* **A676**, 191–202 (1994).
5. P. Orth, and H. Engelhardt, *Chromatographia* **15**, 91 (1982).
6. B. Bendiak, J. Orr, I. Brockhausen, G. Vella, and C. Phoebe, *Anal. Biochem.* **175**, 96–105 (1988).

12 Ion-Exchange Chromatography

Ion-exchange chromatography is used for the separation of ionic or ionizable compounds. The separation principle is an exchange of analyte ions with the counterions to the fixed ions of the ion exchanger. The techniques has been in use for a long time, and some specific analytical techniques based on ion exchange can be viewed as predecessors to high-performance liquid chromatography. A specific example is the amino-acid analyzer, which combined the high-performance ion-exchange separation of amino acids with the derivatization with ninhydrin as a highly sensitive and selective detection technique.

Today, ion exchange is rarely used in mainstream HPLC, which is dominated by reversed-phase chromatography. However, it plays an unequaled role in the separation of biomolecules, specifically, proteins and nucleic acids. Also, it has a natural place in the analysis of inorganic ions. The specific techniques used for the analysis of inorganic ions have acquired the name of ion chromatography.

12.1 RETENTION MECHANISM

Ion exchangers are prepared by attaching either cations or anions to a fixed matrix. To satisfy the need for electroneutrality, the fixed ions are associated with counterions of the opposite charge. These counterions are mobile and can be exchanged by other ions of the same charge, including the analyte ions

(Fig. 12.1). Different ions have affinities different from those of the fixed ions, which can be expressed as an ion-exchange equilibrium constant. The different affinities are the basis of the ion-exchange separation. When analytes with affinities different from those of the ion-exchange resin are injected into an ion-exchange column, those with a higher affinity to the fixed charges migrate through the column more slowly. Retention is reduced by increasing the concentration of competing ions in the mobile phase. Retention and selectivity of the separation can be influenced by the type of competing ion and its charge, and/or by changing the charge of the analyte ions, for example, by changes in pH or by complexation.

Ion exchange can be applied to the separation of organic and inorganic ions, and biomolecules such as amino acids, peptides, proteins, or nucleic acids.

12.1.1   General Ion-Exchange Mechanism

To examine the ion-exchange mechanism in detail, let us start with a cation exchanger in the sodium form. Such an ion exchanger could be formed by the sulfonation of a styrene–divinylbenzene bead. When this ion exchanger is brought into contact with a solution of KCl, an equilibrium is established between the sodium and potassium ions that function as counterions to the fixed sulfonate ions of the ion exchanger and the mobile chloride ions:

$$Na_S^+ + K_M^+ \Leftrightarrow K_S^+ + Na_M^+ \tag{12.1}$$

where the subscripts S and M stand for stationary and mobile, respectively. The ion-exchange equilibrium constant is therefore

$$K_{iex} = \frac{[Na_M^+]}{[K_M^+]} \frac{[K_S^+]}{[Na_S^+]} \tag{12.2}$$

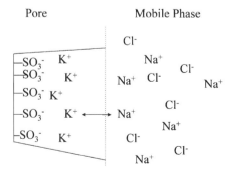

Figure 12.1   Cartoon illustrating the basic principle of ion exchange. The counterions to the fixed ions of the ion exchanger can be exchanged for analyte ions and ions in the mobile phase.

Let us now assume that we have a cation exchanger in the sodium form and a mobile phase containing sodium chloride. We inject potassium chloride; thus the potassium ion is our analyte. As we know from Section 2.3.2, the retention factor of an analyte is defined as the ratio of the number N_S of molecules (or ions) in the stationary phase to their number N_M in the mobile phase:

$$k = \frac{N_S}{N_M} = \frac{V_S}{V_M} \frac{[K_S^+]}{[K_M^+]} \tag{12.3}$$

where V_S and V_M are the volumes of the stationary phase and the mobile phase, respectively. We can substitute the ratio of the concentrations of the potassium ions from Equation (12.2):

$$k = \frac{V_S}{V_M} K_{iex} \frac{[Na_S^+]}{[Na_M^+]} \tag{12.4}$$

At low analyte concentrations, the concentration of the sodium ions in the stationary phase is equal to the concentration of the fixed ion-exchange groups, $[A_S^-]$. Then equation (12.4) becomes

$$k = \frac{V_S}{V_M} K_{iex} \frac{[A_S^-]}{[Na_M^+]} \tag{12.5}$$

This shows that the retention of the analyte is proportional to the concentration of the ion-exchange groups, specifically, the capacity of the ion exchanger, and inversely proportional to the concentration of the competing ion in the mobile phase. The same equation can be derived from other models as well, but the mass-action approach used here is the one that is most natural to most chemists.

We can formulate the equations for the more general case of multiple charges o and p of the analyte cation C^{p+} and the competing ion B^{o+}:

$$pB_S^{o+} + oC_M^{p+} \Leftrightarrow pB_M^{o+} + oC_S^{p+} \tag{12.6}$$

This results in the general equation for the retention factor:

$$k = \frac{V_S}{V_M} \frac{K_{iex}^{1/o} [A_S^-]^{p/o}}{o^{p/o} [B_M^{o+}]^{p/o}} \tag{12.7}$$

The retention factor depends on the concentration of the competing ion in the mobile phase raised to the power of the ratio of charges of the analyte to the competing ion. The same dependence is observed for the capacity of the ion exchanger. If the charge of the competing ion is 1, Equation (12.7) simplifies to Equation (12.8):

$$k = \frac{V_S}{V_M} \, K_{\text{iex}} \, \frac{[\text{A}_S^-]^p}{[\text{B}_M^+]^p} \tag{12.8}$$

A plot of the logarithm of the retention factor versus the logarithm of the concentration of the counterion in the mobile phase (Fig. 12.2) can be used to determine the charge of the analyte. For small molecules or inorganic ions, the charge is known, but for large molecules such as proteins it is not. Therefore, this plot can be used to determine the effective charge with which a protein interacts with the ion exchanger. This effective charge is not the net charge of the entire molecule, but rather a localized surface charge. Consequently, proteins can be retained on an ion exchanger even at their isoelectric point. The effective charge has also been called the *Z number* in the literature [after a formulation of Eq. (12.8) that used Z to designate the effective charge] (1).

The Z number for proteins varies a lot, but typical values are between 3 and 7. This means that the dependence of the retention factor on the concentration of the competing ion (and therefore the ionic strength of the mobile phase) is rather steep. Small changes in ionic strength result in large changes of retention. This makes the isocratic elution of macromolecules rather impractical.

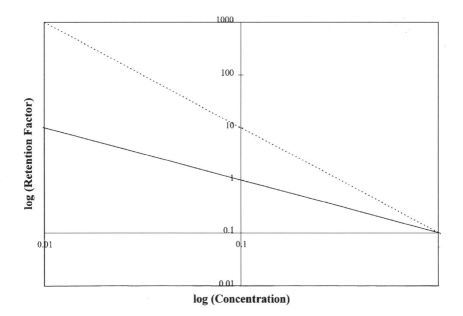

Figure 12.2 Plot of the logarithm of the retention factor versus the logarithm of the ionic concentration for ions with a single charge (solid line) and dual charge (dashed line).

This is the reason why gradient elution is used exclusively in the ion-exchange chromatography of proteins and other macromolecules. The same effect can be found in reversed-phase chromatography.

12.1.2 Behavior of Coions

The treatment presented here is a significant simplification of the actual processes, but nevertheless quite useful for an understanding of the major effects of ion-exchange chromatography. The discussion neglected, for example, the influence of activity coefficients or the swelling and shrinking of the ion exchangers, which is essentially driven by the difference in the ion concentration between the pores of the ion exchanger and the external fluid. We have also up to now ignored the coions, which are the ions that accompany the analyte ions or the counterions to the ion exchanger during their migration with the mobile phase. Coions carry the same type of charge as the ion-exchange sites of the ion exchanger. In the following we will discuss what is happening to the coion and derive the factors that influence the concentration of the coion in the pores.

For a quantitative discussion of the concentration of the coion in the pores, let us assume an ion exchanger with fixed charges A^- in equilibrium with a mobile phase containing the 1-1-electrolyte Na^+Cl^-. On the basis of the preceding definition above, Cl^- is the coion in this case. The equilibrium condition for the equilibrium between the sodium and chloride ions in the mobile phase and in the stationary phase, respectively, can be written as follows:

$$[Na_S^+][Cl_S^-] = [Na_M^+][Cl_M^-] \tag{12.9}$$

This equilibrium is called the *Donnan equilibrium*. As a consequence of the requirement for electroneutrality, it follows that the total concentration of sodium ions outside the pores is equal to the concentration of chloride ions outside the pores. Also, inside the pores, the total concentration of sodium ions is equal to the concentration of the ion-exchange groups and the concentration of chloride ions. Equation (12.9) then becomes

$$([A_S^-] + [Cl_S^-])[Cl_S^-] = [Cl_M^-]^2 \tag{12.10}$$

A little algebra shows that the concentration of the coion in the pores, $[Cl_S^-]$, depends on the concentration of the coion outside the pores as follows:

$$[Cl_S^-] = \sqrt{[Cl_M^-]^2 + \frac{[A_S^-]^2}{4}} - \frac{[A_S^-]}{2} \tag{12.11}$$

If the concentration of the coion in the mobile phase is negligibly small compared to the concentration of ion-exchange groups in the pores, the coions will be nearly completely excluded from the pores. On the other hand, if the situation is inverted and the concentration of the coions in the mobile phase is much larger than the concentration of ion-exchange groups in the pores, the concentration of coions in the pores will be significant and can approach the concentration of the coions in the mobile phase. The consequence is a concentration-dependent exclusion of the coions from the pores. This ion-exclusion effect can be used for the separation of ions of the same charge as the ion exchanger. The chromatographic technique is consequently called *ion-exclusion chromatography*.

12.1.3   Elution Strength Scales

Equation (12.7) showed that the elution strength of a competing ion increases with its charge. But there is also a difference in elution strength within a group of ions of equal charge, reflected in the constant in Equation (12.7). One can establish a sequence of elution strength by measuring the retention factor of an analyte as a function of the type of the competing ion, while keeping the concentration of the competing ion constant. Conversely, one can measure the retention factors of various ions at a constant concentration of a weakly eluting ion.

At equal ionic charge, the elution strength sequence depends primarily on the size of the ion and its polarizability, but it also depends on the properties of the ion exchanger itself. The elution power of cations and anions increases with their increasing interaction with a particular ion exchanger. For cations, the following sequence has been established (2), but on specific ion exchangers the order can change depending on the crosslinking of the resin:

Monovalent cations: $Tl^+ > Ag^+ > Cs^+ > Rb^+ > K^+ > NH_4^+ > Na^+ > H^+ > Li^+$

Divalent cations: $Ba^{2+} > Pb^{2+} > Ca^{2+} > Ni^{2+} > Cd^{2+} > Cu^{2+} > Co^{2+}$
$> Zn^{2+} > Mg^{2+} > Mn^{2+}$

For anions, the following sequence was found, but here also the order can change depending on the hydrophobicity and the crosslinking of the resin:

Monovalent anions: $I^- > NO_3^- > Br^- > SCN^- > CN^- > NO_2^- > Cl^-$
$> HCO_2^- > CH_3CO_2^- > OH^- > F^-$

For multivalent anions, the sequence depends on the pH. Commonly used multivalent ions are sulfate, phosphate, oxalate, and citrate. In addition, carbonate and phthalate are often used in ion chromatography.

For organic ions, additional effects need to be considered, since the ions may interact with the ion exchanger by hydrophobic interaction on top of the ionic interaction. Therefore, the elution strength might be different on ion exchangers based on hydrophobic styrene–divinylbenzene or based on hydrophilic meth-acrylates. This effect influences already the retention of simple inorganic ions such as I^-, SCN^-, and ClO_4^-.

12.1.4   Influence of pH

For compounds whose ionic charge is a function of pH, a change of pH is a powerful tool to manipulate retention and the selectivity of a separation. While pH control is one of many tools in other chromatographic modes, such as in reversed-phase chromatography, it is a primary tool in ion-exchange chromatography. For instance, pH gradients can be used to elute analytes from the column as a function of their pK_a. The best known example of a separation that is primarily driven by a pH gradient is the classic amino-acid analysis. The separation is effected on a cation exchanger (sulfonated styrene–divinylbenzene) using a pH gradient from acidic pH to alkaline pH.

   The principle of the dependence of retention on the pH can be explored without difficulty. Let us look at the example of the separation of a weak acid on a strong anion exchanger. At high pH, the analyte is in the anionic form, and it is retained on the ion exchanger by anion exchange. At low pH, it is protonated and neutral; the anion-exchange mechanism is eliminated, but we will initially assume that there is a finite retention factor even for the uncharged species. The chemical equilibrium can be written as follows:

$$AH \Leftrightarrow A^- + H^+ \tag{12.12}$$

with the equilibrium constant

$$K_a = \frac{[A^-][H^+]}{[AH]} \tag{12.13}$$

The retention factors of acid AH and the anion A- are defined as

$$k_{AH} = \frac{N_{AH,S}}{N_{AH,M}} \tag{12.14}$$

$$k_{A^-} = \frac{N_{A^-,S}}{N_{A^-,M}} \tag{12.15}$$

The combined retention factor k_{A*} of the acid and the anion can be calculated from

1. The definition of the retention factor [Eq. (12.16)]
2. The knowledge that the number of analyte molecules in either the stationary phase or the mobile phase must be the sum of the analyte molecules in the acid form and in the anionic form [Eqs. (12.17) and (12.18)]

3. From the ratio of the acid form to the anionic form [Eq. (12.19)]:

$$k_{A^*} = \frac{N_{A^*,S}}{N_{A^*,M}} \tag{12.16}$$

$$N_{A^*,S} = N_{AH,S} + N_{A^-,S} \tag{12.17}$$

$$N_{A^*,M} = N_{AH,M} + N_{A^-,M} \tag{12.18}$$

$$\frac{N_{A^-,M}}{N_{AH,M}} = \frac{[A^-]}{[AH]} = \frac{K_a}{[H^+]} \tag{12.19}$$

The number of molecules in the stationary phase and in the mobile phase can be derived from this as follows:

$$N_{A^*,M} = N_{AH,M}\left(1 + \frac{K_a}{[H^+]}\right) \tag{12.20}$$

$$N_{A^*,S} = N_{AH,M}k_{AH} + N_{A^-,M}k_{A^-} \tag{12.21}$$

$$N_{A^*,S} = N_{AH,M}\left(k_{AH} + k_{A^-}\frac{K_a}{[H^+]}\right) \tag{12.22}$$

By definition, the retention factor is the ratio of the number of molecules in the stationary phase to the number of molecules in the mobile phase:

$$k_{A^*} = \frac{N_{A^*,S}}{N_{A^*,M}} = \frac{k_{AH} + k_{A^-}\dfrac{K_a}{[H^+]}}{1 + \dfrac{K_a}{[H^+]}} \tag{12.23}$$

From this equation, we can see that at low pH, the retention factor of the analyte is that of the acid form and at high pH, of the anionic form. There will be a rapid transition from one retention factor to the other one when the pH is around the pK_a of the acid. If the retention factor of the acid form is 0, Equation (12.23) simplifies to

$$k_{A^*} = \frac{N_{A^*,S}}{N_{A^*,M}} = \frac{k_{A^-}\cdot\dfrac{K_a}{[H^+]}}{1 + \dfrac{K_a}{[H^+]}} \tag{12.24}$$

Figure 12-3 shows the dependence of the retention factor on pH according to Equation (12.23), assuming that the retention factor of the anionic form itself

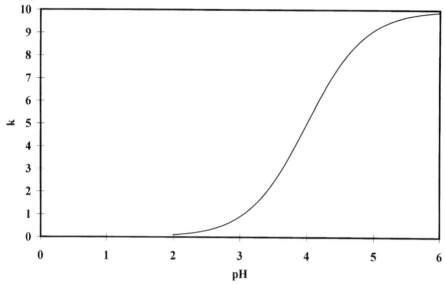

Figure 12.3   pH dependence of the retention factor. The analyte has a retention factor of 10 in its negatively charged form and a retention factor of 0 in its protonated form. Its pK_a is assumed to be 4.

is 10 and that the pK_a of the acid 4.0. At the pK_a of the acid, the retention factor is exactly half the retention factor of the anionic form, since only half of the analyte is in this form. Obviously, this derivatization includes many simplifications, but it can be used as a reasonable model. More complicated relationships can be derived in the same way as outlined in this section.

12.1.5   Ion-Exchange Capacity

The exchange capacity of an ion exchanger is an important measure of its retentivity. But care has to be taken. What counts is not the exchange capacity of a packing per unit mass, but rather per unit packed-bed volume. Otherwise, a comparison between packings of different densities can be misleading. Thus, the exchange capacity of a sulfonated polystyrene packing can be 5 meq/g, but its packing density may be as little as 0.2 g/mL. On the other hand, the exchange capacity of a silica-based quaternary ammonium anion exchanger may be as high as 1.5 meq/g, but the packing density is 0.5 g/mL. Thus the first exchanger has a capacity of 1 meq/mL, and the second one a capacity of 0.75 meq/mL; hence it is very similar for both types. Thus one should be very careful about statements that the capacity of organic ion exchangers is much higher than that of ion exchangers based on silica.

The ion-exchange capacity—properly normalized for a constant bed volume—determines the retentivity of a packing. It also determines the ionic strength of the eluent required for equal retention factors. If a mobile phase with a low conductivity is desired, as in the case of ion chromatography, a resin with a low capacity is preferred. On the other hand, for high retention and high sample capacity, ion exchangers with a high capacity should be used.

Large molecules such as proteins cover a substantially larger area of the surface of the ion exchanger than indicated by the formal surface charge. Therefore the binding capacity of an ion exchanger for proteins is significantly lower than what is expected based on the ion-exchange capacity. It is largely determined by the accessibility of the charge or by the available surface area. The surface area of macroporous ion exchangers is limited by the pore size. A significant improvement of the binding capacity can be obtained by filling the pores of a macroporous packing with a loose network that carries the ion-exchange sites. When properly designed, the network does not hinder diffusion and mass transfer (it can actually accelerate mass transfer by allowing diffusion in the "bound" state), while providing a high ion-exchange capacity at the same time. Commercial examples of this type of ion exchanger are the Hyper-D packings from Sepracor, USA, or the tentacle-type ion exchangers from Merck, Germany. For additional discussion of the protein binding capacity, see Section 12.4.1.

12.1.6 Nonideal Interactions

The discussion of the ion-exchange mechanism in Section 12.1.1 assumed that the only interactions between the analyte and the ion exchanger are electrostatic, specifically, by an ion-exchange mechanism. This is a largely idealized situation. The retention of an analyte may be influenced not only by the desired electrostatic interaction but also, for example, by hydrogen bonding or, more importantly, hydrophobic interaction. Hydrophobic interaction is especially prevalent for ion exchangers based on styrene–divinylbenzene. Even simple inorganic analytes such as the iodide ion can be partially retained by hydrophobic interaction, but it plays a more significant role for organic analytes or macromolecules. Examples are membrane proteins, which contain large hydrophobic regions that interact strongly even with hydrophilic methacrylate-based ion exchangers. Membrane proteins often require the addition of organic solvents to the mobile phase for elution.

The hydrophobic interaction complicates the theoretical treatment of the separation process, but it can be a bonus for the practice of chromatography, since a change in the hydrophobic interaction can be used to manipulate the selectivity of a separation. One can use the addition of organic solvents such as methanol or acetonitrile to accomplish this goal. The newer types of ion exchangers based on macroporous resins are quite tolerant to organic solvents, since swelling and shrinking is minimal.

12.2   ION-EXCHANGE PACKINGS

Although many different ion exchangers can be synthesized, only a few are used in high-performance chromatographic applications. We will cover only those that are available for use in HPLC and ion chromatography.

Ion exchangers can be classified into four general categories: strong and weak cation exchangers and strong and weak anion exchangers. The functional group of strong ion exchangers remains in its charged state over all or most of pH range. The charge of the functional group of weak ion exchangers can be removed by a change in the pH. On one hand, this limits the pH range in which a weak ion exchanger can be used. On the other hand, this feature provides another tool for the elution of analytes. Furthermore, the exchange capacity of weak ion exchangers can be adjusted by changes in the pH.

The functional group of strong cation exchangers is the sulfonic acid group, $-SO_3^-$. It could be either an aromatic sulfonic acid or an aliphatic sulfonic acid.

The functional group of weak cation exchangers is the carboxylate group, $-COO^-$. Most commonly, it is an aliphatic carboxylic acid with a pK_a of ~ 4.5. Aromatic carboxylate groups are slightly more acidic.

Strong anion exchangers contain a quaternary amino group, $-NR_3^+$. This group remains positively charged over the entire pH range. The functional group of weak anion exchangers is a primary, secondary, or tertiary amine. Tertiary amines are most commonly used. The pK_a of these ion-exchange groups is around 9.

12.2.1   Silica-Based Ion Exchangers

Silica is the base material for most commonly used packings in HPLC. Thus it is not a surprise that a wide range of ion-exchange packings are based on silica. Many of these ion exchangers share an advantageous column efficiency with their reversed-phase counterparts. But their general use is hampered by the limited stability of silica at alkaline pH. On the other hand, silica-based ion exchangers neither swell nor shrink.

Caution should be exercised when comparing the ion-exchange capacity of silica-based ion exchangers with the capacity of organic ion exchangers. Ion-exchange capacities are typically quoted in milliequivalents per gram, but this figure is misleading. Silica has a much higher skeleton density (2.2 g/mL) than organic polymers (~ 1 g/mL), which results in a packing density that is a factor of >2 higher for silica compared to a polymeric ion exchanger of equal porosity. Thus the weight-based ion-exchange capacity of a silica-based ion exchanger will be a factor of 2.2 lower than that of an equivalent polymer-based ion exchanger. Ion-exchange capacities of many commercially available silica-based ion exchangers are between 1 and 1.5 meq/g, if a silica with 10-nm pore size is used. Examples are the silica-based ion exchangers Nucleosil SA and SB (Macherey-Nagel, Germany) with an ion-exchange capacity of 1 meq/g. For silicas with a larger pore size, the surface area and therefore the exchange capacity is lower (e.g., $\leqslant 0.3$ meq/g for a 30-nm pore size).

Silica-based ion exchangers exhibit a low hydrophobicity compared with their styrenic counterparts. This can be an advantage if the analytes are organic compounds. The ion-exchange mechanism is not complicated by hydrophobic interaction.

Today, silanes are available or can be synthesized for a one-step surface synthesis of all the different types of ion-exchangers. However, the preparation of some of the older types is likely to involve multiple surface reactions. The details of the preparation of many commercial products is not public knowledge, so it is difficult to know which products are made by which procedure. When multiple reactions are carried out on the surface, a more heterogeneous product is obtained. For example, if the silica is first derivatized with propylamino silane, and the product is subsequently quaternized with methyl iodide, an appreciable amount of tertiary amino groups will remain on the surface. Such an ion exchanger has properties different from those of an ion exchanger that contains purely quaternary amino groups.

To obtain a strong cation exchanger, silica can either be derivatized with an appropriate silane or it can first be reacted with a silane containing a phenyl group, which is then subjected to chlorosulfonation. Even under mild reaction conditions, some fraction of the phenyl silane is split off the silica during the sulfonation reaction.

Silica-based strong anion exchangers can also be obtained directly by derivatization of the silica with the appropriate silane. However, the more common approach is to derivatize the silica first with an amino silane, which is then quaternized by reaction with methyl iodide.

Silica-based weak anion exchangers are obtained by direct silanization and are the precursors to strong ion exchangers. Aminopropyl silanes and diethylaminopropyl silanes are readily available commercially. To increase the exchange capacity, the simple amino group can be replaced with an ethylene diamine or a diethylene triamine function. Silica-based weak cation exchangers with an aliphatic carboxylic acid are available as well.

Silica can be coated with polyethylene imine. The strongly adsorbed imine can then be permanently attached through crosslinking with an appropriate reagent, such as epichlorohydrine. PEI-silica has found many applications, especially as an anion exchanger for proteins and peptides.

As described in Section 6.2.5, a family of ion exchangers can be obtained from polysuccinimide silica: poly(2-sulfoethyl aspartamide) is an aliphatic strong cation exchanger, and polyaspartic acid is a weak cation exchanger. The primary use of this family of products is in the area of protein and peptide separations. These packings exhibit a superior column efficiency compared to their counterparts based on organic polymers.

It needs to be pointed out that silica itself is a cation exchanger of intermediate strength. Very good and reproducible results have been obtained using silica for the separation of tricyclic antidepressants with a mobile phase consisting of a phosphate buffer at pH 7 and 50% v/v acetonitrile (3). Unfortunately, very few application examples have found their way into the literature. Also, alumina can be used for ion exchange (4).

12.2.2 Ion Exchangers Based on Organic Polymers

Both hydrophilic and hydrophobic polymers have served as the matrix for ion exchangers. In the world of HPLC, only those polymers capable of withstanding the high operating pressures of HPLC are used. They comprise mainly poly(styrene–divinylbenzene), which is a hydrophobic resin, and crosslinked polymethacrylate and polyvinylalcohol, both hydrophilic resins.

Styrene–divinylbenzene is one of the oldest base-materials for ion exchangers. Classical ion exchangers were derived from lightly crosslinked resins. These ion exchangers are not very pressure-stable and swell significantly when exposed to an aqueous medium. The newer types, which are more suitable for HPLC, are based on highly crosslinked—most often macroporous—resins. However, the synthetic routes for the modern ion exchangers is the same as for the classic types.

The strong cation exchanger with SO_3^- functional groups can be synthesized conveniently by sulfonation of an HPLC-grade styrene–divinylbenzene resin.

The sulfonation can be carried out by either treatment with chlorosulfonic acid at low temperature, with oleum or with SO3. It can be directed such that every benzene ring in the polymer is sulfonated, resulting in a capacity of about 5 meq/g. Highly crosslinked packings may give a lower degree of sulfonation. But if a low ion-exchange capacity is required, as, for example, in ion chromatography, the sulfonation can be controlled to the desired level.

Styrene-based strong anion exchangers are obtained from a styrene–divinylbenzene-based packing by chloromethylation followed by reaction with a tertiary amine. Weak anion exchangers are prepared through the same route, but ammonia, primary, or secondary amines are used in the second step.

The choice of the amine can have a substantial influence on the selectivity of the separations, even for simple inorganic ions. The elution behavior is a function of the hydrophobicity of the functional group (5,6). As an example, the retention of inorganic anions on pellicular latex-based ion-exchange resins prepared from different amines is shown in Table 12.1 (5).

One route to weak cation exchangers based on styrenic resins also starts from the chloromethylated resin. The chloro group is converted to a cyano group by reaction with KCN, and the cyano group is then hydrolyzed to a carboxylate. Exchangers based on this chemistry are rarely used in HPLC. A more convenient route to weak cation exchangers is through acrylic acid derivatives.

Styrene–divinylbenzene is very hydrophobic, and ion exchangers based on styrene–divinylbenzene maintain a high hydrophobicity. This makes the

TABLE 12.1   Retention Factors for Anions on Ion Exchangers of Different Hydrophobicities

Resin	Analyte						
	F^-	Cl^-	Br^-	NO_3^-	ClO_3^-	SO_4^{2-}	PO_3^{3-}
MDEA	0.06	0.24	0.92	1.1	1.0	.20	.31
DMEA	0.14	1.1	4.5	5.0	4.9	3.0	6.7
TMA	0.30	4.4	19.2	22.5	21.2	51.4	> 100
TEA	0.30	5.8	26.1	55.8	35.7	24.9	> 100

Abbreviations: MDEA, methyldiethanolamine; DMEA, dimethylethanolamine; TMA, trimethylamine; TEA, triethylamine.
Source: Reprinted with permission from H. Small and Plenum Press.

classical preparations unsuitable for the chromatography of proteins. Proteins interact strongly with styrenic resins through hydrophobic interaction, which makes elution difficult and results in a denaturation of the protein. Nevertheless, styrene–divinylbenzene is the base resins for several of the most popular ion-exchange packings for protein separations. Examples are Mono Q or Mono S from Pharmacia, Sweden, or the Hyper-D packings from BioSepra, USA. In the preparation of these resins, the styrene–divinylbenzene resin is first rendered hydrophilic in proprietary processes, and the ion-exchange functions are then added in a secondary step.

More convenient and direct is the synthesis of hydrophilic ion exchangers based on hydrophilic methacrylate resins. Resins designed for aqueous size-exclusion chromatography are an excellent starting point for this route. An example is the strong cation exchanger used in the separation of proteins with the designation sulfopropyl. It is obtained by reacting the alcohol groups on the surface of the hydrophilic beads with sultone, the cyclic ester of γ-hydroxypropyl sulfonic acid. To obtain anion exchangers, these resins can be reacted with epichlorohydrin, and the resulting product can be reacted further with ammonia or an amine. Tertiary amines are used to obtain strong anion exchangers with a quaternary amino group. Primary amines, secondary amines, and ammonia yield weak anion exchangers. The choice of the amine influences the balance of ionic interaction and hydrophobic interaction.

To obtain weak anion exchangers of the DEAE type, a hydrophilic resin is reacted with diethylaminoethylchloride. Weak cation exchangers can be prepared from acrylic or methacrylic acid with an appropriate crosslinker. The structure of the functional group for the DEAE and the SP ion exchangers is shown below:

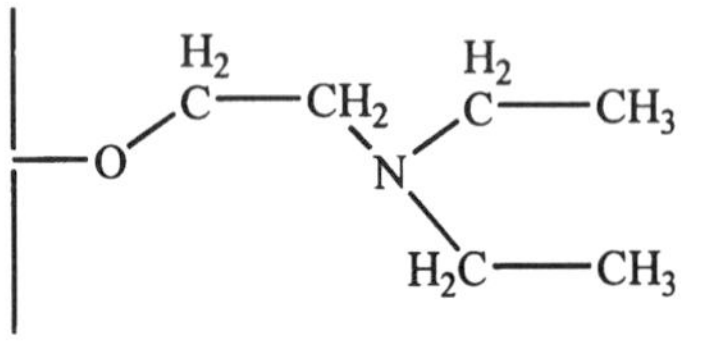

Diethylaminoethyl ion exchanger   Sulfopropyl ion exchanger
(weak anion exchanger)   (strong cation exchanger)

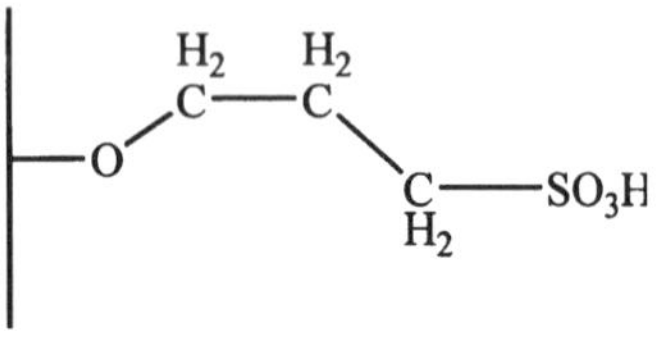

12.2.3   Ion Exchangers for Ion Chromatography

Ion exchangers used for ion chromatography need to have a low capacity (see Section 12.3. There are several ways to accomplish this. One way is to simply derivatize a matrix with a classic technique to a low level of substitution. For example, cation exchangers useful for the ion chromatography of transition metal ions can be obtained by a low-level sulfonation of a styrene–divinylbenzene resin. Commonly used ion exchangers are based on methacrylate resins. The first ion exchangers used for ion chromatography, however, were prepared quite differently; to create an anion exchanger with a low capacity, a high-capacity cation exchanger is coated with small particles of an anion-exchange latex. As a result, one obtains an exchanger with an interactive surface of a low capacity and a noninteracting core. Particles designed this way are known to exhibit good mass-transfer properties, which results in an improved column efficiency.

A special ion exchanger for the analysis of alkali and alkaline-earth ions in a single run is produced by coating silica with a crosslinked poly(butadiene–maleic acid) copolymer (7).

12.3   ION CHROMATOGRAPHY

Ion-exchange chromatography is obviously ideally suited for the separation of inorganic ions. However, it took a while before a suitable HPLC technique became available. The problem was detection. Most inorganic ions do not absorb at the standard wavelengths of the UV detectors used in the early times of HPLC, and the detection with refractive index detectors was too insensitive to be of practical use. Conductivity detection would be natural, but the eluents needed for elution with standard ion exchangers had too high a background conductivity. This problem was solved by Small et al. (8). The technique is very instructive from the standpoint of the design of a chromatographic method.

(*Note*: Originally, the term *ion chromatography* was reserved for the analysis of inorganic ions via the combination of a separation on low-capacity ion exchangers with conductivity detection; today, the term is used for a broader range of separations, including the separation of organic ions, but it escapes a precise definition.)

The basic idea is to use an eluent of a low ionic concentration for the separation and to convert this eluent to a neutral form after the separation. Detection is then carried out with a conductivity detector that now operates against a virtually blank background. Therefore, high sensitivity can be achieved. To accomplish the first goal, a low concentration of elution ions, one simply needs an ion exchanger of low capacity [see Eq. (12.7) or (12.8)]. Exchange capacities of ion exchangers used in ion chromatography can be as low as 0.01 meq/g. To accomplish the second goal, Small and his co-workers used a high-capacity ion-exchange column with special eluents. For the separation of cations, a mobile phase of a dilute (10^{-3} M) acid is used together with a low-capacity cation exchanger. The effluent from the separation column then flows into a second column, called the *suppressor column*, which contains a high-capacity anion exchanger in the OH^- form. The suppressor column binds the anion of the mobile phase, and the released hydroxide ion neutralizes the hydronium ion of the eluent. The analyte cations move through the suppressor column unretained and are eluted into the detector with OH^- ions as counterions against a background of water.

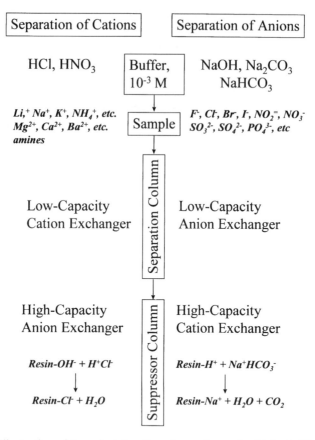

Figure 12.4   Illustration of the principle of ion chromatography. (Adapted from Ref. 9.)

The procedure for the separation of anions is reversed. Eluents are sodium hydroxide, carbonate, or hydrogen carbonate, and the separation takes place in a low-capacity anion exchanger. The suppressor column is a high-capacity cation exchanger in the hydrogen form. It binds the sodium ion and releases hydronium ion, which neutralizes the basic eluent, forming water and/or carbon dioxide. Analyte anions pass through the suppressor column un-retained and elute with H^+ as counterion against a background of water or carbon dioxide dissolved in water. The principle of the process is illustrated in Figure 12.4.

The suppressor column has to be regenerated periodically. Also, the suppressor column contributes to band spreading and limits the performance of the technique. Therefore, alternatives were developed a few years later (10). Instead of a suppressor column, a hollow ion-exchange fiber can be used. If we are separating cations using a dilute acid as mobile phase, we should use an anion-exchange fiber. The anion exchange fiber is immersed in a bath of a dilute solution of OH^- ions, which can permeate the wall of the hollow fiber and neutralize the acidic mobile phase, while the anions of the mobile phase are swept away with the external bath. At the same time, cations cannot penetrate the anion-exchange membrane. The inverse principle holds for the separation of anions. In this case, a cation-exchange fiber is used immersed in a bath of dilute acid, which neutralizes the basic mobile phase. Anions cannot penetrate the anion-exchange fiber and are measured at the detector against a background of water. Therefore the ion-exchange fibers work exactly like a suppressor column, but the regeneration of the suppressor system is continuous. Flat ion-exchange membranes can be used instead of the ion-exchange fibers.

Fiber and membrane suppressors represent an improvement over the classical suppressor columns, but they are not entirely problem-free. Contamination builds up; for example, heavy-metal ions strongly adsorb on the cation-exchange membranes used as suppressors in anion chromatography. Therefore, the membranes also need to be regenerated or replaced in intervals that depend on the application.

The use of suppressor technology is, however, not the only way to do ion chromatography. An alternative technique, called *nonsuppressed ion chromatography*, relies on the fact that the specific conductivity or equivalent conductance of various ions is quite different. Thus one can use a mobile phase of a dilute solution of ions of low equivalent conductance, while the analytes commonly of interest have a high equivalent conductance. Or one can use a mobile phase of high equivalent conductance, while the analytes have a lower equivalent conductance. The technique is also called *single-column ion chromatography* or *ion chromatography with electronic suppression* (as compared to chemical suppression using a suppressor column). As in the classic form of ion chromatography, the separation columns are ion exchangers of a low capacity. Mobile phases of a low equivalent conductance are, for example, potassium hydrogen phthalate at concentrations around 5 mM for the separation of anions. A very effective eluent with a low equivalent conductance is prepared from borate, gluconate, and glycerol (11). This eluent contains an equilibrium mixture of borate esters of gluconate and glycerol. For the separation of cations, dilute (3 mM) nitric acid can be used as mobile phase. Figures 12.5 and 12.6 show examples of single-column anion and cation chromatography. Established and tested methods are available for many common analytical problems of ion chromatography. Nearly all ion-chromatographic methods use isocratic elution.

12.4   ION-EXCHANGE SEPARATIONS OF BIOPOLYMERS

The by far largest application area of high-performance ion-exchange chromatography is the separation of biopolymers: proteins and nucleic acids. Although a special term, *fast protein liquid chromatography* (FPLC), is occa-

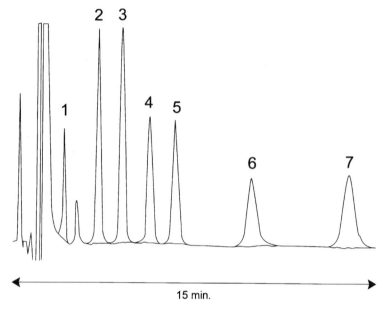

Figure 12.5 Separation of a seven-anion standard with single-column ion chromatography. Column: [*Chromatographic conditions*: IC-Pak Anion HR, 4.6 mm × 75 mm, Waters Corp. Eluent: borate/gluconate. Sample: seven-anion standard: (1) fluoride; (2) bicarbonate; (3) chloride; (4) nitrite; (5) bromide; (6) nitrate; (7) phosphate; (8) sulfate. Detection: conductivity. (Chromatogram courtesy of Waters Corp.)

sionally used for the high-performance separation of proteins, there is nothing intrinsically special about the HPLC separation of biopolymers, which will be covered in the next two paragraphs.

12.4.1 Proteins

All four types of ion exchangers are used in the separation of proteins. The commonly used strong cation exchanger is the sulfopropyl type, abbreviated SP, with a pK_a of ∼2. The weak cation exchanger is designated CM (carboxymethyl), with a pK_a of ∼5. The commonly used weak anion exchanger has a diethylaminoethyl functional group (DEAE), with a pK_a ∼8–9. The strong quaternary ion exchanger is based on quaternary methyl groups and designated QMA.

   In the introductory section we have already discussed a few aspects of the synthesis of ion exchangers designed for the HPLC of proteins. Generally, a larger pore size is required for the chromatography of proteins than for small molecules. Ion exchangers for proteins have a pore size of at least 30 nm, and a more common pore size is 100 nm. The ion-exchange capacity of the

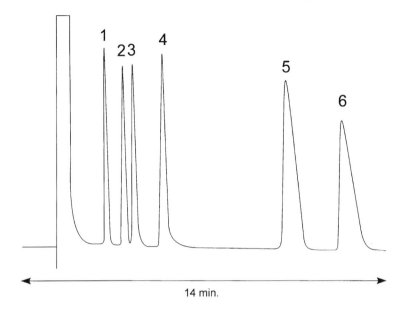

Figure 12.6 Separation of mono- and divalent cations with single-column ion chromatography. [*Chromatographic conditions:* Column: IC-Pak C M/D, 3.9 mm × 150 mm, Waters Corp. Eluent: 0.1 mM EDTA, 3.0 mM HNO_3. Flow rate: 1 mL/min. Sample: (1) lithium; (2) sodium; (3) ammonium; (4) potassium; (5) magnesium; (6) Calcium. Detection: conductivity.] (Chromatogram courtesy of Waters Corp.)

packings with the larger pore size is lower than that of smaller pore size packings, but what counts for both retentivity and loadability is the ion-exchange capacity for the large molecules that are separated with these resins. The capacity is therefore best assessed using proteins as samples under fixed conditions, and is given as the protein binding capacity of the packing. A convenient way to measure the protein binding capacity of a packed column is to pump a solution of a known concentration of protein onto the column under conditions known to bind the protein and measure the 50% point of the breakthrough profile of the protein (Fig. 12.7). After subtracting the volume of the system, including the mobile-phase volume of the column, the amount of protein bound to the column can be calculated. The protein binding capacity is then obtained by dividing the amount of protein bound by the column volume. The volume of the system and the column volume are obtained by carrying out the same experiment with a marker that does not interact with the packing. This marker could be a small molecule or another protein, or the same protein under nonbinding conditions. The protein binding capacity obtained in this manor is called the *dynamic capacity*. It should be noted that the protein binding capacity of a packing depends on the protein. Smaller proteins can enter a larger portion of the pore volume than larger proteins and

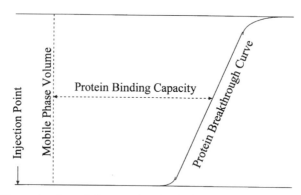

Figure 12.7 Determination of the protein binding capacity. One measures the differ-ence between the elution volume of a noninteracting species and the protein used for the binding study by calculating the difference at the 50% point of the step height. The amount of protein bound is calculated from the concentration of the protein in the feed and this volume.

can therefore explore a larger fraction of the surface of the ion exchanger. Therefore the protein binding capacity is specific for a given protein.

The protein binding capacity of strong and weak anion exchangers can be determined with bovine serum albumin (BSA) in 20 mM Tris/Cl at pH 8.2 (12,13). Similarly, the protein binding capacity of strong and weak cation exchangers can be determined with cytochrome c in 25 mM 2-(N-mor-pholino)ethanesulfonic acid (MES) buffer at pH 5.0.

Typical ion-exchange capacities for ion exchangers designed for protein separations are around 0.1–0.4 meq/mL. The protein binding capacity of most commercially available ion exchangers is similar, for example, 40 mg/mL of bovine serum albumin (BSA) under standardized conditions. Exceptions are the ion exchangers with gel-filled pores (Hyper-D packings from BioSepra) with a protein binding capacity of about 120 mg/mL of BSA, and the Poros ion exchangers by Perceptive Biosystems, USA with an exceptionally low protein binding capacity. Typical protein binding capacities of all four types of ion exchangers are given in Table 12.2.

An important feature of ion exchangers used for preparative protein separations is the ability to clean the packed bed with acidic or alkaline solutions. These procedures are used to removed strongly bound protein from the ion exchangers. Especially a washing protocol using 0.1 N NaOH (pH 13) is commonly applied. Under these conditions, proteins are hydrolyzed into smaller peptides, which are more easily removed from the column. Both the skeleton of the ion exchanger and the functional group should be able to withstand these conditions. Exchangers based on styrene–divinylbenzene are stable over this range. Exchangers based on methacrylate or polyvinylalcohol can be exposed to these conditions for a limited time period without adverse effects.

TABLE 12.2   Ion-Exchange Capacities and Protein Binding Capacities of Some Ion Exchangers for Protein Separations (14)

	Ion Exchanger			
	Protein-Pak Q HR	Protein-Pak DEAE HR	Protein-Pak CM HR	Protein-Pak SP HR
Ion-exchange capacity	0.20 meq/mL	0.25 meq/mL	0.175 meq/mL	0.225 meq/mL
Protein binding capacity	60 mg/mL	40 mg/mL	25 mg/mL	40 mg/mL

The ion-exchange separation of proteins is nearly always carried out using gradient elution. Usually, the pH remains fixed, and the elution strength of the buffer is increased by increasing the ionic strength. Whether a protein interacts with an ion exchanger depends on its surface charge rather than its net charge. Since the surface charge is not known, it is nevertheless a good approach to use the isoelectric point of a protein as a first guide to select a pH for retention. At pH values below the isoelectric point, the net charge of a protein is positive, and therefore, it is retained on a cation exchanger. Conversely, at a pH value above the isoelectric point one expects the protein to be retained on an anion exchanger. However, because the surface charge rather than the net charge is important, this can serve only as a first approximation. Many proteins interact with ion exchangers at their isoelectric point, but close to the isoelectric point the interaction is relatively weak.

The combined effects of ionic strength and pH are used to optimize a separation. One can view pH as the tool to manipulate the selectivity of a separation, while ionic strength is used to change retention. The nature of the buffer or the salt used to increase ionic strength has only a minor influence on retention or selectivity. Therefore, the salt most often used to change ionic strength is simply sodium chloride. In a typical elution protocol, the proteins are bound to the ion exchanger in a buffer of the chosen pH and of low molarity (e.g., 20 mM). They are then eluted using a linear ionic strength gradient over 10–20 column volumes to the same base buffer containing 0.25 to 0.5 M NaCl. This type of protocol is effective for many different applications. Regeneration of the packing between runs can be carried out by flushing the column with 1 M NaCl in the base buffer. Examples of the separation of standard protein mixtures are shown in Figures 12.8 and 12.9.

12.4.2   Nucleic Acids

Ion-exchange packings designed for the separation of nucleic acids are nonporous ion exchangers with quaternary amines as functional groups. While

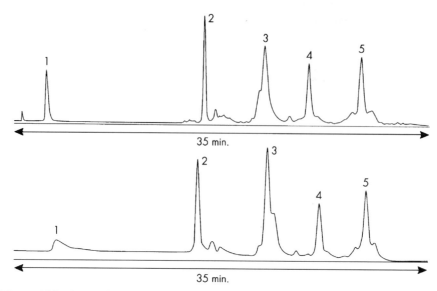

Figure 12.8 Separation of a mixture of protein standards on two strong cation-exchange columns. [*Chromatographic conditions*: Sample: (1) myoglobin; (2) ribonuclease A; (3) chymotrypsinogen A; (4) cytochrome *c*; (5) lysozyme. Columns: *top*—Protein-Pak SP 8 HR Minicolumn, 5 mm × 50 mm, Waters Corp.; *bottom*—Mono S, 5 mm × 50 mm, Pharmacia. Buffer A—20 mM sodium phosphate pH 7; buffer B—buffer A + 1 M sodium chloride. Gradient: 0–50% buffer B in 40 min at 0.4 mL/min. Detection: 280 nm.] (Chromatograms courtesy of Waters Corp.)

separations on more classic ion exchangers are possible, much higher resolution is obtained on non porous packings. The resolution of the *n*-mer from the (*n* − 1)-mer of synthetic oligonucleotides is possible up to a chain length of about 30 bases. Also, with the same nonporous ion-exchangers, the separation of DNA restriction fragments or plasmids of a length of 5000 basepairs is possible. The particle size of these nonporous ion exchangers is smaller than the particle size typical for other ion-exchange applications, typically at 2–3 μm. The higher backpressure of the columns is, however, not an issue, since the nonporous packings are rather resistant to compression. Because of the very high molecular weight of some analytes, the inlet frit of the column should have a pore size and porosity that is large enough to prevent a shearing of the analyte.

As with any technique dealing with large molecules, the standard elution technique is gradient elution. An example is shown in Figure 12.10. The products of a synthesis of a 20-mer phosphorothioate oligonucleotide are separated on a nonporous anion-exchange resin using a salt gradient. A clear separation between the 20-mer and the failure sequences is obtained.

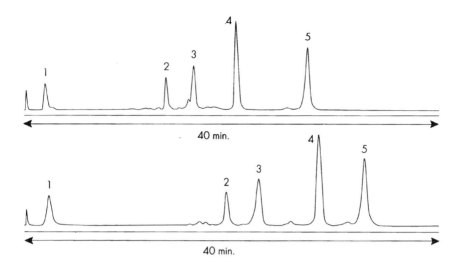

Figure 12.9 Separation of a mixture of protein standards on two strong anion-exchange columns. [*Chromatographic conditions*: Sample: (1) adenosin; (2) carbonic anhydrase; (3) human transferrin; (4) ovalbumin; (5) soybean trypsin inhibitor. Columns: *top*—Protein-Pak SP 8 HR Minicolumn, 5 mm × 50 mm, Waters Corp. *bottom*—Mono S, 5 mm × 50 mm, Pharmacia. Buffer A—20 mM tris/Cl, pH 8.2; buffer B—buffer A + 1 M sodium chloride. Gradient: 0–25% B in 20 min at 0.4 mL/min. Detection: 280 nm.] (Chromatograms courtesy of Waters Corp.)

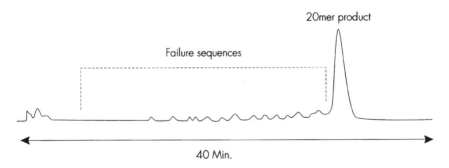

Figure 12.10 Separation of a 20-mer detritylated phosphorothioate oligonucleotide from failure sequences. [*Chromatographic conditions*: Column: Gen-Pak FAX, 4.6 mm × 100 mm, Waters Corp. Eluent A—25 mM Tris HCl, pH 8.0 with 10% acetonitrile; eluent B—eluent A + 2.0 M NaCl. Gradient: 10% B to 100% B, linear over 30 min, then hold at 100% B for 10 min. Flow rate: 0.75 mL/min. Temperature: 65°C. Detection: 260 nm.]

12.5 PRACTICAL ASPECTS

This chapter has shown the broad applicability of ion-exchange chromatography, ranging from the separation of small inorganic ions in ion chromatography to the separation of proteins or plasmids. It is difficult to accommodate all these different techniques under the heading of practical aspects of ion-exchange chromatography. However, there are a few common issues in all types of ion-exchange chromatography, which we will cover first. Then we will look at a few specific topics of the different applications and treat them separately.

In all types of ion-exchange chromatography, the dominating tool to manipulate retention is the ionic strength of the eluent. As we have seen in the theoretical section, the power of the dependence of retention on ionic strength depends on the relative charges of the analyte ions and the eluting ions. For multiply charged ions, the dependence of retention on the ionic strength can be quite strong.

If the ionization of the analyte depends on the pH, it should be tightly controlled. Well-calibrated pH meters are of utmost importance. But the preparation of the buffer itself can also be an issue. Imprecise instructions for buffer preparation can result in reproducibility problems. A 50 mM phosphate buffer, pH 7.0, that is (correctly) prepared by dissolving 50 mmol of NaH_2PO_4 and adding NaOH until the desired pH is obtained is not the same as a buffer prepared from 50 mmol of Na_2HPO_4, to which HCl is added until pH 7.0 is reached. However, in another viable preparation of a 50 mM phosphate buffer, pH 7.0, solutions of 50 mM NaH_2PO_4 and 50 mM Na_2HPO_4 are mixed until the pH of the combined solution is 7.0. This preparation and the first preparation give buffers with identical concentrations of all constituents.

One also should be careful in the specification of the point at which the pH of a buffer is measured. The pH measurement is a determination of the activity of the hydronium ion. Therefore it depends on the ionic strength. One will obtain different pH values in buffers before and after the addition of a neutral salt. Therefore, the specifications "50 mM sodium phosphate buffer, pH 7, with 0.5 M NaCl" and "50 mM sodium phosphate buffer with 0.5 M NaCl, pH 7" are not identical.

Buffers have their maximum buffer capacity at the pK_a of the corresponding acid. If the desired pH is more than ~ 1.5 pH units away from the pK_a of the buffer, the buffering capacity becomes too low. For example, phosphate has no buffering capacity whatsoever at pH 4.5. Also, a solution of ammonium acetate at pH 7 is a salt solution, but not a buffer.

Separations of proteins or nucleic acids are commonly carried out using gradients of increasing ionic strength. It is a good practice to monitor these gradients with a conductivity meter. This provides added information for troubleshooting.

Generally, the regeneration of cation exchangers is accomplished using a strong acid; and the regeneration of anion exchangers, using a strong base. Mildly retained contaminants can often be removed simply with a strong salt solution (e.g., 1 M NaCl). Ion exchangers for protein separations are often cleaned with a strongly alkaline solution (e.g., 0.1 M NaOH). Under these circumstances, the tightly bound protein is hydrolyzed into peptides, which are less strongly bound.

Humic acids, which often accumulate on the column in anion chromatography, can be removed by a combination of organic solvents and an acidic medium. However, these eluents can be used only with the newer generation columns that are compatible with organic solvents. Check the manufacturer's recommendations!

Neutral organic compounds can be retained strongly on ion-exchange resins through hydrophobic interaction. Sometimes this can be prevented by adding organic solvents to the mobile phase. Alternatively, the contamination can be removed by washing the column with organic solvents. Before doing so, you should consult the manufacturer's recommendations about the ion exchanger's compatibility with organic solvents.

Transition-metal contamination of ion exchangers can be removed by washing the column with an EDTA-containing eluent.

REFERENCES

1. N. K. Boardman and S. M. Partridge, *Biochem. J.* **59**, 543–552 (1955).
2. H. Engelhardt, *Hochdruck-Flssigkeits-Chromatographie*, Springer-Verlag, Berlin, 1977
3. U. D. Neue, unpublished results (1979).
4. G. Schmitt and D. Pietrzyk, *Anal. Chem.* **57**, 2247 (1985).
5. H. Small, *Ion Chromatography*, Plenum Press, New York, 1989.
6. R. W. Slingsby and C. A. Pohl, *J. Chromatogr.* **458**, 241–253 (1988).
7. P. Kolla, J. Köhler, and G. Schomburg, *Chromatographia* **23**, 465–472 (1987).
8. H. Small, T. S. Stevens, and W. Baumann, *Anal. Chem.* **47**, 1801 (1975).
9. V. R. Meyer, *Practical High-Performance Liquid Chromatography*, Wiley, Chichester, 1994
10. T. S. Stevens, J. C. Davis, and H. Small, *Anal. Chem.* **53**, 1488 (1981).
11. T. H. Walter and D. J. Cox, "The Composition and Elution Strength of Gluconate/ Borate Eluents for Anion Chromatography," in *Advances in Ion Chromatography*, P. Jandik and R. M. Cassidy, eds., Century International, Medfield (USA), 1990, pp. 179–196.
12. D. M. Dion, K. O'Connor, D. J. Phillips, G. J. Vella, and W. Warren, *J. Chromatogr.* **535**, 127-145 (1990).
13. D. J. Phillips, P. J. Cheli, D. M. Dion, H. L. Hodgdon, A. M. Pomfret, and B. R. San Souci, *J. Chromatogr.* **599**, 239–253 (1992).
14. D. J. Phillips, private communication (1990).

ADDITIONAL READING

H. Small, *Ion Chromatography*, Plenum Press, New York, 1989.
S. Yamamoto, K. Nakanishi, and R. Matsuno, *Ion-Exchange Chromatography of Proteins*, Chromatographic Science Series, J. Cazes, ed., Marcel Dekker, New York, 1988.

13 Hydrophobic-Interaction Chromatography

13.1 Retention Mechanism
13.2 Operating Conditions
13.3 Packings for HIC
13.4 Applications
13.5 Practical Aspects

As hydrophilic-interaction chromatography can be viewed as the extension of normal-phase chromatography to aqueous mobile phases, hydrophobic interaction chromatography (HIC) can be seen as the extension of reversed-phase chromatography to aqueous mobile phases without organic modifiers. Pure water functions as the strong eluent, and retention is increased by adding salts to water. The primary application of the technique is the separation of proteins. Proteins are stable in the salt solutions used as mobile phase in HIC, while they are often denatured in the presence of organic solvents required for reversed-phase chromatography. Therefore, the biological activity of proteins is often maintained in HIC.

As with any technique of retention chromatography of macromolecules, gradient methods are used instead of isocratic methods. The proteins are loaded onto the column using a buffer with a high salt concentration as the starting mobile phase, and are eluted by a gradient with decreasing ionic strength. This is the opposite to the procedure used in ion-exchange chromatography.

The principle underlying retention in HIC has been known for a long time (1) by the name "salting-out chromatography," but it acquired its current name only after the synthesis of specific stationary phases based on soft gels in the early 1970s (e.g., 2–4). Later, in the early 1980s, it evolved into an HPLC technique after the development of packings that were compatible with higher pressures (e.g., 5,6). Today it stands next to ion-exchange chromatography, and size-exclusion chromatography as another important tool for the separation and purification of proteins.

While proteins frequently denature under conditions of reversed-phase chromatography, due to the need for organic solvents for elution and/or the strength of the interaction with the stationary phase, the mild elution conditions and mild hydrophobic interaction of HIC packings allow the elution and recovery of undenatured proteins.

13.1   RETENTION MECHANISM

In hydrophobic interaction chromatography, retention is promoted by high salt concentrations and elution is achieved with low salt concentrations. The packings are very hydrophilic and have only a very mild hydrophobicity. The increase in salt concentration drives the analyte molecules out of the mobile phase and onto the stationary phase.

When an analyte is insufficiently retained on a reversed-phase packing with a mobile phase consisting of water or buffer without any organic modifier, its retention can be increased by adding salt to the mobile phase. Proteins or hydrophilic polymers that are highly soluble in water can be precipitated by dissolving large amounts of salt. Also, small organic molecules can be precipitated from water by the addition of salt. Finally, the interaction and retention of proteins and other analytes with hydrophilic stationary phases can be promoted by the addition of salt to the mobile phase.

All these phenomena have a common denominator. Ions dissolved in water bind the water tightly. The addition of salt to an aqueous solution reduces the amount of free water and promotes any molecular interactions that result in a release of water molecules. At the same time, water—or, more precisely, structured water—surrounds the hydrophobic part of the surface of all molecules dissolved in water. This water is released as these molecules bind to each other or to another hydrophobic surface. Therefore, any interaction of hydrophobic surfaces is enhanced by an increase in the salt concentration. If the molecules interact with each other, precipitation by "salting out" results. If the hydrophobic surface is provided by a packing, we are dealing with HIC.

The water surrounding hydrophobic surface elements is structured water. Therefore, the release of this water from the hydrophobic surface is accompanied by an increase in entropy. Consequently, an increase in temperature often results in an increase in retention in HIC. This is the opposite of any other type of retention chromatography.

Melander, Horváth and co-workers (7–9) have treated this phenomenon quantitatively. They derived an equation of the following form for the dependence of the retention factor on the molal salt concentration m_s in HIC:

$$\ln(k) = \ln(k_0) - \frac{O\sqrt{m_s}}{1 + P\sqrt{m_s}} + \Lambda m_s \qquad (13.1)$$

where k_0 is the retention factor in the absence of salt and O, P, and Λ are constants. The second term represents electrostatic interactions and can be considered to be a "salting-in" factor, describing the increase in protein solubility with increasing salt concentration observed at very low salt concentrations, due to the binding of ions to the protein surface. This term becomes constant at higher salt concentrations. The last term in Equation (13.1) describes the hydrophobic-interaction phenomenon. The constant Λ depends both on the retentive strength of the salt and on the contact area between the analyte and the surface of the packing. The retentive strength of the salt is proportional to the molal surface tension increment of the salt used. This is covered in the next section.

At higher salt concentrations, this equation simplifies to

$$\ln(k) = \ln(k_0) + \Lambda m_s \tag{13.2}$$

This equation predicts a linear relationship between the logarithm of the retention factor and the molal salt concentration, which is indeed commonly observed. Figure 13.1 shows the retention behavior of several proteins on a silica-based polar bonded phase, propylacetamide. The eluents were various concentrations of ammonium sulfate in 0.1 M phosphate buffer, pH 7. The slope of the relationship is 2–2.5 (L/mol) for this stationary phase. For n-butyl- and phenyl-derivatized silica-based phases, values of between 1 and 2 (L/mol) were observed (5).

For some proteins, such as ovalbumin in Figure 13.1, a change in the slope may be observed. This can be attributed to conformational changes of the protein. These structural changes can be observed in gradient chromatography as well (11). If the conformational changes are sufficiently slow, multiple peaks can be observed in the chromatogram for a single compound.

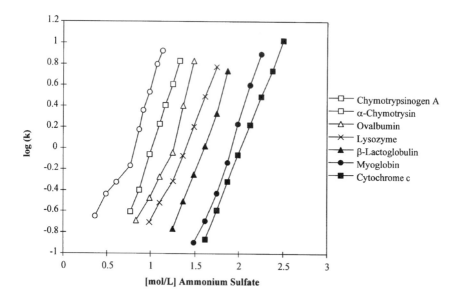

Figure 13.1 Dependence of the retention factor on the molal salt concentration. The stationary phase is propylacetamide bonded to a silica with an average pore size of 25 nm. Mobile phase: ammonium sulfate in 0.1 M phosphate buffer, pH 7. (Reprinted from Ref. 10, p. 3232, by courtesy of Marcel Dekker, New York, 1990.)

13.2   OPERATING CONDITIONS

Since the retention factor is a steep function of the salt concentration, hydrophobic interaction chromatography is normally carried out using gradient elution from a high salt concentration to a low salt concentration. The selectivity of the separation can be influenced by the stationary phase, the type of salt, the temperature, the pH, and other additives to the mobile phase.

The increase in retention at a given concentration depends on both the anion and the cation of the salt. Specifically, the retention increment depends on the molal surface tension increment of the salt. Table 13.1 contains this value for various salts. The higher the value, the greater is the retention at a given concentration. It should be noted that the order in the table parallels the Hofmeister salting-out series (12).

Because of the very high solubility of ammonium sulfate (limit ~ 4 M), it is the most commonly used salt in HIC. However, other salts can be used to modify the selectivity of the separation. The solubility limit of sodium sulfate is about 1.5 M. Sodium chloride is also quite effective because of its higher solubility, and sodium/potassium phosphates are suitable as well. Citrates often yield significant changes in selectivity, but they cannot be used with low-UV detection. Calcium and magnesium salts can have a profound influence on the separation, but the effects depend on the protein and are generally not very predictable.

Another tool for the manipulation of selectivity in HIC is the pH of the mobile phase. As in other forms of chromatography, the choice of the pH can significantly influence the position of the compounds in the chromatogram.

TABLE 13.1   Molal Surface Tension Increment for Various Salts

Salt	Molal Surface Tension Increment $\sigma \times 10^3$ (dyn·g)/(cm·mol)
K_3-citrate	3.12
Na_2SO_4	2.73
K_2SO_4	2.58
$(NH_4)SO_4$	2.16
Na_2HPO_4	2.02
K_2-tartrate	1.96
NaCl	1.64
$KClO_4$	1.40
NH_4Cl	1.39
NaBr	1.32
$NaNO_3$	1.06
$NaClO_3$	0.55

Source: Reprinted from Ref. 12, p. 103 by courtesy of Marcel Dekker, Inc.

However, the effect tends to be smaller in HIC than in ion exchange. This is not unexpected. A change in pH will affect retention only if it affects the neighborhood of the hydrophobic area that binds to the surface. Experiments did not conform with the expectation that retention should be greatest at the isoelectric point of a protein (10).

The most interesting aspect of HIC is the frequent increase of retention with increasing temperature. As mentioned above, this is due to the fact that HIC is an entropy-driven process. This behavior has two advantageous consequences at both high and low temperature. At high temperature, the viscosity of the mobile phase is reduced. Also, lower salt concentrations can be used for comparable retention, which also reduces the viscosity of the mobile phase. The lower viscosity of the mobile phase results in sharper peaks and better separations. Low temperature can be used with advantage for proteins that are labile. On one hand, the low temperature helps to maintain the structure of the protein. On the other hand, the hydrophobic interaction is weakened, which also helps in preserving the tertiary structure of the protein. Therefore, HIC at low temperatures maintains the biological activity of proteins extremely well.

However, the increase in retention with increasing temperature is not always observed. If interactions other than hydrophobic interaction influence retention, the retention behavior as a function of temperature can be more complicated.

Organic modifiers can be used to reduce retention in hydrophobic interaction just as in reversed-phase chromatography. However, organic modifiers with greatly reduced elution strength are used in place of the standard reversed-phase solvents. Typical modifiers used here are glycerol, ethylene glycol, and ethylene glycol monomethylether. They are particularly useful for the elution of very hydrophobic proteins, such as membrane proteins. Nonionic or zwitterionic surfactants can be used for the same purpose. The addition of these organic modifiers or surfactants can also be used to manipulate the selectivity of the separation.

13.3 PACKINGS FOR HIC

The packings suitable for HIC can be classified into two groups: group A comprises packings with a homogeneous surface, and group B contains packings with a low population of hydrophobic functional groups. Group A packings also serve often as stationary phases for aqueous size-exclusion chromatography. In fact, hydrophobic interaction is frequently one of the undesirable side-effects in aqueous size-exclusion chromatography of both proteins and synthetic polymers (see Fig. 8.4 for an example).

Examples of group A packings are the acetamide phase studied by Schön (10) or the ether phases studied by Wu et al. (11). They are silica-based bonded phases, whose structures are given below.

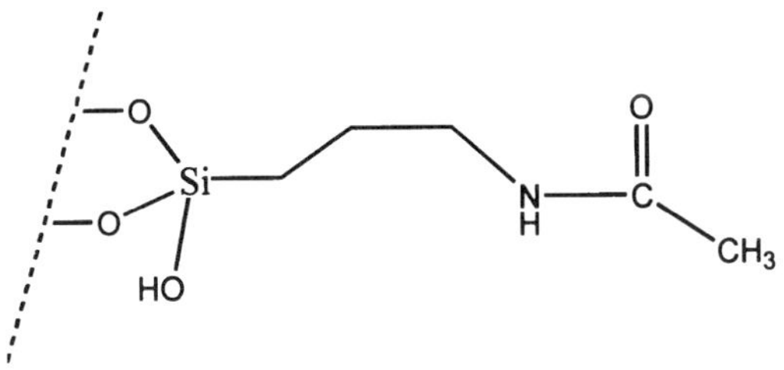

Acetamide bonded phase

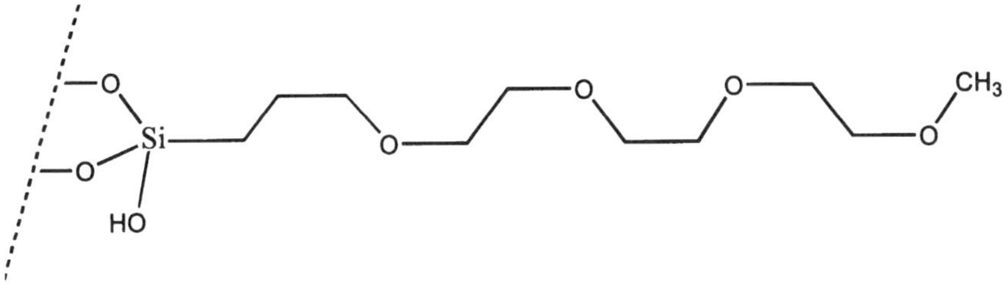

Methylether bonded phase (13)

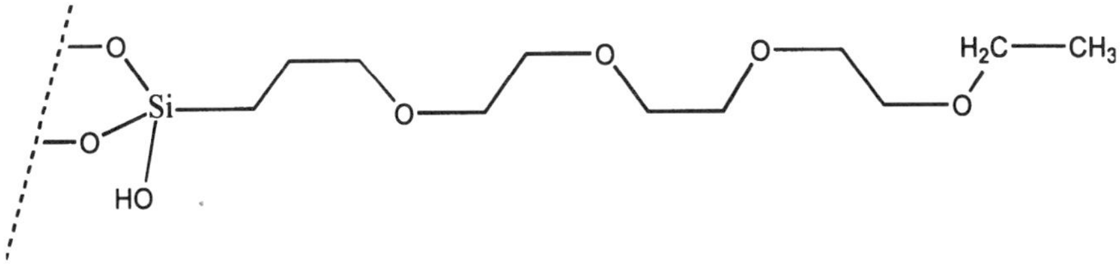

Ethylether bonded phase (13)

Commercially available are silica-based diol phases, which are used for the size-exclusion chromatography of proteins, but also exhibit a weak hydrophobic effect. Similarly, the methacrylate phases used for general aqueous SEC display a mild HIC effect as well. A commercial packing specifically designed for HIC is the Spherogel CAA-HIC from Beckman Instruments, Inc. (CA, USA), which is a polyether phase based on a silica with a pore size of 30 nm.

Group B stationary phases are characterized by a sparse population of hydrophobic groups incorporated into a hydrophilic matrix. The hydrophobic groups used are short aliphatic hydrocarbon chains such as propyl or butyl groups or phenyl groups. The ligand density of these groups is about 2 orders of magnitudes lower than the ligand density for reversed-phase packings. One examples of this type of packing is SynChropak-Propyl from SynChrom, USA. It is based on a silica with a pore size of 30 nm to which a hydrophilic polyamide layer carrying the hydrophobic propyl ligand is covalently attached.

Another example is TSK-Phenyl 5-PW from Toyo Soda, Japan. To prepare this packing, hydrophobic phenyl groups are attached to TSKgel G5000PW, a hydrophilized methacrylate packing used for aqueous size-exclusion chromatography (6).

The influence of the ligand concentration and ligand type on retention has been studied with classical gels. A higher ligand concentration results in a larger retention, as does a longer chain length of the ligand. However, the influence of the design of the stationary phase on the selectivity of the separation is not yet understood (14). Nevertheless, it is known that significant selectivity differences can be observed between different packings. This is contrary to the situation in reversed-phase chromatography, where the selectivity differences for proteins between different stationary phases are sparse. Under HIC conditions, of proteins are in their native conformation, and hydrophobic areas are often found in the folds and crevices of the tertiary structure. It is highly likely that the selectivity differences are related to the ability of the hydrophobic ligands on the surface of the packing to interact with these shielded hydrophobic regions. This expectation is confirmed by a comparison (15) of the retention behavior of TSK-Phenyl 5-PW, SynChropak Propyl and Spherogel CAA-HIC. The study allows the conclusion that the columns carrying a sparse population of hydrophobic groups (type B packings) are more similar to each other than they are to the polyether phase, a type A packing.

13.4 APPLICATIONS

The exclusive application area of hydrophobic-interaction chromatography is the separation of proteins and large peptides. The retention mechanism of hydrophobic-interaction chromatography is orthogonal to ion-exchange chromatography and size-exclusion chromatography. Therefore it can be used effectively in schemes that combine all three mechanisms in the separation of proteins from complex mixtures. The protein binding capacity of HIC packings is comparable to ion exchangers of similar design. However, loading the sample is more tricky than in ion exchange. In order to promote binding, the sample needs to be loaded in a buffer with a high salt concentration. However, because of the salting-out effect, the solubility of proteins in high-salt buffers is limited. This constrains the sample load that can be applied in preparative applications.

If the injection volume is not too large, the sample can simply be diluted with the starting buffer and injected. This is shown in two examples (16). The first example is a separation of human milk whey (Fig. 13.2). In this case the whey was simply diluted 1:1 with the high salt buffer, and the resulting mixture was injected. In the second example (Fig. 13.3), the mouse ascites fluid was first diluted 1:1 with buffer B, then diluted 1:8 with buffer A to bring the salt molarity in the sample close to the composition of buffer A.

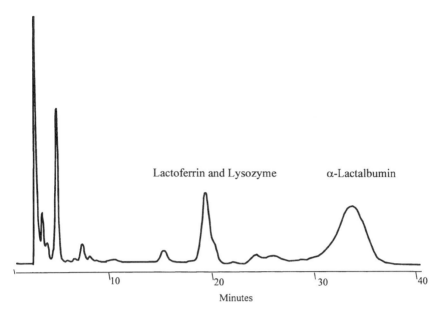

Lactoferrin and Lysozyme   α-Lactalbumin

Figure 13.2 HIC separation of human milk whey. (*Chromatographic conditions:* Column: Protein HIC PH-814, 8 mm × 75 mm. Buffer A—1.7 M ammonium sulfate in buffer B; buffer B—0.1 M sodium phosphate, pH 7.0. Gradient: 0% B to 100% B linear in 15 min. Flow rate: 1.0 mL/min. Injection: 100 μL of mouse ascites fluid diluted 1:1 with buffer A. Detector: UV at 280 nm.) (From Ref. 16.)

A very interesting application study was carried out by Fausnaugh and Regnier (17). The authors investigated the retention behavior of lysozyme isolated from related bird species. Lysozyme is an enzyme with 129 amino acids. Studied were the lysozymes from chicken, Japanese quail, ring-necked pheasant, Peking duck (3 isozymes), and turkey. The authors concluded that the hydrophobic retention was caused by a contact area opposite the catalytic cleft of lysozyme. Plots of retention versus the molarity of ammonium sulfate in a base buffer of 10 mM potassium phosphate, pH 7.0, yielded straight and nearly parallel lines for all lysozymes. This is in agreement with the theoretical expectation that the slope of the line should be a function of the size of the contact area of the analyte with the stationary phase. However, significant differences in the intercept were observed that were caused by amino acid substitutions in the contact area. The intercept increased as the contact area became more hydrophobic. A change in pH of the mobile phase also affected only the intercept but not the slope. The authors concluded that the pH change did not alter the size of the contact area, but that the ionization state of the amino acids in the contact area determined the overall hydrophobicity of the contact area and therefore the intercept. This work represents the clearest demonstration of the principles of HIC available in the literature. It also

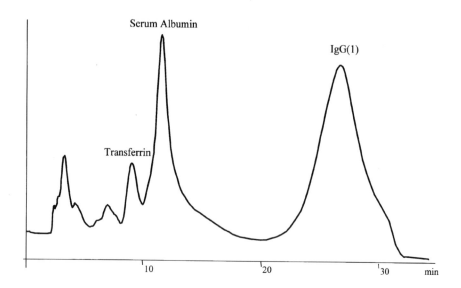

Figure 13.3 HIC separation of mouse ascites (IgG1). (*Chromatographic conditions*: Column: Protein HIC PH-814, 8 mm × 75 mm. Buffer A—1.6 M ammonium sulfate in buffer B; buffer B—0.1 M sodium phosphate, pH 7.0. Gradient: 20% B to 75% B linear in 5 min, hold 75% B for 5 min, then 75% B to 100% B concave in 15 min. Flow rate: 1.0 mL/min. Injection: 200 μL of human milk whey diluted 1:1 with buffer B, then diluted 1:8 with buffer A. Detector: UV at 280 nm.) (From Ref. 16.)

includes a beautiful example of the separation power of HIC. Duck B and C lysozymes differ from each other only by a single amino acid substitution (Pro-Arg in position 79), which is located in the contact area. This single substitution made it possible to distinguish chromatographically between the isozymes: duck C lysozyme with the amino acid arginine was significantly more hydrophilic than duck B lysozyme.

13.5   PRACTICAL ASPECTS

A general difficulty in hydrophobic-interaction chromatography is the application of the sample. If the sample is dissolved in water or a low-buffer solution, only small injection volumes should be applied. Remember that water is the strong elution solvent in HIC! On the other hand, the addition of salt to the sample can result in a precipitation of the protein. However, the addition of salt also allows a larger injection volume. The salt concentration in the sample therefore needs to be carefully optimized. This problem also limits the usefulness of HIC for preparative separations. A possible solution to this problem is the repeated injection of small volumes of sample dissolved in a low-concentration buffer (18).

The high salt concentrations used can present a challenge for the HPLC equipment. Pumps with seal-wash are preferred. If chloride buffers are used, HPLC systems with nonmetallic fluid paths are recommended. It is also advisable to flush the salt solution out of the system when it is not in use.

Temperature and pH can affect the separation and should therefore be controlled. But both the temperature effect and pH effect are smaller than in other forms of chromatography (with the exception of size-exclusion chromatography).

Labile proteins can be chromatographed at low temperature and mild chromatographic conditions to maintain biological activity. The conformation of some proteins can be stabilized by adding Ca^{2+} and Mg^{2+} ions to the mobile phase.

Columns can be washed with organic solvents to remove contaminants. This is not a problem with silica-based HIC columns. The compatibility of polymer-based HIC columns with organic solvents may be limited. Consult the manufacturer's recommendations!

REFERENCES

1. A. Tiselius, *Ark. Kemi. Min. Geol.* **B26,** 1 (1948).
2. S. Shaltiel and Z. Er-el, *Proc. Nat. Acad. Sci.* USA **70**, 778 (1973).
3. B. H. J. Hofstee, *Anal. Biochem.* **52**, 430 (1973).
4. S. Hjertén, J. Rosengren, and S. Páhlman, *J. Chromatogr.* **101**, 281–288 (1974).
5. Y. Kato, T. Kitamura, and T. Hashimoto, *J. Chromatogr.* **266**, 49–54.
6. Y. Kato, T. Kitamura, and T. Hashimoto, *J. Chromatogr.* **292**, 418–426 (1984).
7. W. Melander and Cs. Horváth, *Arch. Biochem. Biophys.* **183**, 393 (1977).
8. Cs. Horváth, W. Melander, and I. Molnar, *Anal. Chem.* **49**, 142 (1977).
9. W. Melander, D. Corradini, and Cs. Horváth, *J. Chromatogr.* **317**, 67–85 (1984).
10. U. Schön, Dissertation, Saarbrücken (1984).
11. S.-L. Wu, A. Figueroa, and B. L. Karger, *J. Chromatogr.* **371**, 3–27 (1986).
12. R. E. Shansky, S.-L. Wu, A. Figueroa, and B. L. Karger, "Hydrophobic-interaction Chromatography of Biopolymers," in *HPLC of Biomolecules: Methods and Applications*, K. M. Gooding, and F. E. Regnier, eds. Marcel Dekker, New York, 1990.
13. N. T. Miller, B. Feibush, and B. L. Karger, *J. Chromatogr.* **316**, 519 (1984).
14. L. Szepesy and G. Rippel, *J. Chromatogr.* **A668**, 337–344 (1994).
15. L. Szepesy and G. Rippel, *Chromatographia* **34**, 391–397 (1992).
16. W. Sawlivich, K. O'Connor, D. Phillips, and G. Vella, poster presentation at the Symposium of the American Society for Biochemistry and Molecular Biology/ America Association of Immunologists, 1990.
17. J. L. Fausnaugh and F. E. Regnier, *J. Chromatogr.* **359**, 131–146 (1986).
18. N. Miller and C. T. Shieh, *J. Liq. Chromatogr.* **9**, 3269–3296 (1986).

ADDITIONAL READING

R. E. Shansky, S.-L. Wu, A. Figueroa, and B. L. Karger, "Hydrophobic-interaction Chromatography of Biopolymers," in *HPLC of Biomolecules: Methods and Applications*, K. M. Gooding, and F. E. Regnier, eds., Marcel Dekker, New York, 1990.

14 Special Techniques

In this chapter several special techniques are covered. We first discuss column coupling, including the use of guard columns. Then we review column switching techniques. This section includes a brief discussion of orthogonal chromatographic techniques. Then, more sophisticated column switching techniques are covered: recycling and the walking column. The last two sections are dedicated to two forms of continuous chromatography. The moving-bed technique is a method of countercurrent chromatography used for binary separations. We will explain the simulated moving bed as a form of column switching. In the final section, we examine how multicomponent separations can be accomplished by continuous chromatography. Although some of these techniques are not HPLC methods in the strictest sense and are used only for certain specific problems of preparative chromatography, a book on column technology would be incomplete without their discussion.

14.1 COLUMN COUPLING

In column coupling, columns of the same type or of different types are connected in series. This technique is standard practice in size-exclusion chromatography, but is only rarely used in retention chromatography. Another common application of column coupling is the use of guard columns.

14.1.1 Principles of Column Coupling

There can be two reasons for the coupling of columns: (1) the increase in raw separation power by increasing the plate count; and (2) the combination of the selectivity of the individual columns to obtain a separation that is not possible using the individual column types alone. Both reasons apply to the use of column coupling in size-exclusion chromatography.

If we add identical columns in series, the plate count of the column bank increases directly with the number of columns. This means that resolution increases proportionally to the square root of the number of columns, while the analysis time increases in direct proportion to the number of columns. To be precise, it should be pointed out that plate counts are not additive, but that the elution volumes and the variances are additive. Therefore the total plate count of any bank of coupled columns should be calculated as follows:

$$N_T = \frac{\left(\sum V_R\right)^2}{\sum \sigma^2} \tag{14.1}$$

The peak variance is measured in volume units. This equation also assumes that there are no additional sources of flow that would dilute the sample. An alternative way to write this equation is as follows:

$$N_T = \frac{\left(\sum V_R\right)^2}{\sum \dfrac{V_R^2}{N}} \tag{14.2}$$

If we analyze this equation, we will find that when we add columns of the same type (i.e., the same elution volume) but different performance (different variance), the plate count of the bank is strongly influenced by the worst column in the bank. This can best be seen by assuming that the retention volumes on all columns are the same, but the plate count is different:

$$N_T = \frac{n_c^2}{\sum \dfrac{1}{N}} \tag{14.3}$$

where n_c is the number of columns in the bank. An inspection of this equation shows that the best plate count is achieved when the plate count of every column is the same. It could easily be that the performance of a column bank is worse than the performance of its best column alone. This phenomenon has to be considered carefully when coupling columns. For example, in the early times of chromatography it was popular to pack precolumns or guard columns with packing material of a large particle size. These types of precolumns can have a deleterious effect on the performance of the system.

If columns of different selectivity are combined, there is always a deterioration in plate count compared to a bank of columns having the same selectivity. Even if all columns in the bank have the same plate count, the optimum column efficiency is achieved only when the retention volumes of all columns

COLUMN COUPLING   263

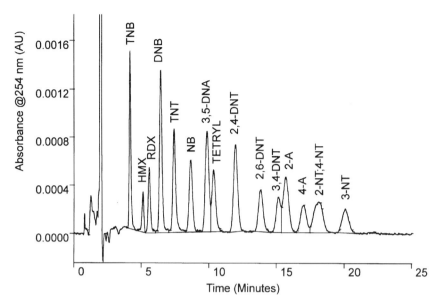

Figure 14.1 Separation of explosives on a column bank consisting of a Waters Nova-Pak CN, 3.9 mm × 150 mm and a Nova-Pak C$_8$ column, 3.9 mm × 100 mm. (*Chromatographic conditions*: Mobile phase: 82% water/18% isopropanol. Flow rate: 1 mL/min. Temperature: 25°C.) (Chromatogram courtesy of E. S. P. Bouvier, Waters Corp.)

are the same. You can examine this for yourself by looking at the right-hand part of Equation (14.4):

$$N_T = N \frac{\left(\sum V_R\right)^2}{\sum V_R^2} \tag{14.4}$$

Despite the fact that the plate count of a mixed bank is inferior to the performance of a bank of columns of the same type, the gain in selectivity obtainable with a mixed column bank may offset this disadvantage. Many examples are found in size-exclusion chromatography, where often only a combination of columns with different pore sizes covers the molecular-weight range that needs to be separated. One of the rare examples of such a separation problem in retention chromatography is shown in Figure 14.1, where the selectivity of a CN column is combined with the selectivity of a C$_8$ column.

14.1.2 Guard Columns

The use of guard columns or precolumns also falls into the category of column coupling. Guard columns are highly recommended for samples that contain constituents that can be strongly adsorbed on the column, such as proteins in serum samples in reversed-phase chromatography. Guard columns work best if the packing in the guard column is identical to the packing in the analytical column. There are two reasons for this. When the guard column contains the same packing as the analytical column, you can blindly assume that whatever packing would usually be strongly adsorbed on the analytical column will now be captured by the precolumn. If you use a different packing, you cannot make this assumption. Second, the band spreading of the guard column is less disruptive, if the two packings are identical. As a matter of fact, you can view a well packed guard column as an extension of the analytical column, which — if well packed — can actually add to the plate count and improve your separation. We always have to keep in mind that guard columns have a limited capacity and need to be replaced on a regular basis. How frequently they need to be replaced depends on how dirty the sample is. So there is no good way of judging the lifetime of a guard column. It really has to be determined empirically.

In some circumstances the guard column may be used to remove a particular type of constituent from the sample. For example, one can use ion exchangers to selectively remove ionic compounds from a sample. This can be used to eliminate small amounts of interference. Another example could be the use of an anion exchanger in the hydroxide form to neutralize samples from dissolution tests. When you use a guard column in this way, you have to be even more aware of its limited capacity.

Precolumns can also be used to condition the mobile phase. In this case they are usually operated in a position upstream from the sample injection point. For example, a silica column loaded with a large amount of amine modifier can be used to equilibrate the mobile phase with the amine modifier. This technique has been employed in carbohydrate separations on silica columns that are dynamically coated with an amine modifier. Silica columns have been used to saturate the mobile phase with silica in an attempt to prolong the column life of silica-based columns in alkaline mobile phases. That this technique actually works has, however, not yet been demonstrated unambiguously.

In high-pressure gradient applications, reversed-phase precolumns have been used to remove contaminants from the more aqueous mobile phase, which otherwise would give rise to a change in the baseline during the gradient. In this case the precolumn is located between the pump delivering the aqueous component of the gradient and the point where the gradient is mixed. The column can be regenerated once in a while by flushing it with the stronger mobile phase. This technique is quite effective when high-quality HPLC-grade water is not available.

Precolumns can be used to increase the concentration of analytes present in low concentrations. The most effective implementation of the trace enrichment technique uses column switching. In a reversed-phase application, a large sample volume is pumped onto the precolumn with a dedicated sample pump.

The precolumn is then switched to a position upstream of the analytical column, and the analytes are eluted off the precolumn by the mobile phase used for the separation. While the HPLC analysis is carried out, a second precolumn can be used to enrich the next sample.

14.2   COLUMN SWITCHING

At the end of the last section we already mentioned an application that is most efficiently carried out using column switching. However, the most powerful applications of column switching take advantage of different separation mechanism in the different columns. The less correlated the separation mechanisms are, the more powerful the column switching technique can be.

Assume that we are interested in the quantitation of a single compound in a very complex matrix. It may be difficult or even impossible to separate the compound of interest from all constituents of the matrix by any single chromatographic technique. In this case, the combination of orthogonal (= uncorrelated) separation mechanisms significantly increases the separation power.

For the sake of simplicity, we assume that we are running a gradient separation with a run time t_g. In this case, all peaks can be assumed to have the same width w_g measured in the same units as the gradient run time. If we have n_m number of compounds in the sample matrix, and if they are randomly distributed over the chromatogram, the number of compounds n_i that interfere with the quantitation of our analyte is simply

$$n_i = n_m \frac{2w_g}{t_g} \tag{14.5}$$

For typical gradient HPLC applications, the peak : width gradient time ratio is in the order of $\frac{1}{50}$. Therefore, if the sample matrix contains about 1000 compounds and if our assumptions hold, we can expect around 40 compounds to interfere with the quantitation of our target analyte.

If we now take all the compounds that elute in the retention-time window of our analyte and subject this mixture to a completely orthogonal second gradient separation, the 40 compounds will be distributed evenly over the entire gradient window, and the chances of an interference are significantly reduced. In practice, a little bit of optimization work would eliminate any remaining interferences, and the analyte can be quantitated without difficulty. In a one-dimensional chromatographic system, this task would be virtually impossible.

In practice, it is not easy to design truly orthogonal column switching techniques. For example, the retention on different reversed-phase packings is highly correlated, and one gains little from column switching between different reversed-phase columns. In other cases, the mobile phase needed for one dimension is incompatible with the separation technique for the second dimension. Probably the best example of an orthogonal separation is the combination of ion exchange with reversed-phase chromatography. Prerequisite is that either the analyte or a portion of the interferences is ionizable and can be made to interact with the ion exchanger.

As an example, a urine sample containing a basic drug can be injected onto a cation-exchange column using a mobile phase with a low concentration of buffer. The separation is carried out by increasing the ionic strength, and the fraction containing the analyte is switched onto a reversed-phase column, where it accumulates on the top of the column. The reversed-phase column is subsequently eluted with a gradient from buffer to organic solvent.

It is not always necessary that the separation techniques are orthogonal for all constituents of the sample. A special sample cleanup technique combines the use of internal-surface reversed-phase packings (see Section 10.4.3) with standard reversed-phase packings. Internal-surface reversed-phase (ISRP) packings have an external noninteracting surface and a reversed-phase surface in the small pores. Proteins, which are excluded from the small pores, can interact only with the external surface and the surface in the large pores. Therefore they pass through this packing unretained, while small molecules are retained in the small pores through a reversed-phase mechanism. This allows us to remove proteins from a sample that otherwise would clog a standard reversed-phase column. In a typical application of this technique, a serum sample is injected into the internal-surface reversed-phase column, while the separation column is off line. After the proteins have been eluted from the ISRP column, this column is positioned upstream of the analytical column and the low-molecular-weight analytes are separated by reversed-phase chromatography. This method combines the advantages of an ISRP column with the advantages of using a standard reversed-phase column for the actual analysis.

We have discussed above how column switching can be used to cut out a single retention window in a complex sample and subject the remaining compounds to an orthogonal analysis. If this is done with many retention windows, one approaches two-dimensional chromatography. An example could be as follows. A complex mixture of peptides is separated via a salt gradient on an ion-exchange column. The effluent from the ion-exchange column is directed onto a reversed-phase column, where the peptides enrich on the top of the column via hydrophobic interaction. In regular intervals, the reversed-phase column is replaced with another one. After all fractions have been collected onto reversed-phase columns, these columns are themselves eluted one after the other with a reversed-phase gradient. Thus one obtains a complete fractionation of the sample by two orthogonal retention mechanisms. This is akin to two-dimensional thin-layer chromatography. Unfortunately, this elegant and very powerful technique is also very complex and time-consuming. Therefore it has not been used to solve real application problems.

14.3   RECYCLING AND WALKING COLUMN

In this section, we will discuss two techniques that are used to obtain the resolution of a long column or a large column bank using only one or two columns. They are applied to separate a small number of compounds, usually two, that elute very closely to each other (low selectivity). One technique is called *recycling*; the other technique is known as the *walking column*.

In the case of recycling, a single column is used. The mobile phase carrying the pair of compounds that need to be separated is redirected to the pump and pumped back into the column. Every new pass through the column increases the distance between peaks, while the peak width increases only with the square root of the number of passes through the system. Thus the number of plates and the resolution continuously increase with each pass. This can be repeated for quite a while, but there is a limiting factor. Once the combined peak width (in length units) of the band that we are trying to resolve approaches the column length, we either need to shave off the sides of the peaks or terminate the protocol. If the extracolumn band spreading were negligible, a very large number of plates would be achievable by this technique. Every pass through the system would add the number of plates achievable with a single column alone. Unfortunately, the band spreading caused by recirculating the sample through the pump can be quite severe if the peak width is small. Therefore, an alternative technique is more effective.

This alternative technique uses two columns instead of one and is called the *walking-column technique*. It works as follows. The columns are arranged in a bank with switching valves between the columns. The sample is injected onto the bank. Once the peaks of interest have completely entered the second column, the first column is switched downstream of the second column; thus their sequence is reversed. The switching is repeated once the sample has migrated from column 2 back to column 1 and the original sequence is restored. The advantage of this technique is the significant reduction of the band spreading in the switching valves compared to the band spreading in pump heads. With this technique, the increase in the number of plates with each switching of the columns is very close to the number of plates achievable with a single column. As above, the limit is reached when the combined width of the bands that we are trying to resolve approaches one column length. In principle, one can repeat this cycle several hundred times with a set of high-performance columns, and several million plates can be reached with this procedure. In practice, this technique is mostly of academic interest.

The recycling techniques have been used for preparative chromatography using lower-resolution columns packed with larger particle sizes. The application is usually a binary separation. In this case, it is possible to extract the sides of the bands during each run to obtain the pure compounds. This technique is called *peak shaving*. The center of the band is recycled as usual. Thus the load on the column continuously decreases, and the width of the band can be controlled. With this method it is possible to obtain high-resolution preparative separations with inexpensive, low-resolution columns. Of course, nothing is free, and this technique is quite time-consuming.

A variant of this procedure has been commercialized under the name Cyclo-Jet (Merck, Germany). The same column switching technique is used as in the walking-column technique, but fresh sample is injected into the center of the sample band any time it passes from one column to the other. Also, the edges of the band are collected as described in the peak-shaving technique mentioned above. Semicontinuous binary separations are possible with the Cyclo-Jet technique that are reminiscent of the simulated moving-bed technique described in the next section, but the technique is not as solvent efficient as the simulated moving bed.

14.4   SIMULATED MOVING BED

The simulated-moving-bed technique (1,2) is a preparative technique for binary separations, specifically, the separation of two compounds or the separation of one compound from the rest of the world. In the second case, it is necessary that the peak of interest elutes either first or last in the chromatogram. The simulated moving bed is the chromatographic technology that uses the packing and the solvent most efficiently. This fact, combined with the fact that it is a binary separation technique, makes it ideal for the preparation of chirally pure compounds from an enantiomeric mixture at industrial scales.

The concept of the moving-bed chromatography is rather simple. A binary sample is injected into a column, in which both the eluent and the "stationary" phase, the packed bed, are moving. See Figure 14.2 for an illustration! The eluent moves from left to right, and the packed bed moves from right to left. The sample is separated into its components. The velociticities of the mobile phase and of the packed bed are adjusted such that the midpoint between both

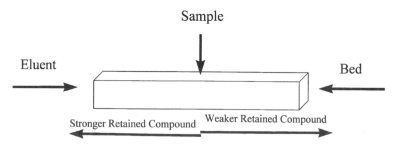

Figure 14.2 Separation principle of the moving-bed chromatography. The eluent moves from left to right, the packing moves from right to left, and the injection point of the binary sample remains stationary in the center. The velocities are adjusted such that the stronger retained compound moves with the packing and the weaker retained compound moves with the eluent.

sample constituents remains stationary. This means that the compound with the lower retention factor will move into the direction of the eluent, that is, from left to right, and the second compound will move with the packed bed into the opposite direction. It also means that the injection point remains stationary. Thus we can continuously inject sample at this point and one compound will move toward the right and the other toward the left. Somewhere downstream on both sides we elute each component in its pure form.

The actual implementation is more difficult to understand. A train of columns arranged in a circle is used. In principle, four columns are sufficient, but typical trains contain 8–16 columns arranged into four zones. Eluent is fed in on one side, and in regular intervals the columns are incremented in the direction opposite to the mobile-phase stream. The sample is continuously fed into the center of the train, and the separated compounds are collected at both upstream and downstream of the feed point.

The principle of the simulated moving-bed chromatography can be examined from two different viewpoints: as a theoretical moving bed or as a column switching technique. From the standpoint of optimization, it is best to view it as a theoretical moving bed. From the standpoint of understanding, a discussion of the simulated moving bed as a column switching technique is more fruitful.

We will first simplify the operation by assuming that the flow rate of the sample feed is negligible compared to the eluent flow. With this simplification, the principle of the operation can be understood by examining just the two center columns of the train, at the feed point of the sample. We will assume that the mobile phase is pumped just through these two center columns of the train and that the sample is fed into the stream between both columns. To simplify the discussion here, we will also pretend that the load on the columns is low enough to stay within the linear range of the isotherm (=constant retention factor), and that we can neglect dispersion. As we will see, the second assumption is not a very severe constraint, and we can get a very good impression for the operation of the system. But in practice the system is complicated mainly because it is operated in overload mode, where the retention factor changes with the sample concentration. Initially, we will also ignore the problem of eluting the compound from the column at the top of the train.

Let us suppose that we want to separate a mixture of a blue dye and a red dye. The blue dye has a lower retention factor than the red dye. Thus the blue dye will move into the same direction as the mobile phase, and the red dye will move in the direction of the bed. For quantitative purposes, we assume that the blue dye migrates $\frac{5}{4}$ column lengths during one time interval, while the red dye moves $\frac{3}{4}$ column length.

There is a time period between the start of the operation and the point where the steady state is reached. To understand the operation, we will discuss this transition state in detail. At the beginning, we are continuously feeding sample into the mobile-phase stream that enters column 2 and a frontal analysis profile develops in column 2, consisting of a first front of blue dye followed by a slower moving front of red dye mixed with blue dye. As we continue feeding the sample into column 2, some pure blue dye will at some point exit the column and will be collected. Once that happens, we can switch the train; thus column 2 becomes column 1 and a fresh column replaces column 2. The point for switching the train should be selected somewhere between the point that blue dye has broken through and red dye has not yet reached the exit of column 2. If we switch too early (and always switch at the same time interval), the blue dye will accumulate in column 2 and we will never be able to collect it. Also, if we switch the train at the point where the red dye has just reached the exit of column 2, the red dye will accumulate indefinitely in column 2. Therefore we will switch the train right in the middle. At that time, the red dye will have moved down $\frac{3}{4}$ of the length of column 2 and the blue dye would have moved down $\frac{5}{4}$ of the length of the column, if that were possible. The evolution of the concentration profile is shown in Figure 14.3.

Let us now first look at what is happening at column 1 after we switched the train. Column 1 is full of blue dye, and the red dye has migrated partly down the column. As we pump mobile phase into this column, pure blue dye will first be eluted and pumped into column 2, where it is mixed with new

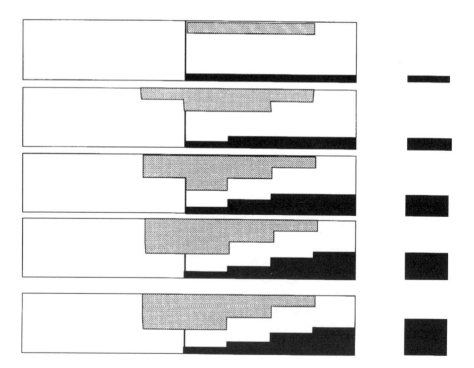

Figure 14.3 Evolution of the concentration profile in a simplified simulated moving bed. See the text for a detailed discussion!

sample. Then, a quarter of a time interval later, a band of a mixture of blue dye and red dye will elute and be pumped into column 2. And finally, a band of pure red dye develops at the top of the column, migrating slowly down the column. At column 2, sample is continuously added to this stream, increasing the concentration of blue dye 2-fold. Also, as the red dye breaks through from column 1 to column 2, there is soon an increase in the concentration of the red dye as well. Now we have a band with double the concentration of blue dye and double the concentration of red dye. Once the band of pure blue dye reaches the outlet of column 2, we collect it. Then, we switch the column train at exactly the same time as before, that is, before any red dye reaches the end of column 2. At the time of the switch, all blue dye from column 1 has been pumped into column 2, and there is only pure red dye left in column 1. After the switch, this column, which is now column 0, is taken out of the train and set aside for extraction of the pure red dye.

After the switch, the process repeats, with the difference that the bands of blue and red dye in column 1 are at double the concentration than before. The evolution of the concentration train is shown in Figure 14.3 for the next switches. In our simple example, after three switches a constant amount of red dye is removed at the top with every column that is taken out of the train, and after five switches a constant concentration of blue dye is eluted at the bottom of column 2, and a steady state is reached.

In true simulated-moving-bed applications, the feed flow rate is not negligible compared to the mobile-phase flow rate. As a matter of fact, an important feature of simulated-moving-bed applications is the fact that the dilution of the sample stream is kept to a minimum. So let us add a substantial feed flow. We can increase the flow in column 2 (downstream of the feed point) maximally to a flow that just avoids a breakthrough of the red dye from the bottom of column 2. In our example this is $\frac{4}{3}$ of our original flow. In column 1, upstream of the feed point, we can decrease the flow rate down to a value that allows us to just remove all blue dye from column 1 in a single time interval. In our example, this is $\frac{4}{5}$ of the original flow. The difference between the flow coming from column 1 and the flow in column 2 is made up by the feed flow rate: $\frac{4}{3} - \frac{4}{5} = \frac{8}{15}$. Thus we can add a feed flow of over $\frac{1}{2}$ the eluent flow rate and still keep the system in balance. Of course, the concentration profile for the dyes becomes more complicated. It also means that it takes longer for the system to reach steady state.

In addition, we had collected the entire eluent stream that emerged from column 2. In a complete simulated moving bed, we want to recover some of the eluent and pump it back into the system. In order to do this, we pump a portion of the eluent stream carrying the blue dye into column 3. Pure solvent exits column 3. The flow that can be pumped through column 3 in a given time interval without letting the blue dye break through cannot exceed $\frac{4}{5}$ of the original flow rate. Therefore we need to remove $\frac{8}{15}$ of the original flow rate between columns 2 and 3. The fact that column 3 carries blue dye with it as it is switched upstream and becomes column 2 increases the concentration of blue dye in column 2 and ultimately in the *raffinate*, as the stream of purified blue dye is generally called.

Let us now consider what happens to the columns at the top end of the train that contain the purified red dye. To completely extract the red dye from this column, we need a flow rate that is large enough to remove all red dye from a column in a single time interval. In our example this means that the flow rate needs to be at least $\frac{4}{3}$ of the original flow, since the red dye was moving at a rate of $\frac{3}{4}$ of one column volume at the original flow rate. Since we need a flow of $\frac{4}{5}$ of the original flow in column 1 and a flow of at least $\frac{4}{3}$ of the original flow in column 0, we can extract a flow of $\frac{8}{15}$ of the original flow rate between column 0 and column 1. As a matter of fact, the general name for this stream is the extract. Also, the fact that we pump back $\frac{7}{16}$ of the flow into column 1 increases the concentration of red dye inside the system as well. Since we are recovering an eluent flow of $\frac{4}{5}$ of the original flow from the bottom of the train, the makeup flow at the top of the train is also $\frac{8}{15}$ of the original flow.

We can now examine the balance of all flows in a complete simulated moving bed (Fig. 14.4). Mass balance requires that the sum of the makeup flow and the sample feed flow equals the sum of the raffinate and extract flow. Also, in a steady state, the amount of dye entering the system in any given time interval must equal the amount of dye exiting the system. Therefore we can calculate the dilution of the dyes from the ratio of the extract or raffinate flow to the sample feed rate. In our case, the ratio is 1. Therefore *no dilution* of the sample took place. This is quite different from any other chromatographic technique.

In a real system, the boundaries are not infinitely sharp and we have to adjust the various flow rates to take this fact into account. This would decrease the sample feed rate and increase the extract and raffinate flows. Therefore it decreases the efficiency of the system, but it does not change the fundamental principle of the operation.

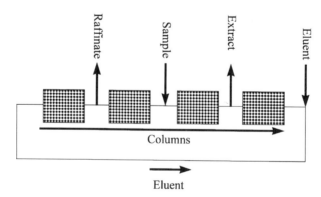

Figure 14.4   Flow paths in a simulated moving bed.

In practice, more columns are kept in the train, which brings the system closer to the ideal of a true moving bed. The major complication in actual applications is the very high sample concentration, which brings the operation out of the linear range of the adsorption isotherm. In this case, the operation is best studied and optimized through a computer simulation, which takes into account the competitive isotherms and the dispersion. Nevertheless, the simple picture presented here is a good first cut toward an understanding of the simulated moving-bed technology. It should be pointed out that the number of theoretical plates required for applications of the simulated moving bed can be rather low. Especially for applications outside the linear range of the isotherm, column efficiency is only of a secondary concern.

At the time of this writing, there is a renewed interest in simulated-moving-bed technology, and further advancements have been made. Of special interest is the use of supercritical fluids as eluents.

14.5 CONTINUOUS CHROMATOGRAPHY

One of the most interesting aspects of column technology is the attempt to generate a continuous chromatographic separation of multiple components. This is of special interest for preparative chromatographic separations. The principle of the technique is relatively straightforward, but the technical implementation is quite difficult.

To understand the principle, let us imagine a rectangular bed of a finite length and depth, but of unlimited width. The mobile phase is flowing through this bed everywhere at the same rate. Let us assume that we inject the sample continuously onto a point at the top of the bed, while we move the bed slowly in a direction perpendicular to the eluent flow. This principle is demonstrated in Figure 14.5. Every longitudinal section of the column acts, in principle, like an individual column, and the peaks migrate down this section at different speeds, just as in an ordinary column. Also the band spreading is akin to the band spreading in an individual column. However, since we are continuously moving the injection point, we are continuously moving into a fresh column section, while the separation continues to proceed in the old column sections. Thus the separation that happens in individual columns with time is transformed into a separation along the axis in which the injection point is moving. At some point the separated bands reach the bottom of our infinite column. The points where they reach the bottom are offset from the injection point by a constant distance, which depends on the migration velocity of the band and the speed with which the injection point is moved. Therefore, we can collect the separated compounds at collection points that move at the same speed and into the same direction as the injection point. In this way a continuous separation is achieved.

Injection point

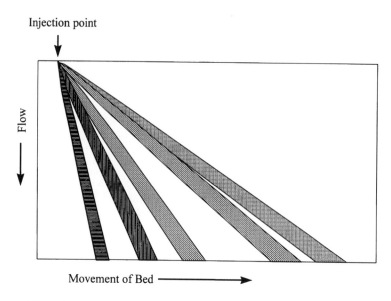

Flow

Movement of Bed ⟶

Figure 14.5 Principle of continuous multicomponent chromatography. The bed is moving from left to right in a direction perpendicular to the mobile-phase flow.

To generate an infinitely wide chromatographic bed is difficult and may not be cost-effective. Therefore one of the practical implementations of this technique uses a chromatographic bed in the form of an annulus (3). Mobile phase is continuously pumped through the annular bed from the top to the bottom, and the sample feed points and the collection points are continuously moved around the annulus. The moving of the injection point and the collection points is accomplished through valves. This technique is quite practical, but it appears that the preparation of a uniform annular bed is technically difficult. To my knowledge, there are no commercial implementations of this technique.

The annular moving bed can be simulated by a large number of columns that are moved around in a circle. As above, the injection point is moved from column to column in regular intervals, and the collection points for the separated compounds are also moved around. In this way, a continuous separation of several different compounds is possible. The number of columns that are needed depends on the retention factor of the last compound and how often we are switching the columns. In this approach, we can even use gradient elution, at least in the form of step gradients. In practice, we need a pump for every column to assure a reliable operation of the system. This increases the cost of such a system, but otherwise this technique is quite practical and assumes only that all columns are reasonably similar to each other. However, separation problems where several different compounds need to be purified from the same feed stream are quite rare, therefore this technique is also of limited practical interest.

REFERENCES

1. F. Charton and R.-M. Nicoud, *J. Chromatogr.* **A702**, 97–112 (1995).
2. R.-M. Nicoud, *LC-GC Intl.* **5**(5), 43–47 (1992).
3. J. M. Begovich, C. H. Byers, and W. G. Sisson, *Sep. Sci. Technol.* **18**, 1167 (1983).

15 Preparative Chromatography

The major difference between analytical chromatography and preparative chromatography lies in its purpose. In analytical chromatography, the goal of the separation is information; in preparative chromatography, it is purification. Therefore, preparative chromatography has a set of performance criteria, which are different from the performance criteria of analytical chromatography. Examples are load, purification factor, and production rate. Since the remainder of this book has been dedicated to analytical chromatography, we will approach the subject of preparative chromatography using the knowledge of analytical chromatography as a base and treat preparative chromatography as a departure from this base. For an in-depth discussion of preparative chromatography, specialized textbooks should be consulted (1).

15.1 PRINCIPLES

Since the goal of preparative chromatography is the production of large quantities of purified compounds, a large sample volume or a large mass of sample or a combination of both is usually applied to the column. This leads to additional effects not commonly encountered in analytical chromatography, such as shifts in retention times and additional broadening of the peaks. In the discussions in the remainder of the book, we tacitly implied the absence of such "overload" effects. For example, we assumed that the concentration of the analytes remains within the linear portion of the adsorption isotherm, while preparative chromatography is usually carried out in the nonlinear portion of the isotherm. Preparative chromatography is therefore also sometimes called *nonlinear chromatography*.

Nonlinear effects significantly complicate the theoretical treatment of chromatography. Retention times, peak widths, and peak shapes are all a function of the concentration of the sample constituents. At high sample load, the contributions of the nonlinear effects completely overshadow the effects encountered in analytical chromatography. A rigorous treatment of these contributions is mathematically complex and is still the subject of intense research. In the majority of cases, analytical solutions to the question of retention times and peak shapes are not available and the answers are obtained by computer simulations. Nevertheless, one can get a good impression of the phenomena involved in nonlinear chromatography by making some simplifying assumptions, for which rigorous solutions are possible. The entire phenomenon can then be viewed as a combination of the simple phenomena. A similar approach has been used by Knox and Pyper (2).

In the following discussion, we will examine the factors that influence peak width and peak shape of a single peak in preparative chromatography. There are three distinct factors, which for the purpose of simplification we will treat separately.

- The first factor is the band spreading due to the chromatographic process itself. The same mechanisms are in effect as in linear chromatography. These phenomena have been discussed in Section 2.2.

- The second factor contributing to the peak shape and peak width is the injection volume of the sample. The effect is the same as the extracolumn bandbroadening encountered in analytical chromatography. It is well understood and can be treated rigorously.

- The third factor stems from the departure of the adsorption isotherm or distribution coefficient from linearity. This factor can be easily understood by assuming an ideal column free from the other band-spreading effects, and a single-component isotherm. This simplification is quite valid for high-resolution columns.

15.1.1 Influence of the Injection Volume

The influence of large injection volumes can be treated like any other extracolumn band-spreading effects. The band spreading inside the column is independent of the band spreading outside the column. Thus we can simply add the variance of the injection volume to the variance of the band contributed by the column:

$$\sigma^2_{v,\text{total}} = \sigma^2_{v,\text{injection}} + \sigma^2_{v,\text{column}} \qquad (15.1)$$

The variances σ_v^2 are expressed in volume units. This relationship holds even for different shapes of the bands. For example, the injection band at large injection volumes is rather a plug, while the shape of the band in the column may approach a Gaussian distribution. The variance of a plug injection is given as

$$\sigma_{v,\text{plug}}^2 = \frac{V_{\text{plug}}^2}{12} \tag{15.2}$$

While the variance of the combined contributions can be calculated easily, this relationship does not give us any information on the shape of the band emerging from the column. At low injection volumes, when the variance of the column dominates the band spreading, we obtain a band that departs very little from the Gaussian peak shape of analytical chromatography. On the other extreme, if the injection volume is completely dominating the band spreading, we obtain a nearly rectangular peak shape with rounded edges due to the influence of the band spreading of the column. The shape of the rounded edges are the same as those obtained in frontal chromatography; thus they can be represented well by the integral over the Gaussian distribution. In between, the peak shape resembles a Gaussian distribution with a flattened top.

The position of the beginning of the band remains constant and at the same retention time as the band obtained from a very small injection volume. We can view the injection of a large injection volume as an uninterrupted series of small injections. In this way we understand that the center of a band obtained at a large injection volume is displaced to longer retention times (and volumes). More specifically, the center of the band is located at the retention volume of the band obtained with a negligible sample volume plus $\frac{1}{2}$ of the injection volume. This is shown in Figure 15.1.

15.1.2   Influence of Mass Overload

The treatment of mass overload is significantly more complicated than the treatment of the influence of a large injection volume. To get an understanding of the fundamental effects that influence peak shape and peak width, we will make the simplifying assumption that the dispersion in the column can be neglected. This means that we assume that the column plate count is very high, even infinite. This assumption is actually not that far-fetched for the high-performance columns in use today. This simplification allows us to get an understanding of the principles of nonlinear chromatography using only simple algebra and simple calculus as tools. The derivation is adapted from Reference 3.

Let us consider a column without dispersion, in which the equilibrium between mobile phase and stationary phase is instantaneous. The mass balance

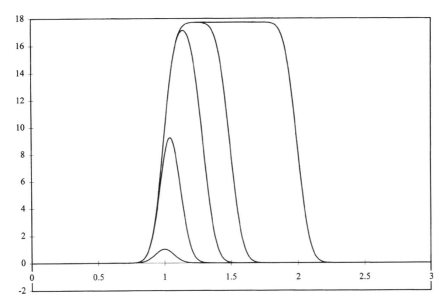

Figure 15.1   Evolution of peak shape with increasing injection volume.

for the sample in the column-length element dL is then

$$V_M \frac{dc_M}{dt} + V_S \frac{dq_S}{dt} + V_M u \frac{dc_M}{dL} = 0 \tag{15.3}$$

where the subscripts M and S stand for the mobile and stationary phases, respectively; V is the volume of the phase; c_M is the sample concentration in the mobile phase; q_S is the surface concentration in the stationary phase; and u is the mobile-phase velocity.

This equation can be rearranged:

$$\frac{dc_M}{dt} \left(1 + \frac{V_S}{V_M} \frac{dq_S}{dc_M} \right) + u \frac{dc_M}{dL} = 0 \tag{15.4}$$

The second factor in the last term in parentheses is merely the derivative of the adsorption isotherm, and the factor preceding it is the phase ratio.

Further rearrangement yields the migration velocity of a sample band of the concentration c_M:

$$\frac{dL}{dt} = \frac{u}{\left(1 + \dfrac{V_S}{V_M} \dfrac{dq_S}{dc_M} \right)} \tag{15.5}$$

Consequently, the retention factor k for this band is

$$k = \frac{V_S}{V_M} \frac{dq_S}{dc_M} \qquad (15.6)$$

Now, we can calculate the shape of the peak at the column exit from the adsorption isotherm. The shape of the peak is simply a plot of the concentration in the mobile phase versus the retention factor, which is the relationship stated implicitly in Equation (15.6).

We will use the Langmuir adsorption isotherm for the following calculation; it is a good description of the adsorption isotherms typically encountered in HPLC:

$$q_S = \frac{O c_M}{P + c_M} \qquad (15.7)$$

where O is the surface concentration at saturation and P is a constant, defining the slope of the isotherm at low surface concentration.

The first derivative of the isotherm is given below:

$$\frac{dq_S}{dc_M} = \frac{OP}{(P + c_M)^2} \qquad (15.8)$$

Both the Langmuir isotherm and its first derivative are shown in Figure 15.2. With Equation (15.7), the retention factor can be expressed as a function of the concentration in the mobile phase:

$$k = \beta \frac{OP}{(P + c_M)^2} \qquad (15.9)$$

where β is the phase ratio. The chromatogram is described by this relationship. It starts with a sharp step, followed by a long tail. The idealized chromatogram is shown in Figure 15.3. It is compared to an actual chromatogram obtained with increasing sample load in Figure 15.4. The end of the peak is the retention factor at very low sample load. As the sample load increases, the peak front moves toward shorter retention, but the tail of the peak follows a common envelope determined by the first derivative of the adsorption isotherm.

We can calculate the variance of such a peak without difficulty. The general formula is

$$\sigma_k^2 = \frac{\int_{k_{max}}^{k_o} c_M (k - k_{ave})^2 \, dk}{\int_{k_{max}}^{k_o} c_M \, dk} \qquad (15.10)$$

where σ_k is the standard deviation in retention-factor units, and is therefore

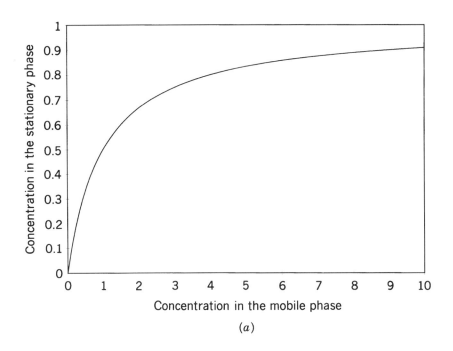

(a)

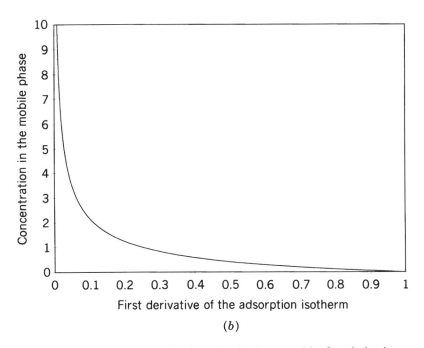

(b)

Figure 15.2   The Langmuir adsorption isotherm and its first derivative.

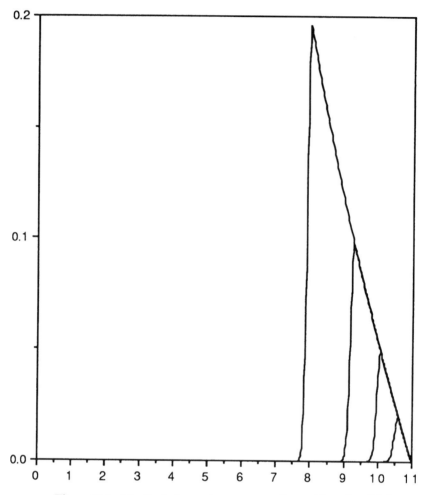

Figure 15.3   Idealized chromatogram according to Equation (15.8).

dimensionless; k_0 is the retention factor at low concentration; and k_{max} is the retention factor of the peak maximum. The average retention factor k_{ave} can be calculated as follows:

$$k_{ave} = \frac{\int_{k_{max}}^{k_0} c_M k \, dk}{\int_{k_{max}}^{k_0} c_M \, dk} \tag{15.11}$$

For a Langmuir isotherm, the average retention factor is

$$k_{ave} = \frac{k_0}{6} \left(3 \frac{k_{max}}{k_0} + 2 \sqrt{\frac{k_{max}}{k_0}} + 1 \right) \tag{15.12}$$

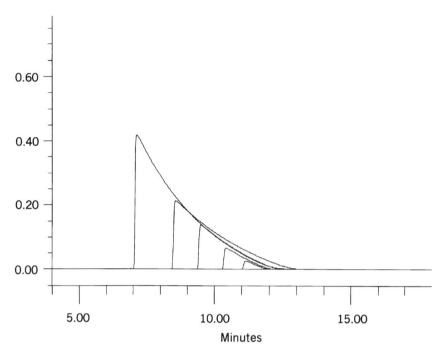

Figure 15.4 Actual chromatograms obtained with increasing sample load. (*Chromatographic conditions*: Sample: ethacrynic acid, 43 mg/mL. Injection: 5–100 μL. Column: 7 μm SymmetryPrep C_{18}, 3.9 mm × 150 mm, Waters Corp. Mobile phase: 25% acetonitrile/75% 20 mM phosphate buffer, pH 7.0. Flow rate: 0.7 mL/min.) (Chromatograms courtesy of M. Z. El Fallah, Waters Corp.)

and the variance of the peak is

$$\sigma_k^2 = \frac{k_0^2}{180} \left(15 \frac{k_{max}}{k_0} + 18 \sqrt{\frac{k_{max}}{k_0}} + 7 \right) \left(\sqrt{\frac{k_{max}}{k_0}} - 1 \right)^2 \qquad (15.13)$$

The variance is expressed here in dimensionless form, in units of the retention factor. We can use this equation to estimate the peak variance contribution from mass overload. The total mass injected can be obtained by integration over the peak:

$$m = V_M P k_0 \left(\sqrt{\frac{k_{max}}{k_0}} - 1 \right)^2 \qquad (15.14)$$

It can also be expressed as a fraction of the total mass m_{sat} that can be loaded onto the stationary phase at saturation:

$$\frac{m}{m_{\text{sat}}} = \frac{m}{V_S O} = \left(\sqrt{\frac{k_{\text{max}}}{k_0}} - 1 \right)^2 \tag{15.15}$$

This ratio is also called the *loading factor* L_f. It uniquely determines the peak shape. Conversely, one can obtain the loading factor from an overloaded peak by simply measuring the retention factors at the beginning and at the end of the peak. This allows an objective comparison of the loadability of different packings to each other. Of course, the underlying assumption is a common isotherm type, here a Langmuir isotherm.

The discussion of the peak shape in this section assumed that the isotherm of the compound of interest is not influenced by the presence of the other constituents of the sample. This is true only for very large concentration differences between the peak of interest and the other constituents of the sample. When the concentrations of two or more compounds fall into the same order of magnitude, the competitive nature of the isotherms gives rise to significant complications in the peak shapes of the compounds. A treatment of this situation is outside the scope of this book, and a more advanced book on preparative chromatography, such as Reference 1, should be consulted.

15.1.3   Influence of Column Dispersion

In the previous section we assumed that the dispersion of the column was negligible. However, dispersion does exist, and the factors that govern it are the same as in linear chromatography, namely, without mass overload. Therefore they can be assessed by the same equations as those derived in Section 2.2. Using the van Deemter equation, we can derive the variance due to dispersion effects as follows:

$$\sigma_L^2 = L \left(A + \frac{B}{u} + Cu \right) \tag{15.16}$$

where σ_L is the standard deviation of the peak in length units. Typical coefficients of the van Deemter equation are given in Section 2.2.

We have now derived the variance contribution due to volume overload, mass overload, and dispersion. In order to obtain the effect of the combined contributions, all we need to do is add up the variances, making sure that they all have the same dimension.

15.1.4   Combined Effect of Injection Volume, Mass Overload, and Dispersion

In the following we will examine the combined effects of the injection volume, the mass overload, and the column dispersion. A commonly used graphic representation for this purpose is a plot of plate count versus injected mass or injected volume. The relationship necessary for this is the law of the additivity of the variances, as applied, for example, in Equation (15.1). The plate count is simply a dimensionless representation of the variance. In the case of volume overload, it can be written as

$$N_{\text{total}} = \frac{V_R^2}{\sigma_{v,\text{total}}^2} = \frac{1}{\dfrac{1}{N_{\text{column}}} + \dfrac{\sigma_{v,\text{injection}}^2}{V_R^2}} \tag{15.17}$$

As we can see from this equation, the influence of the injection volume is normalized by the retention volume of the peak. We can therefore obtain a unique plot for the influence of the injection volume on column performance, if we plot the total plate count obtained with various columns of different efficiencies against the injection volume divided by the retention volume. (For a given chromatographic separation with a constant retention factor, this is the same as normalizing the injection volume by the column volume.) The resulting relationship is plotted in Figure 15.5. As one can see, at high-volume overload all curves merge into a single curve, independent of the inherent performance of the column. This means that if the resolution requirements of the separation can tolerate a large injection volume, one can obtain the same separation on a low-resolution column as on a high-resolution column.

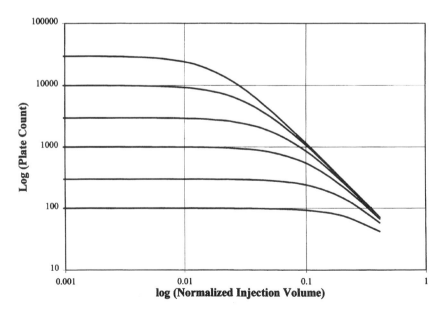

Figure 15.5 Plot of the logarithm of the plate count versus the logarithm of the reduced injection volume for columns of different plate counts.

Low-resolution columns packed with larger particles are less expensive than high-resolution columns.

A similar relationship can be demonstrated for mass overload. We can define a plate count under conditions of mass overload, N_{mass}, for the ideal column described in Section 15.1.2:

$$N_{mass} = \frac{k_{ave}^2}{\sigma_k^2} \qquad (15.18)$$

where k_{ave} and σ_k are the same as defined in Equations (15.12) and (15.13), and their dependence on the injected mass (or, better, the loading factor) can be derived from Equation (15.15). If the law of the additivity of the variances can be applied to the combination of the contributions to the peak variance from dispersion and from mass overload, we can write

$$N_{total} = \frac{L^2}{\sigma_{total}^2} = \frac{1}{\dfrac{1}{N_{column}} + \dfrac{1}{N_{mass}}} \qquad (15.19)$$

(*Note:* It can be argued that dispersion effects are kinetic in origin and the mass overload effects are thermodynamic in origin and that this justifies the assertion that the law of the additivity of variances can be applied.)

The resulting relationship between plate count and loading factor is plotted in Figure 15.6 for columns with various plate counts. The family of curves obtained is similar to the family of curves obtained for volume overload. At high load, all curves merge into each other. The limiting line is defined by the adsorption isotherm. For mass overload the same holds as for volume overload — at high overload, the same separation can be obtained on a low-performance column as on a high-performance column.

There is one important difference between mass overload and volume overload. In mass overload, the plate count drops by one decade, when the mass load is increased by one decade. In volume overload, the plate count drops by two decades when the volume load is increased by one decade. In practice this means that it is desirable to work with concentrated solutions of the sample.

It is also instructive to examine Figure 15.6 from a global standpoint. In the part of the curves that are parallel to the x-axis, band spreading is dominated by dispersion and can be predicted well with the tools of linear chromatography. In the part of the curves that approaches asymptotically the solution obtained for ideal chromatography, the peak shape, especially the tail of the peak, can be predicted fairly well by assuming a total absence of dispersion. Only in the intermediate range, which is fairly limited, do we need to use more

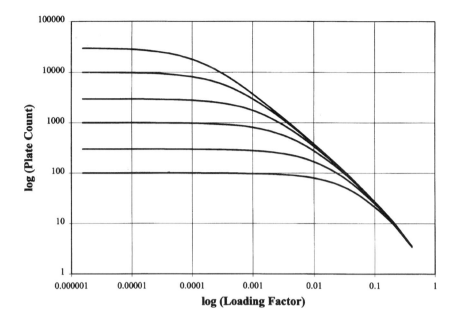

Figure 15.6 Plot of the logarithm of the plate count versus the logarithm of the normalized mass load for columns of different plate counts.

sophisticated tools to predict the peak shape. This in itself is justification for the treatment chosen in this section.

The types of curves described by the relationships shown in this section have been measured by many investigators. In Figure 15.7 the theoretical curve of an overload plot based on a Langmuir isotherm is compared to measurements obtained on a silica column, to which samples with increasing concentration were applied at a constant injection volume. As one can see, agreement between the measured loadability curve and the theoretical curve is quite good.

The load at which the column plate count declines by 10% compared to the plate count at very low load has been interpreted in the literature as the loadability of a packing. Figure 15.6 teaches us that this point is not a characteristic of the packing, but is a function of the plate count of the column. Therefore this definition of loadability commonly found in the literature has to be rejected. Unfortunately, this definition has also lead to the common misperception that the loadability of packings with a small particle size is inferior to the loadability of packings with a large particle size. This is not the case. At equal plate count, the loadability of a packing with a small particle size is identical to the loadability of the same packing with large particle size.

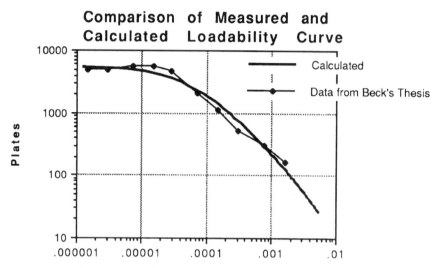

Figure 15.7   Comparison of a measured and a calculated mass loadability curve. (The data were obtained from Ref. 4.)

15.2   TECHNOLOGIES FOR LABORATORY-SCALE PREPARATIVE CHROMATOGRPHY

Standard steel columns of increasing diameter are available for preparative chromatography. A scaling of a separation by a factor of about 100 is possible with columns and instruments designed for the laboratory. Thus the purification of gram quantities is still within the reach of laboratory-scale chromatography.

The packing of larger-diameter columns is technically more demanding than the packing of smaller-diameter columns, especially if high-performance columns packed with small particles are needed. However, as we have seen in Section 15.1.4, because of the column overload of a typical preparative separation, one can most often use lower-performance columns packed with larger particle sizes. These columns are easier to pack and provide a lower backpressure than columns packed with analytical-grade packings. Also, the cost of columns packed with a larger particle size tends to be lower than the cost of columns packed with smaller particles. However, you want to make sure that the surface chemistry of the packings in the preparative column is identical to the surface chemistry in your scaling column. Otherwise, the scaling experiments (see below) need to be repeated with the preparative packing.

A major problem of large-diameter steel columns is their tendency to develop voids at the column top due to a collapse of the bed. The void space is detrimental to the separation. Also, the collapse of the bed may create large nonuniformities in the packed bed itself, which also is detrimental to the separation. Therefore, technologies have been developed that are designed to prevent this phenomenon: active compression of the packed bed by either radial compression or axial compression. These patented technologies have significantly improved the reliability of preparative chromatography. They have been discussed in more detail in Section 3.4. In the current context, it should only be mentioned that the design of radially compressed columns allows the easy stacking of several cartridges. This can be used with advantage to tailor the amount of packing to the amount of sample that needs to be separated. Figure 15.8 shows a comparison of the loadability of a single

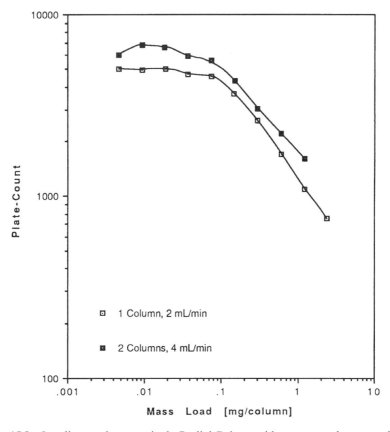

Figure 15.8 Loading study on a single Radial-Pak cartridge compared to a stack of two cartridges in series. The loadability of the stack was 2 times higher than the loadability of the single column. The run time was the same, since the flow rate was doubled for the stack.

radially compressed cartridge to a stack of two cartridges. As expected, the loadability of the stack is roughly 2-fold higher than the loadability of the single cartridge. There was no increase in run time, since the flow rate was increased by a factor of 2 as well. Therefore the production rate of the stack was 2-fold higher than the production rate of a single cartridge.

In the next section, we discuss the scaling of a preparative separation from one column size to another.

15.3   PRACTICAL ASPECTS

In most cases, the practice of preparative chromatography differs substantially from the practice of analytical chromatography. In analytical chromatography, the problem is solved when an adequate separation has been developed. In preparative chromatography, the separation is only the starting point. We now need to determine, how much sample can be loaded onto the column until the separation is lost. Then we need to determine what column we need to obtain the desired product in the desired quantities.

15.3.1   Loading Studies

As a general rule of thumb for small molecules, a broadening of a peak due to mass overload is observed at a load of 0.1–1 mg of sample per milliliter of packed bed. However, if the separation is not critical, the mass load can be up to a factor of 10 higher. Also, if we want to separate a major compound from impurities, and if all impurities elute behind the parent peak, a significantly larger load can be applied.

In practice, the loadability of a column has to be studied empirically. It is also preferable to perform the loading study on a small column. This way, we do not waste a large amount of sample. Once the maximum load has been established on a small volume column, it should be scaled to a larger column capable of delivering the amount of pure compound that is desired.

The loading study is performed by steadily increasing the mass of sample injected. This is best performed with a sample that is dissolved in mobile phase at the highest concentration possible. From the standpoint of volume overload, it would be desirable to dissolve the sample in a solvent that is a weaker eluent than the mobile phase. Unfortunately, in most cases the solubility of the sample in such a solvent is lower than in the mobile phase. Conversely, a solvent that is a stronger eluent than the mobile phase usually leads to additional distortion of the peak.

A good approach to the loading study is to increase the amount injected by factors of 2. This is fine enough to find the point of maximum load but coarse enough to cover a large concentration range quickly. It is seldom necessary to perform measurements on the chromatogram. A visual inspection is often sufficient. An example of such a study is shown in Figure 15.9. The amount of sample, buspirone, was increased by factors of 2. One can see the loss of resolution in front of the peak as the amount of sample is increased beyond 2.4 mg for the scaling column. One also can see that the peaks behind the major peak are barely affected by the overload of the major compound, even up to a load of over 19 mg.

In many practical cases, the load can be increased significantly beyond the point where the resolution between the peaks is visually lost. This is due to displacement effects. In such a case, the high concentration of the major compound displaces the peaks that elute in front of this compound. The only way to determine whether this effect is taking place is to collect fractions and analyze the fractions. This is, unfortunately, a lot of work.

15.3.2   Scaling Laws

Once we understand how much sample can be loaded on a small column, we need to scale the separation to a larger column. All scaling laws assume a priori that the column chemistry remains constant. They also assume that the intrinsic performance (plate count) of the column to which the separation is scaled is either irrelevant (due to significant overload) or comparable to the scaling column. In most practical applications, the overload is so large that we can ignore the column performance.

All scaling of all parameters of the separation should be in proportion to the column volume. The sample load, in terms of both mass load and volume load, should be scaled in proportion to the column volume. Also, flow rates should be scaled in proportion to the column volume, at least when the scaling column and the preparative column contain the same particle size. Unfortunately, this is not always possible. Chromatographic instruments designed for preparative chromatography usually have a lower pressure capability than analytical instruments. In this case, a reduced flow rate is used, and one simply needs to accept the longer run times. Even if the flow rate cannot be scaled in proportion to the column volume, the load should still be scaled in proportion to the volume.

If the separation is a gradient separation, the same rules apply—the gradient volume should be scaled in proportion to the column volume. If we can scale the flow rate in proportion to the column volume without running into pressure limitations, we can also keep the times in the gradient table.

There is a small problem when scaling gradients, but with a well-designed separation it can usually be ignored. There is a delay between the time when the gradient is mixed and the time when the gradient reaches the column inlet. This delay is due to the volume of instrument parts between the point where the gradient is mixed and the column inlet. This volume is appropriately called the *gradient delay volume*. The effect of the delay volume is a time span during which the separation is carried out isocratically. This can affect the elution times of all peaks in the chromatogram and the position of peaks that elute early in the chromatogram. If the preparative separation is carried out on the

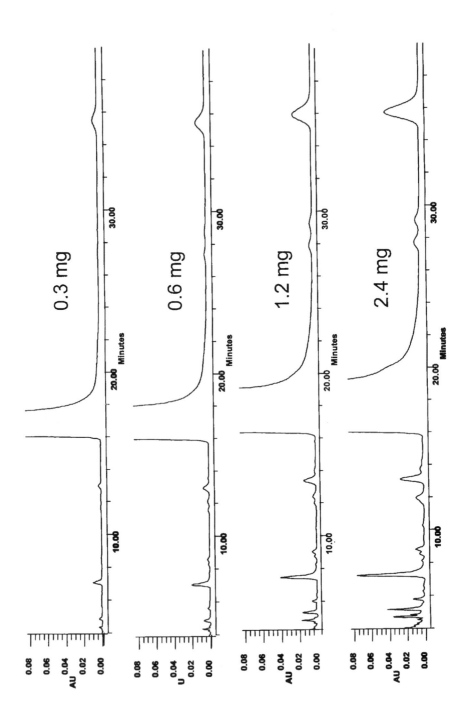

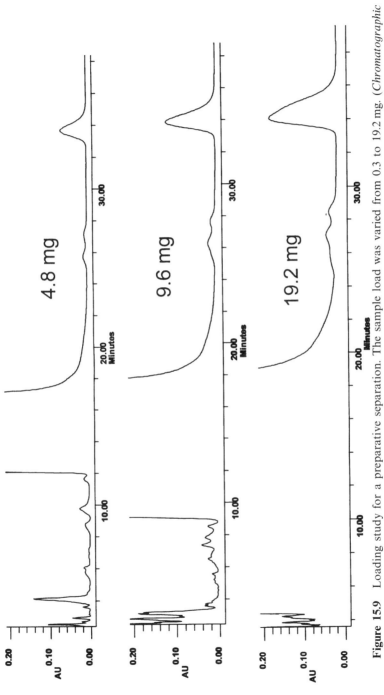

Figure 15.9 Loading study for a preparative separation. The sample load was varied from 0.3 to 19.2 mg. (*Chromatographic conditions:* Sample: buspirone. Column: 7 μm SymmetryPrep C_{18}, 3.9 mm × 150 mm, Waters Corp. Mobile phase: 28% acetonitrile/72% 0.18% triethylenetetramine acetate buffer, pH 7.0. Flow rate: 1 mL/min.) (Chromatograms courtesy of M. Z. El Fallah, Waters Corp.)

same instrument, the isocratic delay time will be less for the scaled-up separation than for the scaling separation. This effect is one of the major reasons for differences in the chromatography during the scaling of gradient separations. If this is detrimental to the separation on the preparative column, an additional isocratic time should be incorporated at the beginning of the gradient to simulate the gradient delay time observed with the scaling column. Unfortunately, this also results in an additional consumption of solvent, which is not desirable.

The formula for calculating this additional delay time t_d is given below, assuming that the same instrument is used for the scaling study and the preparative chromatography:

$$t_d = \frac{V_d}{F_{prep}} \left(\frac{V_{prep}}{V_{sc}} - 1 \right) \tag{15.20}$$

where V_d is the delay volume of the instrument, F_{prep} is the preparative flow rate, V_{prep} is the volume of the preparative column, and V_{sc} is the volume of the scaling column. The delay volume of an instrument can be measured by programming a step gradient from a solvent to another solvent with a small amount of a UV absorber (e.g., 1% acetone in methanol) and measuring the time from the start of the step gradient to the time that 50% of the step is observed in the detector. The time is multiplied by the flow rate to obtain the gradient delay volume. No column is connected to the instrument during this experiment.

15.3.3   Particle Size

If we scale a separation according to these rules, we expect the separation to be similar in resolution, but not necessarily in run time, due to instrument constraints. At the beginning of Section 15.3.2 we stated that we assume that the intrinsic column performance plays only a minor role in the separation due to column overload. This also implies that we can use a larger particle size for the preparative separation. The use of larger particles has two advantages: (1) lower pressure drop and (2) lower cost. However, there are some limitations to the use of larger particles, which depend on the overall resolution requirements of the separation. For most practical cases, a change in particle size by a factor of 2–3 (=a change in column efficiency 10) is tolerable. But larger changes in the intrinsic column performance should be checked carefully. In such a case it is better to obtain a scaling column packed with the same particle size as the target preparative column and to perform a loading study on this particle size.

Finally, you want to make sure that the surface chemistry of the preparative packing is identical to the surface chemistry of the packing used for the loading study. Ask the manufacturer for advice!

REFERENCES

1. G. Guiochon, S. Golshan-Shirazi, and A. Katti, *Fundamentals of Preparative and Nonlinear Chromatography*, Academic Press, Boston, 1994.
2. J. H. Knox and H. M. Pyper, *J. Chromatogr.* **363**, 1 (1986).
3. J. F. K. Huber and R. G. Gerritse, *J. Chromatogr.* **58**, 137 (1971).
4. W. Beck, dissertation, Universität des Saarlandes (1977).

ADDITIONAL READING

G. Guiochon, S. Golshan-Shirazi, and A. Katti, *Fundamentals of Preparative and Nonlinear Chromatography*, Academic Press, Boston, 1994.

16 HPLC Methods Development*

> The pattern juggler lifts his hand;
> the orchestra begin.
>
> — King Crimson, *In the Court of the Crimson King*

16.1   INTRODUCTION

HPLC method development, and more so, method validation, is an essential part of today's analytical work. HPLC has become a method of choice in the pharmaceutical laboratory, overshadowing most other analytical techniques. One can hardly imagine a drug introduced into the market today without the use of HPLC throughout the process of its discovery, development, and manufacturing. From basic research to the release of the final product, modern HPLC is interlaced in the chain of analyses more than any other technique. Its contribution is therefore unequivocal, and in a field as critical as human health, it adds to the quality of life. It withstands the costly yet necessary scrutiny of regulatory authorities. The application of HPLC to the analysis of drugs for human consumption is well on its way to a full worldwide harmonization of governmental regulations. HPLC derives its strength from its versatility, the ease with which it can be learned and applied, its ruggedness, and the relative ease with which it can be interfaced with a wide variety of detection schemes.

*This chapter was contributed by M. Zoubair El Fallah, LabMetrix Technologies, Inc.

Also, steady enhancements, especially the adaptation of other detection techniques, continuously add to its flexibility. The advent of mass spectrometry as a powerful detector significantly improves the applicability of modern HPLC, and NMR is on its way to become just another HPLC detector.

In this chapter, we will discuss the principles involved in the development of an HPLC method. For the majority of users, the final step of method development is the validation of the HPLC method. It is a time-consuming and laborious work, but it is a necessary safeguard that does make sense, by ensuring the specificity, ruggedness, and robustness of a given analytical technique, demonstrating its merits and defining its limits. This ultimately ensures the quality of the method. A well-thought-out method development strategy can generate a significant amount of information up front during the method development process, alleviating the amount of work necessary for the method validation process, and therefore saving both time and money.

16.2   THEORETICAL CONSIDERATIONS

No matter what the underlying chromatographic principle is, high-performance gradient elution chromatography always involves an increase of solvent strength with time, which is achieved by modifying the composition of the mobile phase that enters the column. The technique is used to accelerate the migration of compounds, and therefore to decrease their retention until they exit the column. The theory of gradient elution is fundamental to a successful method development strategy, since we will use gradient elution extensively to examine retention behavior, predict initial isocratic mobile-phase composition, or optimize gradient parameters. Gradients can also be employed to make a primary selection among different choices of chromatographic variables and guide us to the best parameters necessary to accomplish a given separation. Therefore, it is important that we discuss first the fundamentals of gradient chromatography. Snyder and co-workers (1) have developed the linear-solvent-strength theory in great detail and contributed significantly to its acceptance as the most widely used theory in reversed-phase gradient HPLC. Their most original contribution was the creation of fundamental descriptors for gradient elution chromatography that were analogous to parameters commonly used in isocratic elution mode. Readers who desire to learn more about their approach may consult one of their numerous books, reviews, or articles (2–5).

16.2.1   Linear-Solvent-Strength Theory

A linear-solvent-strength gradient is generated when the composition of the mobile phase changes with time such that the logarithm of the *hypothetical* retention factor k_h for a given component decreases linearly with time:

$$\ln(k_h) = \ln(k_a) - b_g \frac{t}{t_0} \tag{16.1}$$

where the hypothetical retention factor k_h is the retention factor of a solute assuming that had it been eluted isocratically using the mobile phase present at time t at the position of the solute in the column, k_a is the retention factor of a solute at the beginning of the gradient, and t_0 is the time the mobile phase requires to travel down the column. The gradient steepness b_g is a very important descriptor of the gradient.

If the gradient steepness is identical for all components in the injected mixture, this type of gradient offers a number of specific advantages:

- The band spreading is constant for all peaks throughout the chromatogram.
- The resolution and the effective plate count (see below) are similar for early- and late-eluting peaks.

In addition, gradients in general offer the following advantages compared to isocratic chromatography:

- Increased sensitivity, specially for late-eluting peaks
- More regular band spacing along the chromatogram

16.2.2   Characteristic Parameters in Linear-Solvent-Strength Gradients

The simple dependence of retention on time in linear-solvent-strength gradient chromatography makes it possible to obtain exact expressions for retention time and peak broadening as well as resolution. This enables us to analyze these parameters and use this information for the optimization of the gradient conditions. In the following, the expressions for gradient retention time, peak width, and resolution are discussed.

16.2.2.1   Retention Time   The retention time in gradient elution chromatography t_g is related to the gradient steepness factor b_g according to the following expression:

$$t_g = t_0 + \frac{t_0}{b_g} \ln(1 + k_a b_g) \tag{16.2}$$

In general, the initial retention in the starting mobile phase is very high, and this relationship simplifies to

$$\frac{t_g - t_0}{t_0} = \frac{\ln(k_a)}{b_g} \tag{16.3}$$

The left-hand side of this equation is analogous to the retention factor in isocratic elution mode.

16.2.2.2 Peak Width Two separate effects influence the peak width in gradient elution chromatography. On one hand, the migration velocity of the peak at the column exit is determined by the solvent composition at the moment when the peak leaves the column. Therefore, the peak width is to a first approximation the same as the peak width obtained isocratically at the solvent composition at the peak apex during the gradient run. However, with the progressive increase in solvent strength in gradient elution, an additional phenomenon plays a role in defining the ultimate peak width at the column outlet. During the elution process, a compression of the band occurs because the rear side of the peak, located in the stronger eluent, migrates faster than its earlier-eluting part. Therefore, the rear part of the peak is always in the process of catching up with the front part of the peak. This peak compression is characterized by a factor, G, which depends on the gradient steepness and on the plate count. The influence of this band compression increases as the gradient steepness parameter b_g increases. A side effect of this compression phenomenon is the fact that peaks that tail in isocratic mode become more symmetric in gradient elution. Peak compression may be used to enhance the detectability of compounds present in low concentrations.

If N is the theoretical plate count observed in isocratic elution, the peak standard deviation, σ_t (in time units) at the column outlet after a linear-solvent-strength gradient is given by

$$\sigma_t = G \frac{t_0}{\sqrt{N}} \left(1 + \frac{k_a}{1 + k_a b_g} \right) \tag{16.4}$$

For high values of k_a, the retention factor at the beginning of the gradient, we simply obtain

$$\sigma_t = G \frac{t_0}{\sqrt{N}} \left(1 + \frac{1}{b_g} \right) \tag{16.5}$$

This relationship illustrates that the peak width depends only on parameters related to the gradient (b_g, G, t_0) and column performance (N), and is independent on the initial retention (as long as $k_a > 10$, which is the case for most peaks in the gradient chromatogram).

16.2.2.3 ***Resolution***   The resolution factor plays an important role in chromatography as it expresses the quality of the separation between adjacent peaks. If t_1 and t_2 are the retention times of a peak pair, and σ_1 and σ_2 are their respective standard deviations, the resolution is defined by the following equation:

$$Rs = \frac{t_2 - t_1}{\sigma_2 + \sigma_1} \tag{16.6}$$

In the case of compounds that are highly retained at the start of the gradient, the combination of Equations (16.3) and (16.5) leads to

$$Rs = \frac{\sqrt{N}}{4} \frac{1}{G(1 + b_g)} \ln\left(\frac{k_{a,2}}{k_{a,1}}\right) \tag{16.7}$$

where $k_{a,2}$ and $k_{a,1}$ are the retention factors of the two compounds at the beginning of the gradient. When these values do not differ greatly, which is the case for closely related compounds, Equation (16.7) can be rewritten

$$Rs = \frac{\sqrt{NQ_g^2}}{4}\left(\frac{k_{a,2}}{k_{a,1}} - 1\right) \tag{16.8}$$

where Q_g is as given by the following relationship:

$$Q_g = \frac{1}{G(1 + b_g)} \tag{16.9}$$

Therefore, in a linear-solvent-strength gradient, NQ_g^2 is the equivalent to the theoretical effective plate count in isocratic chromatography. It is evident that the effective plate count and resolution in a gradient can be enhanced by decreasing the parameter b_g, the gradient steepness. However, this option for improving the resolution leads necessarily to an increase in analysis time (see Section 16.3). Hence, we are confronted with the classic compromise between resolution and analysis time, a well-known fact in isocratic chromatography.

However, it is possible to maximize the parameter NQ_g^2 at constant analysis time by varying the flow rate (=linear velocity). In this case, one simultaneously changes the plate count N and the gradient steepness b_g. Because of the influence of the factor Q_g, one obtains an optimum at linear velocities higher than those that correspond to the maximum plate count in isocratic chromatography (see also the section on gradient optimization).

It should be pointed out that the analysis shown here is valid only if the gradient steepness parameter is the same for the neighboring peaks under consideration. If this were not the case, no simple general conclusions can be drawn for the dependence of resolution on gradient parameters.

16.2.3 Relationship between Solvent Strength and Mobile-Phase Composition

The time dependence of the mobile-phase composition required to realize a linear-solvent-strength gradient depends on the retention mechanism. In the case of reversed-phase chromatography, a linear relationship between the logarithm of the retention factor and the volume fraction φ_B of an organic modifier B has been widely observed (2,3) in aqueous binary mobile-phase systems:

$$\ln(k) = \ln(k_w) - S_r\varphi_B \tag{16.10}$$

where k_w is the solute retention factor in pure water, while S_r is a parameter directly related to the strength of the organic solvent and the nature of the solute. Larger S_r values lead to a faster decrease in the retention factor as the organic volume fraction is increased. For small molecules, S_r values for methanol and acetonitrile are generally in the range of 7–10. In fact, several studies have demonstrated that S_r depends on the nature of the solute. The most important factor is the molecular weight of the solute. The larger the molecular weight, the larger is the factor S_r. Biomolecules are known to exhibit very high S_r values. Note that some authors may express Equation (16.10) in log instead of ln (natural logarithms), and therefore S_r values from their work are 2.3 times smaller than those reported here. The same holds for the parameter b_g.

The combination of Equations (16.1) and (16.10) can be used to obtain the modifier composition profile needed to generate a linear-solvent-strength gradient in reversed-phase chromatography:

$$\varphi_B = \frac{\ln(k_w) - \ln(k_a)}{S_r} + \frac{b_g}{S_r}\frac{t}{t_0} \tag{16.11}$$

Therefore, the volume fraction of the mobile phase needs to increase linearly with time in reversed-phase chromatography.

When $\Delta\varphi_B$ is the difference between the mobile-phase composition at the beginning and the end of a gradient program and t_G is the gradient duration, the gradient steepness factor, b_g, can be expressed using the gradient parameters according to the following relationship:

$$b_g = \Delta\varphi_B S_r \frac{t_0}{t_G} = \Delta\varphi_B S_r \frac{V_0}{Ft_G} \tag{16.12}$$

where V_0 is the column void volume and F the mobile-phase volumetric flow rate. Alternatively, $S_r\Delta\varphi_B$ is related to the solute retention factors, k_a and k_b, in the initial and final mobile phases, respectively:

$$b_g = \frac{V_0}{Ft_G}\ln\left(\frac{k_a}{k_b}\right) \tag{16.13}$$

Equations (16.12) and (16.13) are very important, since they easily can be used to predict the influence of any operational parameter on the steepness factor, b_g, and therefore on the analysis time, efficiency, and resolution. However, they are based on the validity of Equation (16.10). It has been shown that some deviations occur for some compounds and chromatographic systems (6), especially when retention is not governed solely by hydrophobic interaction. This is, for example, the case when the solutes are strongly basic and the stationary-phase acidity is high. Nevertheless, it is always possible to modify the form of the mobile-phase variation with time in order to maintain the applicability of the linear-solvent-strength theory [Equation (16.1)]. As we have seen above, this type of gradient offers a considerable help in the fundamental understanding of the retention behavior of the solutes and in the optimization of a separation.

16.2.4   Measuring the Quality of Separation between Adjacent Peaks

Many criteria can be used to assess the quality of a separation. They include measures of the separation between adjacent peaks, analysis time, peak spacing, and the efficient use of the chromatographic space. In the following paragraphs, we will discuss several criteria that can be used to assess the quality of the separation between adjacent peaks, which is the most fundamental yardstick for the quality of a separation.

16.2.4.1 Resolution   The resolution factor Rs, also simply called *resolution*, is the most popular criterion used to measure the quality of the separation between adjacent peaks. Its definition is given by Equation (16.6). It has the advantage of being simple (7). The criterion $Rs > 1$ is fulfilled when the two tangents at the peak inflection points cross at or below the baseline, or when the peaks are "baseline-separated." Whether this criterion is fulfilled can easily be assessed by visual inspection without the need of an actual measurement. In addition, the resolution factor has the advantage of been directly related to the fundamental parameters of retention, efficiency, and selectivity. The relationship linking these parameters together is widely used to predict their influence on resolution.

The resolution factor is usually estimated from the peak retention times and widths observed in a chromatogram of a mixture of solutes. However, in a rigorous way, a more accurate estimation requires the separate injection of the individual compounds. This is particularly true for closely eluting peaks. It has been established that the retention times measured at the peak apex, and more so the peak widths, are different if the measurement is made on individually injected solutes or on the peaks in a mixture. This difference is more pronounced when the peak shape cannot be described simply by a Gaussian profile, and where the center of gravity of the peak does not correspond to the peak apex (8). Nevertheless, the chief drawback of the resolution factor Rs is the fact that it does not take into account the relative peak height (9) of the adjacent peaks. It is quite possible that one may exceed what is considered to be good resolution without being able to see a separation. The phenomenon becomes more severe as the difference in peak height increases. This fact is well known and has been studied through simulations (10). Therefore, for a given peak height ratio there is a critical resolution Rs^* that must be achieved in order to first see the presence of two peaks and then to perform accurate quantitation. This fact has significant practical implications, for example, in pharmaceutical analysis for impurity profiling in the drug substance or product. In this case, it is necessary to characterize and set up limits for impurities present in concentrations as low as 0.1% of the main drug product, which corresponds to a peak height ratio of $\sim 1 : 1000$. One must keep in mind that the minimum resolution requirement often reaches 2.0 in such a situation (see below).

16.2.4.2   Other Separation Criteria

Several criteria for measuring separation quality have been proposed (11). We can cite here the purity content by Glueckauf (12), the enrichment factor by Boyde (13), the extent of separation by Rony (14,15), and the parameter p introduced by Kaiser, alternatively called "Auflösung" (= resolution) (16) or separation power (17). This criterion, which is easily estimated from a graphic construction varies between 0 and 1, and increases as the resolution between the peaks increases. It is sensitive to the peak-height ratio and to the peak shape. It has also been used in several methods development procedures as a criterion for optimizing chromato-graphic response functions (18). Another criterion, the valley to peak ratio V/P, which is somewhat similar to Kaiser's factor, has been proposed by Christophe (19).

In a comparative study of different criteria used to measure separation quality, Debets et al. (20) concluded that each one of these criteria suffered from a specific limitation. Nevertheless, it seems that criteria that are sensitive to the relative peak height and peak shape performed better in assessing overall chromatographic performance.

16.2.4.3   The Discrimination Factor

El Fallah and Martin (21) introduced a separation factor called the discrimination factor d_0, which uses the height of the smaller peak in the pair h_p and the height at the valley h_v to measure the extent of separation between adjacent peaks (see Fig. 16.1). This factor is similar to the criteria introduced by Kaiser (p) and Christophe (V/P), since it depends on the ratio of the height of the valley to the peak height. It varies between 0 and 1 and increases when the separation improves. It is calculated using the following formula:

$$d_0 = \frac{h_p - h_v}{h_p} \tag{16.14}$$

This factor is easily calculated from the chromatogram. Contrary to Kaiser's factor p, it does not require any graphic construction. Therefore, it can be

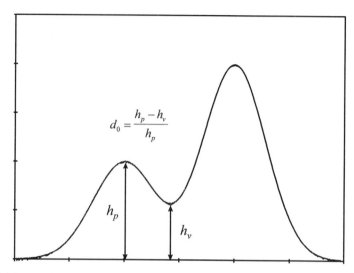

Figure 16.1   Explanation of the discrimination factor d_0. Measured is the ratio of the distance between the apex of the smaller peak and the valley to the height of the smaller peak

calculated easily by computer-based data acquisition and reporting systems. More importantly, it is sensitive to the peak shape and the relative height of the adjacent peaks. When the discrimination factor d_0 approaches 1, an increase in the distance between peak pairs results in a very slow increase of the factor. This has a significant advantage in method development. Criteria of separation quality, which approach a limiting value once we reach a baseline resolution, are a good representation of the goal of our separation. This is not the case with the resolution Rs, which increases indefinitely with increasing separation between the peaks. However, this additional increase affects only the total analysis time, with no real enhancement in the method. In addition, the required resolution Rs needed to achieve a satisfactory separation depends on the relative height of the peaks. Therefore, the optimal condition for the separation is to achieve a discrimination factor close to unity, rather than to reach or exceed a given resolution. The discrimination factor d_0 is a better measure of separation quality than the resolution Rs.

Since Rs is the factor that is most widely used to measure separation quality, it is worthwhile to compare it to the discrimination factor. For peaks exhibiting the same height, and assuming that the observed retention times are very close to the retention times of the individual components (22), the combination of Equations (16.6) and (16.14) leads to the following relationship:

$$d_0 = 1 - 2\exp(-2Rs^2) \qquad (16.15)$$

This relationship is only an approximation, and the deviation increases as the peaks become less resolved. It was used to estimate the resolution at which the valley between peaks disappears ($d_0 = 0$). Equation (16.15) yields $Rs = 0.59$ for $d_0 = 0$ (23), while the correct value should be 0.5 (24). This discrepancy is actually due to the fact that Equation (16.15) leads to a significant error when the peaks are barely resolved (i.e., $d_0 < 0.1$). Figure 16.2 shows the dependence of the critical resolution Rs^* needed to observe a valley between two adjacent peaks on the relative peak height (or peak height ratio) of these peaks, h_{rp}. We clearly see that the resolution necessary to distinguish a peak valley increases significantly when the relative peak height increases. This is why it is important to define the required resolution on the basis of the expected concentrations of the analytes. This is especially true when the method development is performed using standards in concentrations that do not reflect the concentrations in real samples. From the upper curve of Figure 16.2, we see that if one needs to separate peaks of a height ratio of 1 : 1000, which is the case in pharmaceutical laboratories, a resolution of more than 1.2 is needed just to observe the existence of the smaller peak.

Figure 16.3 shows the dependence of the resolution Rs on the discrimination factor d_0 for different values of the relative peak height h_{rp}. Of course, the

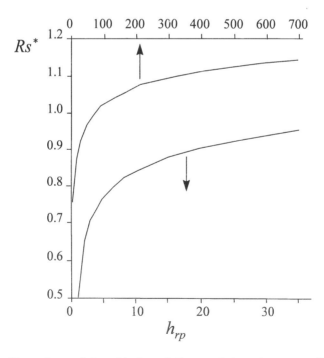

Figure 16.2 Dependence of the critical resolution needed to observe a valley between two adjacent peaks on the relative height of the peaks. The lower curve belongs to the lower axis and the upper curve, to the upper axis. The lower axis is a 20-fold expansion of the upper axis.

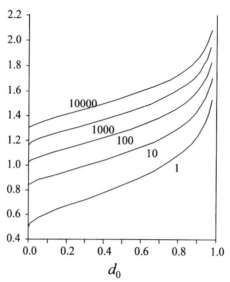

Figure 16.3 Dependence of the resolution Rs on the discrimination factor d_0 for different values of the relative peak height h_{rp}.

resolution increases with d_0. First, it increases rapidly for low values of d_0, then almost linearly with it, and then it approaches asymptotically a vertical line at $d_0 = 1$. We can clearly see that for a 1:1000 peak height ratio, a resolution of 2.0 is necessary to get an adequate valley between the peaks for reliable quantitation.

The discrimination factor has also been used in error analysis in quantitative HPLC (22). Since resolution is not sensitive to relative peak height and the error in quantitation depends strongly on the location of the valley between the peaks, d_0 is a valid parameter for characterizing errors in quantitative analysis. The error can be expressed using the fractional area F_a and fractional height F_h for the smallest peak i (the one for which the error is the highest):

$$F_a = \frac{A_{p,i}}{A_{p,i0}} \tag{16.16a}$$

$$F_h = \frac{h_{p,i}}{h_{p,i0}} \tag{16.16b}$$

where $A_{p,i}$ and $h_{p,i}$ are the area and height of compound i, respectively, as measured from the chromatogram of the mixture, and $A_{p,i0}$ and $h_{p,i0}$ are respectively the area and height of compound i when injected separately. The fractional values are equal to 1, if no quantitation error occurs. F_a and F_h were calculated as a function of the discrimination factor d_0, and for several peak height ratios. Figures 16.4a,b show this dependence.

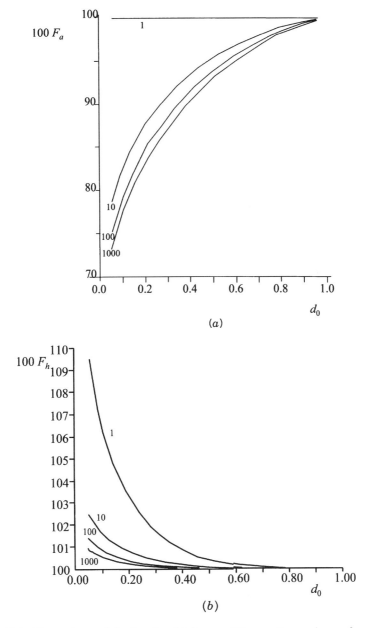

Figure 16.4   Dependence of the fractional (*a*) area of the smaller peak as a function of the discrimination factor d_0 for the different values of the relative peak height indicated near the curves and (*b*) height of the smaller peak as a function of the discrimination factor d_0 for the different values of the relative peak height indicated near the curves. The *y* axis is $100 \times F_h$.

It is clear that, for a given peak height ratio, the quantitation errors in both height and area increase as d_0 decreases. The error is positive for the height and negative for the area. However, the comparison between Figures 16.4a and 16.4b reveals that for a given relative height and discrimination factor, the absolute error of the height measurement decreases while the absolute error of the area measurement increases, when the relative peak height increases. We also observe that for low values of d_0 and relative peak height, the error made by measuring the area is smaller than the error made by height measurement. However, when the peak height ratio increases, the situation is reversed. The choice between area and height measurements for quantitation is therefore dependent on the relative peak height. For peaks with similar peak height, area measurements perform better, whereas height measurements have a clear advantage when the peak heights are significantly different.

These recommendations, however, are based on Gaussian peak profiles, and do not take into account other factors, such as noise, flow-rate fluctuations, and the detection mode, which may have a significant influence on the measurements of area and height.

16.2.5   Basis of Mobile-Phase Selectivity in HPLC

One of the most powerful features of HPLC stems from the fact that the eluent is not merely a transport medium, it rather contributes significantly to the mechanism of separation. Retention as well as selectivity arise from the combined action of mobile and stationary phase on the solutes. Because of its distinct physicochemical characteristics, a given solvent interacts in a specific manner with the solutes. Its capacity to donate or accept protons or to induce a dipole moment defines the nature of its interaction with the solutes in solution. Slight differences in these interactions complemented by stationary-phase action are often enough to provide the desired selectivity in HPLC. Solvent classification based on their elution strength or "polarity" represents a classic area of investigation into the fundamentals of retention mechanisms in HPLC (25–27), and the chapter is not yet closed on its development and refinement (28–30). Here we will present a practical and comprehensive way to exploit the effect that the nature of the solvent has on retention and selectivity in reversed phase HPLC.

There are many empirical scales that describe solvent polarity. Snyder (27) proposed a selectivity triangle that groups solvents into eight groups according to the contribution of proton acceptor, donor, and dipole characters to their overall strength. This approach is similar to that developed by Rohrschneider, who attempted to classify GC stationary phases as well as volatile solvents (31). Each solvent is characterized by three contributing parameters: a proton acceptor, X_e; a proton donor, X_d; and a dipole parameter, X_n. For a given solvent, the sum of the X values is 1. A solvent polarity parameter, P', was defined as the sum of the logarithm of the retention factors of test solutes (ethanol, dioxane, and nitrobenzene). Using this definition, P' for mixed solvents can be estimated using the following relationship:

$$P' = \sum_i \varphi_i P'_i \qquad (16.17)$$

where P'_i is the polarity of solvent i and φ_i its volume fraction in the solvent mixture. Mixtures exhibiting the same P' values are called *isoeluotropic*, and are supposed to provide the same retention range. This description of polarity has its own limitations (29,32).

This polarity scale was first developed for adsorption chromatography, and it can be extended to reversed-phase chromatography by assigning 0 to the P' value of water while neat methanol, acetonitrile, and tetrahydrofuran have P' values of 2.6, 3.2, and 4.5, respectively (33). This amounts to a significant reduction of the original selectivity triangle, which is a consequence of the fact that miscibility with water is a prerequisite for the selection of a solvent in reversed-phase HPLC.

Another more rigorous solvent classification is based on spectroscopic measurements of test solutes that are sensitive to the physicochemical characteristics of a solvent. The transition energy E_T of the maximum absorption of a solvatochromic indicator is related to solvent properties. This approach defines the so-called solvatochromic theory (34–36). The most widely used indicator is 2,6-diphenyl-(2,4,6-triphenyl-N-pyridino)-phenolate, and the scale associated with it is denoted as E_T (30). The drawback of this scale is that it responds to changes in both dipolarity/polarizability and hydrogen-bonding strength (37). An empirical multiparameter scale developed by Kamlet et al. (37,38), called the *Kamlet–Taft scale*, provides a differential contribution from solvent dipolarity/polarizability, hydrogen-bond donating (acidity) and accepting (basicity). It seems to be the most meaningful way to describe solvent effects in HPLC.

In practice, there is only a limited choice of solvents for optimizing selectivity in reversed-phase HPLC. Methanol, acetonitrile, and tetrahydrofuran generally provide enough selectivity changes to deal with most samples, and the use of pH may amplify these differences in selectivity as well. It is helpful to draw some general rules that can help in replacing one solvent with another without affecting retention significantly. We found that two sets of relationships provide reasonably accurate transfer rules from one solvent to another. The first set was proposed by Schoenmakers et al. (39):

$$\varphi_{\text{MeCN}} = 0.32\varphi_{\text{MeOH}}^2 + 0.57\varphi_{\text{MeOH}} \qquad (16.18a)$$

$$\varphi_{\text{THF}} = 0.66\varphi_{\text{MeOH}} \qquad (16.18b)$$

while the second set of transfer rules was proposed by Haddad et al.:

$$\varphi_{\text{MeCN}} = 0.90\varphi_{\text{MeOH}}^2 - 0.21\varphi_{\text{MeOH}} + 0.126 \qquad (16.19a)$$

$$\varphi_{\text{THF}} = 0.66\varphi_{\text{MeOH}} - 0.15 \qquad (16.19b)$$

The set of rules proposed by Haddad seems to perform better with ionizable compounds.

On the basis of our experience and other data in the literature, THF is the solvent of choice when one wants a drastic change in selectivity. Unfortunately, its UV cutoff point is relatively high, and aqueous mixtures exhibit a high viscosity. However, the addition of a few percent of THF to the mobile phase is enough to obtain significant changes in selectivity. Yet, one must be concerned about the ruggedness of the method, as a slight variation in THF content can produce considerable shifts in retention times.

16.3   METHOD DEVELOPMENT STRATEGY

We will now proceed to discuss method development strategies. First, we will contemplate the choice of isocratic or gradient methods. Then we will develop an efficient method development strategy. The focus of the method development is the analysis of low-molecular-weight ionizable compounds by reversed-phase HPLC.

16.3.1   Considerations in the Choice of Isocratic versus Gradient Elution

The choice between an isocratic and a gradient method as the target of method development is probably one of the most controversial questions. As we will see in the following discussion, the answer to this question depends primarily on the sample and matrix to be analyzed. For relatively simple samples, gradient elution is easy to optimize. However, it has the following limitations and constraints:

- It requires an instrument with gradient capabilities.
- Because of inherent differences in the instrumentation, method transfer as well as validation is more problematic than for isocratic methods.
- It cannot be used with refractive index detectors.
- Mobile phases need to be sparged or degassed using an on-line vacuum degassing system because of bubble formation when mixing water with organic solvents.
- The baseline may change during the analysis, when UV absorbing modifiers are used.

The latter problem drastically limits the choice of solvents and modifiers. However, when sensitivity is needed for the detection of solutes present at low levels, gradient chromatography certainly has a clear advantage over the isocratic mode. Nevertheless, the decision to choose one mode over the other has to be made in a rational way. In the following, we will discuss several criteria that will help us in this choice.

In isocratic mode, the peak capacity of a column, P_c, is the theoretical number of peaks that can be separated with a given resolution within a given analysis time. In isocratic chromatography, it is given by the following relationship (40,41):

$$P_c = 1 + \frac{\sqrt{N}}{4Rs} \ln(1 + k_{max}) \tag{16.20}$$

where N is column efficiency, Rs is the resolution between consecutive peak pairs, and k_{max} is the retention factor of the last-eluting peak. The peak capacity increases with column efficiency, and of course with time. Efficiency is the predominant factor determining the peak capacity, while the increase of peak capacity with time is rather slow. However, since in isocratic mode later-eluting peaks are more diluted than early-eluting peaks, one cannot rely indefinitely on longer retention times to increase peak capacity and therefore separate more peaks. For a Gaussian peak profile, and assuming that band spreading is caused only by the column, the concentration c_{max} at the peak apex is related to parameters of retention and column performance according to the following relationship (42):

$$c_{max} = c_0 \frac{V_i}{V_M} \sqrt{\frac{N}{2\pi}} \frac{1}{1 + k_{max}} \tag{16.21}$$

where c_0 is the solute concentration in the sample matrix, V_i is the injection volume, and V_M is the column void volume. For the last peak to be detected or quantified, the signal-to-noise ratio of the last peak, sn_l, is given according to the following relationship:

$$sn_l = sn_0 \frac{V_i}{V_M} \sqrt{\frac{N}{2\pi}} \frac{1}{1 + k_{max}} \tag{16.22}$$

where sn_0 is the signal-to-noise ratio that one would have obtained for the solute l if it would have been injected directly into the detector without dilution through the column. It depends on the concentration of the solute l in the original sample and its detector response.

If the signal-to-noise ratio of the last-eluting peak must exceed a certain value, say, 10, for quantitative analysis, it is possible to calculate the maximum number of peaks that can, in theory, be separated under isocratic elution conditions by combining Equations (16.20) and (16.22):

$$P_c = 1 + \frac{\sqrt{4}}{4Rs} \ln\left(\frac{sn_0}{sn_l} \frac{V_i}{V_M} \sqrt{\frac{N}{2\pi\gamma}}\right) \tag{16.23}$$

As one might expect, this number increases with the column efficiency, and to a lesser extent with the injected mass ($sn_0 V_i$). However, in practice, one has a requirement to quantitate or at least to detect a certain number n_m of components in the sample matrix, and needs to know what is the concentration and response requirement sn_0 for the last-eluting peak that still gives a signal of 3 times the noise (for detection) or 10 times the noise (for quantitative determination) at the column outlet. We can answer this question by using the following relationship:

$$sn_0 = sn_l \frac{V_M}{V_i} \sqrt{\frac{2\pi}{N}} \exp\left[\frac{4Rs(n_m - 1)}{\sqrt{N}}\right] \qquad (16.24)$$

Figure 16.5 shows the requirements for sn_0, the concentration and response of the last-eluting solute in the sample matrix, for detection or quantitation of the peak, as a function of the number of peaks in the mixture. In theory, if the response of the last peak in the original mixture is 20 times the noise, one can detect this peak among 42 peaks and perform its quantitation in a mixture of 27 peaks, all separated with a resolution of 2. These numbers are quite high, but they are misleading since they suggest that isocratic elution could solve most separation problems. One has to remember that even after the most successful optimization, the majority of the chromatographic space is empty (43), while the numbers derived above assume that the peaks are perfectly lined up with a constant resolution between all peak pairs. This is, of course, unrealistic.

Let us examine the impact of the existence of unoccupied chromatographic space on the concentration/response requirement, sn_0. Since an additional

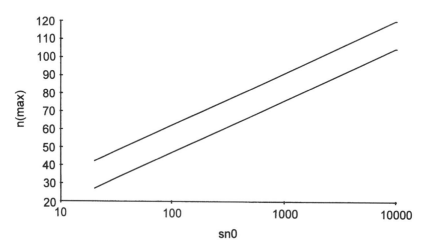

Figure 16.5 Relationship between the maximum number of peaks that can be detected (upper line) and quantitated (lower line) and the concentration and response requirement sn_0 of the last eluting analyte in the sample ($Rs = 2$, $N = 10,000$).

increase in retention does not lead to a proportional increase in the number of peaks that can theoretically be resolved, the fraction of unused chromato-graphic space to the total retention time, γ_t, is not proportional to the fraction of the theoretical number of peaks resolved in the used chromatographic space, γ_n. The latter increases linearly with the logarithm of γ_t, and the slope and origin of this variation are dependent on the total retention time (or k_{max}), according to the following relationship:

$$\gamma_n = O(k_{max}) + P(k_{max}) \ln(\gamma_t) \qquad (16.25)$$

Figure 16.6 shows the variation of γ_n as a function of γ_t, for k_{max} values of 25, 50, and 100. If one assumes that only a fifth of the chromatographic space is used ($\gamma_t = 0.2$), we find that γ_n is 0.47 and 0.43 for k_{max} values of 25 and 50, respectively. This means that the number of peaks that can be resolved theoretically within the given retention space decreases by more that 50%. Note that γ_t can be used as a criterion to measure the success of optimization procedures, since ideally it should approach 1 when the whole chromatographic space is used while maintaining adequate resolution between consecutive peak pairs.

If we now go back to the example above where sn_0 was assumed to be 20 times the noise, we find that isocratic elution can be used to detect up to 21 peaks and quantitate about 13 peaks. This calculation assumes a column with 10,000 theoretical plates and a minimum resolution of 2 between peak pairs. Of course, as the initial signal-to-noise ratio of the last peak increases, the number of peaks one can possibly resolve increases, as discussed above.

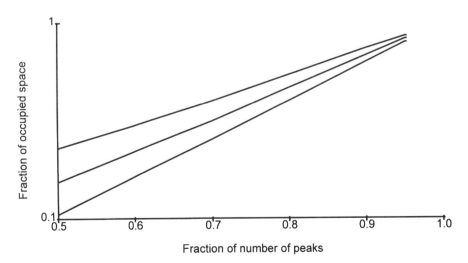

Figure 16.6 Dependence of the number of resolved peaks γ_n on the fraction of occupied chromatographic space γ_t for values of the maximum retention factor k_{max} of 25, 50, and 100.

In summary, we can draw some general conclusions about the usefulness and limits of the isocratic elution mode:

1. Although isocratic elution is preferred because of its ruggedness, ease of method transfer and validation, and the use of cost-effective instrumentation, it may have severe limitations in sensitivity. These limitations depend ultimately on the complexity of the sample matrix, specifically, the number of components to be analyzed as well as their relative concentrations and response factors.

2. In practice, the concentration of minor components in the sample matrix is low, as in the case of impurity profiling in the drug product and/or substance. If one assumes a response of 1 AU for the parent drug peak, impurities that have an initial response of 10^{-3} AU need to be quantitated. With current UV detectors, the noise is around 5×10^{-5} AU. This means that the initial signal-to-noise ratio sn_0 of an impurity with 0.1% relative concentration is around 50. In principle, the maximum number of peaks (if one assumes γ_t around 1) that can be separated with a resolution of 2 between consecutive peak pairs is 38 under these conditions. However, as discussed above, this number is more realistically around 18 peaks for a column of 10,000 plates, when taking into account the existence of unoccupied chromatographic space.

3. One has to emphasize that it is possible to increase the separation power of isocratic elution by carefully optimizing the injection conditions (i.e., the injection volume), the sample concentration, and the composition of the sample solvent. Equation (16.23) suggests that the number of peaks that can be separated using isocratic elution increases logarithmically with the injection volume. However, because of column saturation and/or excessive loss in resolution, the injection volume also cannot be increased indefinitely. A search for the optimum column volume is of utmost importance in enhancing sensitivity. Injection volume considerations will be discussed below.

In contrast to the isocratic elution mode, typical gradient profiles have the advantage of providing far superior peak capacity per unit time. The peak capacity $P_{c,g}$ in a linear-solvent-strength gradient is given by (43)

$$P_{c,g} = 1 + \frac{\Delta t}{4\sigma_t Rs} \tag{16.26}$$

where Δt is the analysis time for which the peak capacity is calculated and σ_t is the peak standard deviation in time units at the given gradient operating conditions, namely, flow rate and gradient steepness parameters [see Equation (16.5)]. This increase in peak capacity is in direct proportion to analysis time, which is due to the fact that the peak width remains approximately constant throughout the gradient. This property makes gradient techniques very useful when high sensitivity is required as in the quantitation of low-level impurities or when complex matrices need to be resolved. Gradient elution should be used in the separation of mixtures that exhibit a wide range of polarity, which will certainly save a considerable time over isocratic chromatography. Also in the case of biopolymers (peptides and proteins), where retention is governed by an "on/off" mechanism, gradient elution is the only acceptable mode of chromatography.

On the basis of Equation (16.26), the peak capacity can be increased by increasing retention time. However, this approach is not very attractive since the analysis will suffer from lengthy retention time and higher detection limits. However, a careful look at the influence of the operating parameters suggests that the peak capacity per unit time will exhibit an optimum at a particular flow rate. This fact will be discussed in the section dealing with the optimization of gradients.

16.3.2 Developing Isocratic Separations

To a large extent, method development is still conducted on the basis of trial and error, although software packages are commercially available that are designed to assist rational method development (see below). The resistance that these software packages have encountered is most likely due to the fact that chromatographers prefer to rely on their visual judgment and experience rather than delegate the process of optimization and chromatographic quality assessment to a sophisticated software program. In addition, one also has to invest time in understanding the theory, assumptions, and limitations that are built into the software, along with the experimental design approach and the methodology used to assess separation quality. A software package is most likely to follow a sound experimental design and decision making; however, knowledgeable chromatographers may require fewer experiments to reach the same optimum as that of a computer-assisted optimization (44). A practical approach to method development would take advantage of the chromatographer's intuition and experience as well as the theory of chromatographic retention and selectivity.

When multivariate optimization is not carried out, meaning that each parameter is optimized sequentially, one has to address the most significant parameter first. This is the operating parameter that produces the largest changes in selectivity. Therefore, it is important to identify the relative importance of parameters that affect retention and selectivity. It is understood that the nature of the column, namely C_{18} versus cyano or phenyl, or even within C_{18} from different manufacturers, is very important in defining selectivity, but it is rather expensive to start the optimization by systematically changing the nature of the stationary phase. Perhaps having two columns with selectivities as different as possible is acceptable, and changing the ligand nature may be used within an overall strategy. However, mobile-phase parameters are easier to change. For a given column, the selectivity and separation quality can be affected by several factors:

- Eluent composition and nature of the solvent(s)
- Mobile phase pH
- Ionic strength
- Temperature

The relative importance of these parameters is solute-dependent, and knowledge of the matrix to be analyzed must be used when available. In pharmaceutical applications, most reversed-phase columns were generally used at acidic mobile phases (i.e., pH 3.0). This stems from the fact that, until recently, silica-based reversed-phase columns exhibited a strong affinity for basic compounds, making it difficult to carry out the separation at neutral pH values, where the activity of surface silanols is very strong and often not reproducible. This lack of reproducibility is caused by a wide, and often bimodal, energy distribution of the free silanols. Some free silanols, although not abundant on the silica surface, may exhibit a strong interaction energy that is characterized by higher retention and easy saturation of these sites. The combination of highly active silanols and the heterogeneity of their distribution leads to a strong peak tailing and retention shifts that are unacceptable when method ruggedness and low detection limits are needed. Often, silanol shielding additives, primarily small aliphatic amines, are a requirement for the use of neutral pH mobile phases, which often is the only way to achieve the desired selectivity.

Recently, high-purity silica became commercially available. For stationary-phases bases on such a silica, the problems mentioned above are minimized, making it possible to use the whole working pH range of silica-based packings (pH 2–8). Since the majority of pharmaceutical compounds are ionizable and/or exhibit basic character, pH is often the most powerful operating parameter, and should be investigated first. Another compelling reason for optimizing the mobile phase pH first is that its change does not need to affect all the compounds in the mixture. If only some of the compounds in the sample matrix are sensitive to pH changes while others are not, significant selectivity changes occur readily. Moreover, changes produced by modifying the mobile phase strength change the retention of all compounds in the same direction, while pH changes may lead to an increase in the retention of some compounds, leaving others unaffected, while increasing the retention of another class of compounds. For ionizable compounds, the silanols and their ionization states, through pH changes, may also play a major role in achieving the desired selectivity. Ionic interactions as well as hydrogen bonding between silanols and solutes can amplify the changes produced by modification of the mobile-phase pH, and therefore induce additional degrees of freedom for manipulation of the separation. This ability to easily affect retention and selectivity is of utmost importance in developing HPLC methods for pharmaceutical compounds.

Therefore, for ionizable compounds, the variation of pH is a primary tool for manipulating the selectivity of a separation.

For ionizable compounds, the ionic strength of the buffer is also an important parameter to consider at an early stage of the optimization. One should scout its effect by using low and high buffer concentrations. In other cases, where the goal is to separate structurally related compounds, such as positional isomers, increased retention is often the key to a successful separation, and it is advisable to investigate solvent composition and selectivity first.

We will focus here on the case where at least a portion of the sample comprises ionizable compounds, which is the case for most pharmaceutical samples. Since it is a priori difficult to predict the initial isocratic mobile-phase composition that would give adequate retention for all peaks, it is best to perform initially a slow gradient run with a relatively weak starting mobile phase (e.g., 10% methanol), to ensure that early eluting peaks are retained. This is especially important when dealing with unknown mixtures, for example, degradation products. However, to investigate the effect of pH on the separation, one needs to perform two gradient runs: one at a low pH value (around 2) and another at neutral pH (around 7). These two runs must have otherwise the same gradient operating conditions (initial mobile phase and gradient steepness). These gradient runs are inspected, and on the basis of these two runs, the best mobile-phase pH is selected. This selection involves the judgment of the chromatographer and is based on the overall quality of the separation, including the resolution and the peak spacing for all compounds of interest.

Once the pH is selected, and assuming that an isocratic elution can separate the mixture at hand (see discussion above), the gradient retention data can now be used to derive the appropriate isocratic composition. One way of performing this task is to assume that the slope S_r of the dependence of the logarithm of the retention factor on solvent composition is known. Although this slope does depend on the nature of the solute and its molecular weight, it is adequate to assume a value of 8 for small molecules as a first approximation. Since one of the important objectives of optimization is the total analysis time, the retention factor for the last compound should not exceed 10 for simple matrices. This requirement may be adjusted depending on the nature of the sample. Therefore, one needs to find out what organic modifier composition leads to the desired retention factor, and therefore total analysis time.

The necessary information can be obtained for every analyte i from its corrected gradient retention time $t_{gc,i}$:

$$t_{gc,i} = \frac{1}{S_r \beta_g} \ln[1 + S_r \beta_g t_0 k_a(\varphi_{B,0})] \qquad (16.27)$$

where $\varphi_{B,0}$ is the mobile phase fraction at the beginning of the gradient, S_r is the assumed slope of the relationship between the logarithm of the retention factor and the solvent composition, and β_g is the rate of change of the mobile-phase composition. The corrected gradient retention time is calculated from the measured gradient retention time t_g as follows:

$$t_{gc} = t_g - t_0 - t_d \qquad (16.28)$$

where t_0 is the retention time of an unretained peak and t_d is the gradient delay time of the instrument. When low-pressure gradient systems are used, there is a delay in the delivery of the gradient to the column head. The delay volume, also called *dwell volume*, can be estimated using a step gradient from reservoir A to B, where B contains a solvent spiked with an appropriate UV absorber (i.e., acetone) while solvent A is the neat solvent.

Using the assumed S_r value for the last peak, one can estimate its retention factor in the neat aqueous phase $k_{w,L}$ with the following equation:

$$k_{w,L} = \frac{1}{S_r \beta_g t_0} [\exp(S_r \beta_g t_{gc,L} + S_r \varphi_{B,0}) - \exp(S_r \varphi_{B,0})] \qquad (16.29)$$

where $t_{gc,L}$ is the corrected gradient retention time for the last-eluting peak. We can now estimate the mobile-phase composition φ_B that gives us the desired retention factor k_L for the last peak with the following simple relationship:

$$\varphi_B(k_L) = \frac{1}{S_r} \ln\left(\frac{k_L}{k_{w,L}}\right) \qquad (16.30)$$

Now we can use this mobile-phase composition experimentally. This method usually works very well, but small deviations can occur, which can be corrected easily. An alternative approach that does not rely on an assumed value for S_r is discussed below.

Another requirement that must be met here is that the mobile-phase composition derived from Equation (16.30) must also lead to enough retention for the first-eluting solute F. We first calculate its retention factor $k_{w,F}$ in neat aqueous mobile phase analogous to the procedure shown for the most retained compound, assuming the same S_r value. Then we can calculate its retention factor k_F at the desired mobile-phase composition as follows:

$$k_F = k_{w,F} - S_r \varphi_B(k_L) \qquad (16.31)$$

The ratio k_L/k_F is also an indication of whether isocratic elution is appropriate for the separation. Typically, if the ratio is above 15, it will be difficult to separate all peaks isocratically with a sufficient sensitivity, and gradient elution may need to be used.

The relationships shown above can be easily used in a spreadsheet program to quickly derive the initial isocratic mobile-phase composition.

An alternative to the preceding methodology for deriving isocratic retention data is to perform an additional gradient run, with a different gradient slope, but at the selected pH. This procedure eliminates the need to assume a certain slope of the dependence of the logarithm of the retention factor with mobile-phase composition. For each compound, this dependence is determined by two independent parameters, k_w and S_r [see Eq. (16.10)]. The combination of two sets of retention times for two gradient runs with different slopes $\beta_{g,1}$ and $\beta_{g,2}$, but with the same starting organic fraction $\varphi_{B,0}$, allows us to calculate both k_w and S_r for each solute. The combination of two sets of Equation (16.27) leads to the following relationship, which can be numerically solved for S_r:

$$\beta_{g,2}\exp(t_{gc,1}S_r\beta_{g,1}) - \beta_{g,1}\exp(t_{gc,2}S_r\beta_{g,2}) = 0 \qquad (16.32)$$

Once S_r has been obtained, k_w is calculated according to Equation (16.29).

The summary of the experimental strategy to obtain an initial isocratic separation for subsequent optimization is as follows:

1. Perform two gradient experiments at two significantly different pH values, for example, pH 7 and pH 2 (phosphate buffers) using the same gradient!

2. Inspect the separations obtained and select the better separation for the determination of the initial isocratic conditions!

3. Calculate the mobile-phase composition that will result in an acceptable retention time under isocratic conditions! This can be done by assuming a value (e.g., 8) for the slope S_r of the dependence of the logarithm of the retention factor on the solvent composition. Alternatively, a third gradient experiment can be performed with a different gradient steepness. In the latter case, one obtains both S_r and the intercept k_w for each solute.

4. Perform the isocratic separation under the calculated conditions and examine the chromatogram! If the separation is not yet satisfactory, solvent selectivity should be used to further improve the separation.

Examples of this strategy are shown in the next section. The solvent selectivity optimization is discussed subsequently in Section 16.3.2.2.

16.3.2.1 Binary Mobile Phases The application of this strategy is demonstrated here for the analysis of two antidepressant drugs and their metabolites. Figure 16.7 shows the gradient separation of amitriptyline, doxepin, and their respective metabolites, obtained at pH 2.3 and 7.0. The pH reported here is that of the aqueous mobile phase. The resulting pH of aqueous–organic mixtures is known to depend on the organic content. On the basis of solvatochromic theory (45), Barbosa and Sanz-Nebot (46) calculated the pH of acetonitrile–buffer mixtures, for several buffer systems. For acetate buffers, the pH increases by 35% when 50% w/w of the aqueous solution is replaced by acetonitrile.

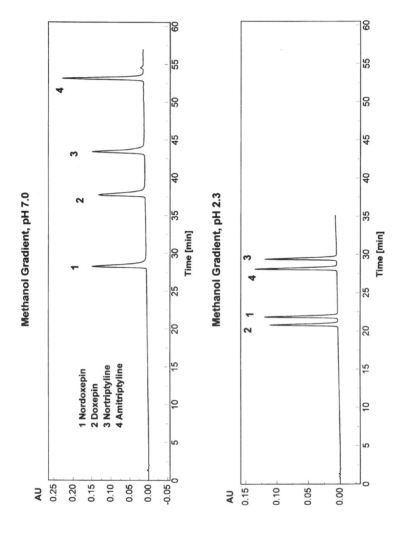

Figure 16.7 Gradient separations of four tricyclic antidepressants at acidic and neutral pH (pH 2.3 and 7.0, respectively). Note the significant selectivity differences at both pH values. (*Chromatographic conditions:* Column: Waters 5 μm Symmetry C$_{18}$; 3.9 mm × 150 mm. Temperature: 35°C. Gradient: from 20% MeOH/80% 20 mM potassium phosphate buffer to 80% MeOH/20% buffer in 60 min. Flow rate: 1 mL/min. Detection: UV at 254 nm.) (Chromatograms courtesy of Waters Corp.)

However, the pK_a values for the buffer and the solutes are also affected to almost the same extent. This means that the state of ionization of the solute is not affected, in general, by the change in the apparent pH due to the addition of the organic solvent.

Note that the peaks are quite symmetric even at neutral pH. Since these are basic compounds, the neutral pH leads to higher retention times. At acidic pH we observe a reversal in the elution order, and the peaks are bunched in pairs, and with a wide space between the pairs. It is clear that an isocratic mobile phase at pH 7.0 will more likely lead to a better and more rugged separation than at acidic pH.

Once the pH is selected, an additional gradient run is needed to derive isocratic retention data. Since we had significant retention at a solvent change rate of 1%/min, it is appropriate to run the next gradient at 2%/min, but with the same starting composition (20% methanol). Using the gradient retention times in combination with the equations 16.32 and 16.29 allows us to estimate S_r and k_w values for each one of these compounds. Note that the dwell volume of the HPLC system as well as the column holdup volume need to be estimated, or better measured, in order to carry out the calculation. Table 16.1 shows the values of the retention parameters for each solute in the mixture.

It is interesting to note that the tertiary amines doxepin and amitriptyline have similar S_r values, while the secondary amines nortriptyline and nordoxepin also have similar S values.

Using the data in Table 16.1, one can now plot the dependence of $\ln(k)$ versus the mobile-phase composition φ_B and choose the appropriate composition leading to an adequate separation of all peaks. Alternatively, one can set up a requirement for total analysis time and check whether the separation of all peaks is possible within this analysis time. It is also important to plot the dependence of $\ln(k)$ versus φ_B because this allows us to detect the occurrence of any selectivity reversal (crossing lines). Figure 16.8 shows the dependence of $\ln(k)$ on the methanol volume fraction. The horizontal line corresponds to a retention time of 10 min for the last-eluting peak. It intercepts the line of the last-eluting solute at 71% methanol. We can also see that an isocratic eluent with 71% methanol would lead to an adequate separation of all solutes. This is evident when the selectivity factors for adjacent peak pairs are calculated (2.0, 1.3, and 2.1). With today's columns these selectivities are expected to lead to a baseline resolution for all peak pairs. The chromatogram at 71% methanol and pH 7 is shown in Figure 16.9.

TABLE 16.1   Retention Parameters Obtained from Gradient Experiments

Solute	Nordoxepin	Doxepin	Nortriptyline	Amitriptyline
k_w	230	1224	1326	6186
S_r	8.39	9.40	8.44	9.63

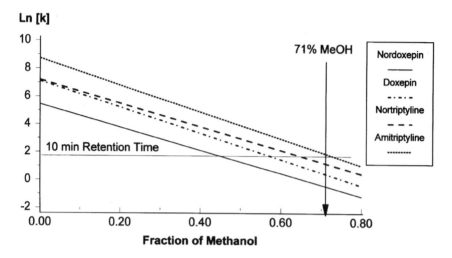

Figure 16.8 Dependence of the logarithm of the retention factor on the volume fraction of methanol.

First, note that the retention time for the last peak is 11 min versus the predicted 10 min. Second, the experimental selectivity values for adjacent peak pairs are 2.4, 2.1, and 2.0. These differences between predicted and experimental values may be due to one or the combination of several reasons:

- The linear model [$\ln(k)$ vs. φ_B] is only an approximation. This is especially true for ionizable compounds as they may exhibit a mixed-mode retention.
- The organic modifier adsorption to the stationary phase does affect the retention time of the samples, as demonstrated by Martin (47) when the modifier adsorption behavior is linear and studied by El Fallah and Guiochon (48) with acetonitrile showing that the gradient applied to the column experiences a delay as well as an increase in its steepness with time.
- Errors in estimating the dwell volume affect significantly the retention-time predictions for early-eluting peaks.
- There is an inherent proportioning error in the mobile phase delivered to the column, both during gradient and final isocratic composition. This error may be as high as $\pm0.5\%$. However, this error can be reduced to less than 0.15 with high-precision HPLC instruments of the latest generation.

Nevertheless, the differences between experiment and prediction are rather small, and the strategy outlined above results in a rapid determination of a reasonable isocratic retention window.

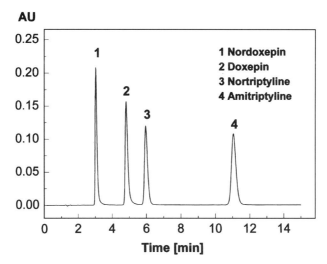

Figure 16.9   Final isocratic separation of the four tricyclic antidepressants. (*Chromatographic conditions*: Column: $5\,\mu m$ Symmetry C_{18} 3.9 mm × 150 mm. Temperature: 35°C. Mobile phase: 71% MeOH/29% 20 mM potassium phosphate buffer, pH 7.0. Flow rate: 1 mL/min. Detection: UV at 254 nm.) (Chromatograms courtesy of Waters Corp.)

16.3.2.2 Solvent Selectivity and Ternary Mobile Phases   When the first isocratic experiment does not lead to a satisfactory result, one should now explore the selectivity differences between different organic modifiers. This is the next logical step after selecting the pH. The use of differences in mobile-phase selectivity is demonstrated in the following example, where the separation of chlordiazepoxide and its metabolites is examined.

Keeping with our methodology, two gradient runs were performed initially at two different pH values (pH 2.0 and pH 7.0). Figure 16.10 shows the chromatograms obtained at these pH values with acetonitrile as the organic modifier. There are significant differences in the selectivity of the separation between neutral and acidic pH. At pH 7, nordiazepam elutes last; while at pH 2 it coelutes up front with norchlordiazepoxide. Demoxepam and oxazepam elute at nearly the same time independently of pH, while chlordiazepoxide moved to lower retention. This demonstrates again the prominence of pH as a tool to effect the separation of ionizable solutes. Although two of the peaks were not affected by pH, there are significant changes in the relative position of the peaks in the chromatogram. As a result of better peak spacing, and the fact that all peaks were resolved, pH 7 was selected as the pH for further work.

A second gradient was run at pH 7 to obtain the retention parameter S_r and k_w for all compounds. From the values obtained for the last-eluting peak, a composition of 32% acetonitrile was estimated to lead to an analysis time of

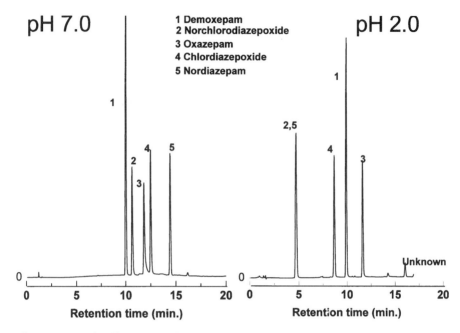

Figure 16.10 Gradient separations of chlordiazepoxide and its metabolites at acidic and neutral pH (pH 2.0 and 7.0, respectively). Note the significant differences in the selectivity of the separation. (*Chromatographic conditions*: Column: 5 μm Symmetry C_{18}, 3.0 × 150 mm. Temperature: ambient. Gradient: from 20% MeCN/80% 20 mM potassium phosphate buffer to 60% MeCN/40% 20 mM buffer. Flow rate: 0.59 mL/min. Detection: UV at 240 nm.) (Chromatograms courtesy of Waters Corp.)

15 min. The chromatogram at this composition is shown in Figure 16.11. We can notice two problems with this separation. First, there isn't enough resolution between peaks 1 and 2. This may be solved by decreasing the acetonitrile content. More important is the fact that the oxazepam peak tails, since this may impact the method sensitivity and ruggedness. The occurrence of strong tailing with acetonitrile is a known phenomenon (49). Since acetonitrile does not provide hydrogen-bonding properties, the silanols on the silica surface are readily available to interact with an analyte. Similar observations can be made with tetrahydrofuran. Methanol, on the other hand, because of its strong hydrogen bonding, helps shield the silanols, and therefore gives better peak shape for analytes that interact strongly with silanols. One solution would be to add a slight amount of methanol to the mixture, for example, 5%. However, since binary mobile phases are preferred, one should try first methanol as the only modifier. On the basis of the isoeluotropic transfer rules discussed above, 59% methanol should lead to approximately the same analysis time as 32% acetonitrile. Figure 16.12 shows the separation obtained with 59% methanol. Peaks 2 and 3 coelute, and conse-

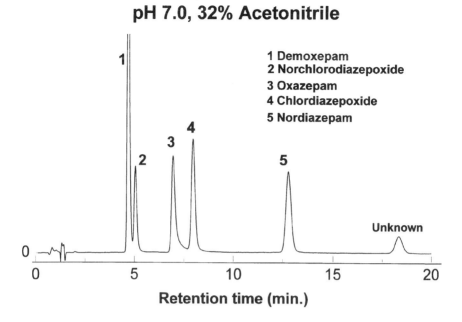

pH 7.0, 32% Acetonitrile

1 Demoxepam
2 Norchlorodiazepoxide
3 Oxazepam
4 Chlordiazepoxide
5 Nordiazepam

Retention time (min.)

Figure 16.11 Separation of chlordiazepoxide and its metabolites. First isocratic experiment with 32% acetonitrile at pH 7.0. Note that the resolution between the first and second peak is not yet satisfactory. (*Chromatographic conditions*: Column: 5 μm Symmetry C_{18}, 3.0 × 150 mm. Temperature: ambient. Mobile phase: 32% MeCN/68% 20 mM potassium phosphate pH 7.0. Flow rate: 0.59 mL/min. Detection: UV at 240 nm.) (Chromatogram courtesy of Waters Corp.)

quently methanol alone cannot be used. However, a careful comparison between the isoeluotropic mobile phases based on acetonitrile and methanol suggests that their combination is likely to lead to a satisfactory separation. In fact, peaks 1 and 2, which were poorly separated in the acetonitrile-based mobile phase, are well resolved with methanol; while peaks 2 and 3, which coelute with methanol, are well resolved with acetonitrile. The logical next step is to proportionally mix acetonitrile and methanol. The chromatogram obtained using this ternary mobile-phase composition is shown in Figure 16.13*a*. As expected, we obtain baseline resolution for all peaks. However, the total analysis time is larger than our target, which was 15 min. The analysis time can be reduced by increasing the solvent strength. This can be done by increasing the total organic content while keeping the ratio of methanol to acetonitrile constant (no selectivity change), or by increasing either methanol or acetonitrile. This decision has to be made rationally. Increasing the total organic content while keeping the ratio of methanol to acetonitrile constant would result in the decrease in resolution between peaks 2 and 3, which is not

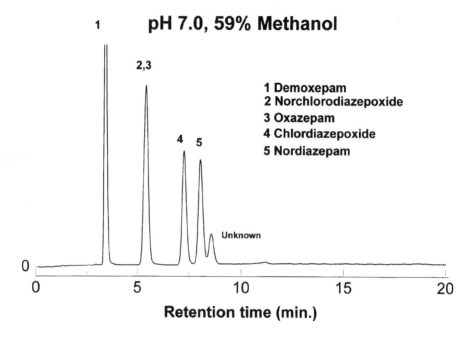

Figure 16.12 Separation of chlordiazepoxide and its metabolites with 59% methanol at pH 7. This mobile-phase composition was calculated to give about the same retention time as 32% acetonitrile. A comparison between Figures 16.11 and 16.12 reveals the solvent selectivity differences between the methanol-based and the acetonitrile-based mobile phase. Other chromatographic conditions as in Figure 16.11. (Chromatogram courtesy of Waters Corp.)

acceptable. However, we know that acetonitrile provides plenty of resolution between peaks 2 and 3. Therefore an increase in acetonitrile is not likely to spoil the resolution of this critical pair. In contrast, with methanol as organic modifier these peaks coelute. Therefore, it is undesirable to try to increase its content. Since typically an increase of 5% v/v of the organic modifier reduces retention by a factor of 2, a 7% v/v increase in the acetonitrile content should result in an analysis time of about 10 min. The final chromatogram is shown in Figure 16.13*b*.

The following summarizes the continuation of the experimental strategy, by using solvent selectivity after pH selection to fine-tune the separation:

5. Calculate isoeluotropic mobile phases using Equation (16.18) or (16.19)!

6. Perform isocratic experiments with these mobile phases! They will reveal selectivity differences of the different solvents for your solutes.

7. Fine-tune the separation by making use of the observed selectivity differences by using combinations of different organic modifiers!

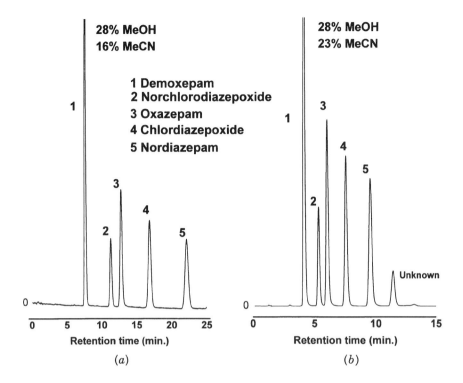

Figure 16.13 (*a*, *b*) Separation of chlordiazepoxide and its metabolites using combinations of acetonitrile and methanol as organic modifiers. Chromatographic conditions as in Figure 16.11, except the concentration of the organic modifiers is as indicated on the graph. (Chromatograms courtesy of Waters Corp.)

16.3.2.3 Separation of Totally or Partially Unknown Mixtures   The analysis of totally or partially unknown mixtures is one of the most challenging tasks in HPLC. Typical applications include the development of a stability indicating method for the separation of degradation products of the bulk drug or the drug product, impurity profiling to identify process related contaminants and/or degradation products, and the analysis of environmental samples such as soil or water extracts. In such applications, resolution alone is not an adequate criterion for measuring the success of the optimization of the separation. Interferences between peaks will often lead to an overlap that cannot always be detected from the peak shape. Therefore it is important to use a detection scheme that can detect coelution, for example, photodiode array detection, or ultimately, mass spectrometry.

For these types of problems, the objective of the optimization needs to be clearly stated. In the case of the separation of impurities or degradation products, one may need to separate all components that are present above a

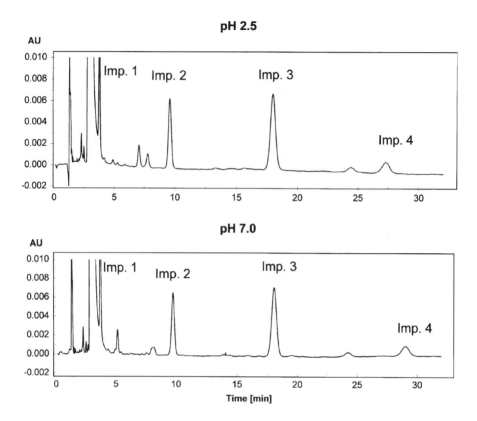

Figure 16.14 Initial chromatograms of the impurities of AZT at low and neutral pH. The off-scale peak next to the peak labeled IMP 1 is the AZT peak. (*Chromatographic conditions*: Injection: 100 μL of 0.5 mg/mL AZT solution. Column: 5 μm Symmetry C$_{18}$ 3.9 mm × 150 mm. Temperature: 45°C. Mobile phase: 33% methanol/67% 10 mM potassium phosphate buffer at the indicated pH. Flow rate: 1 mL/min. Detection: UV at 268 nm. (Chromatograms courtesy of Waters Corp.)

certain level, for instance, 0.1% of the parent drug. Here one has to make sure that minor components (<0.1% of the response of the drug peak) do not interfere with either the drug itself or its major impurities. Smaller impurities below 0.1% response may give rise to a response that is above the 0.1% level if coelution occurs. This, of course, may lead to the false interpretation that this "unknown" with a response above 0.1% needs to be monitored, while in fact it is the result of the coelution of peaks that are below the 0.1% threshold. Typically, the goal of the optimization is to determine limit values for the response of each major impurity, and one has to make sure that interferences from impurities that are below the 0.1% level do not influence the quantitation of the impurities that one needs to monitor. In other applications, one may only need to separate a specific compound or a group of compounds from a matrix, as in the detection of an herbicide and its metabolites. Often, a sample preparation to concentrate the analyte of interest and remove interference is carried out before HPLC analysis.

In all cases, one has to make sure that the compounds of interest are free from interferences from either major or minor components in the mixture.

Let us examine how one can efficiently develop a method for the separation of low-level impurities in a drug product. The example chosen is the separation of impurities in the drug AZT (azidothymidine). The goal of the method is to separate AZT and its four major impurities ($>0.1\%$ response) from each other and from interferences. Figure 16.14 shows the initial separations obtained with methanol at acidic and neutral pH. It is obvious that a pH change does not lead to a significant change in selectivity for the compounds of interest (the peaks labeled IMP1-4 in the chromatograms). However, the retention of minor impurities that do not need to be measured, but may interfere with AZT or its

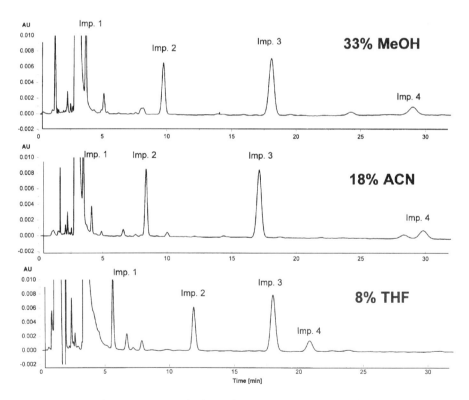

Figure 16.15 Chromatogram obtained for AZT and its impurities in the three isoeluotropic mobile phases at pH 7.0. Only THF is able to separate impurity 1 from the drug peak. Chromatographic conditions as in Figure 16.14 except as indicated in the figure. (Chromatograms courtesy of Waters Corp.)

major impurities, is greatly affected by pH. So again pH may play a major role in the optimization of the method. Since impurity 1 nearly coelutes with the AZT peak, one has to investigate selectivity changes based on the type of the organic modifier. Figure 16.15 shows the separation obtained with the three isoelutropic mobile phases. Using acetonitrile as organic modifier does not give any advantage over methanol as modifier. The separations obtained with both solvents are quite similar to each other, except that with acetonitrile impurity 4 is not well resolved from a minor impurity. However, the separation with tetrahydrofuran as organic modifier shows a significant change in selectivity. In particular, impurity 1 is now well resolved from the drug peak. Nevertheless, the purity-check algorithm of the photodiode array detector showed that the peak of impurity 1 is contaminated, i.e., we have not yet achieved a satisfactory separation. The next step is to mix ternary mobile phases, and we selected THF and methanol to do so, since acetonitrile may cause the coelution of impurity 4 and the minor impurity eluting next to it. One has to use the combination of resolution and peak spectral homogeneity as a criterion for judging the quality of the separation. After a few additional steps of fine-tuning, a satisfactory separation was obtained for AZT and its major impurities from each other and from matrix interferences. Figure 16.16 shows the optimized chromatogram.

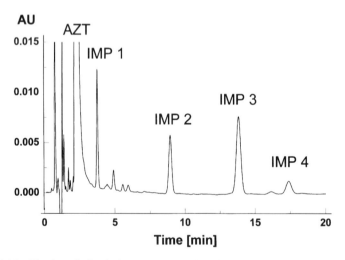

Figure 16.16   Final optimized chromatogram showing the separation of AZT and its major impurities from each other and from matrix interferences. (*Chromatographic conditions*: Injection: 100 μL of 0.5 mg/mL AZT solution. Column: 5 μm Symmetry C_{18} 3.9 mm × 150 mm. Temperature: 45°C. Mobile phase: 6% MeOH/6% THF/88% 10 mM potassium phosphate buffer, pH 2.5. Flow rate: 1.7 mL/min. Detection: UV at 268 nm. (Chromatograms courtesy of Waters Corp.)

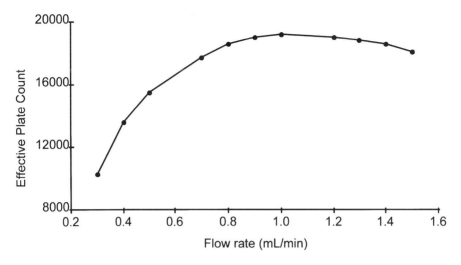

Figure 16.17 Effective plate count versus flow rate. (*Chromatographic conditions*: Gradient from 59% methanol to 95% methanol in 40 min. Column: HiBar C_{18} Lichrosorb C_{18}, 5 μm, 4.6 mm × 250 mm. The sample is pentachlorophenol.)

16.3.3 Developing Gradient Separations

When the mixture is too complex to be resolved using an isocratic method, a gradient elution method needs to be developed. However, before any attempt to develop a gradient method is made, an evaluation of mobile-phase contaminants is necessary. Experienced users are well aware of the fact that organic contaminants from either aqueous or organic mobile phases can be detrimental to achieving rugged gradient methods. The appearance of ghost peaks in the chromatogram is a serious problem, and determining their origin is of utmost importance. One should use the following methodology. First, equilibrate the column at high aqueous composition for enough time to concentrate impurities present in the mobile phase. Then execute a relatively fast gradient to as much organic as possible (high concentrations of an inorganic buffer may precipitate at high organic content, especially with acetonitrile). From this experiment one can estimate the organic solvent composition necessary to wash away these contaminants. A blank gradient must always be run before the start of the analysis. It goes without saying that high-purity HPLC solvent must be used. High-purity water is available from most HPLC solvent suppliers, but the methodology presented above is a cheaper way to solve the problem. Often a fast wash after the actual gradient program is advisable and can prolong column lifetime. Equilibration between consecutive runs must be kept to a minimum to reduce the concentration of contaminants. The subtraction of blank gradients is available with most computer-based data systems, but excellent retention time reproducibility is a must.

One attractive way of developing a gradient method is to specify the total analysis time and then optimize the flow rate to achieve maximum peak capacity. In fact, the effective plate count, NQ_g^2, exhibits a maximum at a certain flow rate. This strategy is equivalent to optimizing the b_g parameter when linear-solvent-strength gradients are performed in reversed-phase chromatography. The optimum b_g parameter is sensitive to both the particle size of the packing and the nature of the solvent. The mobile-phase composition at the start of the gradient can be selected such that the retention factor of the first-eluting peak is around 2. Figure 16.17 shows the dependence of the effective plate count of pentachlorophenol as a function of the flow rate when total gradient run was set to 40 min.

The optimum flow rate for the separation of chlorophenol reaction products was about 1 mL/min for a gradient time of 40 min. If, in such an optimization scheme, the total gradient time turns out to be too long, the gradient time can be reduced and the flow rate increased such that the product of gradient time and flow rate is kept constant. This amounts to keeping the gradient parameter b_g constant.

The effect of organic modifier and pH should be investigated at an early stage of the gradient before trying to optimize the gradient parameters, as a different solvent or pH may actually lead to a better separation. Ternary solvents may be useful when a given pair of peaks is not or poorly resolved with one solvent while the second solvent shows enough resolution for that pair. The strategy for obtaining the dependence of $\ln(k)$ on the organic composition can be applied to gradient runs as well. This information can then be used to optimize the gradient, since retention under gradient conditions can be easily calculated once these parameters have been determined for the critical peak pairs. One can then plot the dependence of selectivity or resolution versus either the starting-mobile phase composition or the gradient slope, and may locate an area where the critical pair or pairs are resolved.

16.3.4 Computer-Assisted Method Development

The application of chemometric techniques to HPLC optimization is too broad a subject to be covered comprehensively in this chapter. We will only outline the major developments and products in this area. If more details are needed, readers can consult some excellent textbooks that deal with this subject (50,51).

These techniques and software programs can be grouped into two categories. The first category includes simulation programs that use retention data and/or structural formulas to predict either the chromatogram at selected chromatographic conditions or the optimum separation conditions based on the selected goal of the separation. The second category includes pure chemometric techniques where models are not necessarily defined, but rather the retention data of a selected set of runs is subsequently used to estimate the optimum conditions for the separation.

Among the simulation packages, the most widely used is the DryLab software from LC Resources (52). This simulation package allows the user to browse the solvent composition scale in order to locate the optimum, after providing retention data from either two gradients or two isocratic runs. The optimum mobile-phase composition can be easily estimated by examining the resolution map of the critical pair (the pair with the least resolution) generated using these two initial runs. Retention modeling is based on the linear-solvent-strength theory. The linearity of the dependence of the logarithm of the retention factor on the volume fraction of the organic modifier is assumed. However, a quadratic model can also be selected when deviation from linearity are observed, that is, when the experiments show a strong deviation from the predictions generated using the linear model. Besides binary mobile-phase composition, other parameters can be varied as well. The effects of a third mobile-phase modifier, temperature, pH, buffer concentration, linear gradient, and multistep gradient conditions can all be investigated. In the case of buffers and additives, ternary mobile-phase compositions, and pH, a third experimental run is necessary to model retention. The effect of column parameters can also be investigated by the user. DryLab has been widely used by chromatographers, and errors in retention predictions are generally less than 5%.

HiPac (53) from Phase Separation is another commercially available software package. In several aspects this software is similar to DryLab, but its most important feature is that it can estimate the optimum mobile-phase conditions for the separation of the mixture at hand or only a selected number of peaks. Recently, ChromSword (commercially available from Merck, Germany) was introduced (54). It uses a retention model based on solvophobic theory. The input for this package can be the structural formulas of the solutes, the combination of structural formulas and retention data from a single run, or retention data from two runs. Data from additional runs are incorporated into the model, and prediction accuracy below 3% can be achieved under these circumstances.

As examples for purely chemometric programs, the sequential simplex method and the modified simplex method are widely used in optimizing mobile-phase conditions in HPLC (55–57). Other chemometric techniques include factorial design (58,59), search point optimization method (60), reduced factorial design based on partial least-squares regression (61,62) and a mixed-level orthogonal array design (63–65).

16.4   ENHANCING SENSITIVITY

Good sensitivity is an important characteristic of a good method. During the process of optimization, one has to keep in mind that sensitivity requirements have to be met, and this consideration has to be built into the method development strategy. Earlier, we discussed some theoretical limitations of isocratic elution in achieving the desired sensitivity, but mainly from the standpoint of matrix complexity. Here, we will discuss the chromatographic parameters that affect sensitivity. Column geometry (internal diameter and length), column performance (peak shape and efficiency), and thermodynamic parameters such as retention and temperature, as well as injection conditions, all influence the sensitivity of an assay. Sensitivity is usually expressed using the limit of detection (LOD) and the limit of quantitation (LOQ), which are, respectively, the lowest concentration that can be detected and reliably quantitated at a given signal-to-noise ratio. These parameters are related to the concentration at the peak apex c_{max}, which depends on the chromatographic parameters as described by the following Equation:

$$c_{max} = \frac{4}{\varepsilon_t \pi \sqrt{2\pi}} \frac{\sqrt{N}}{L d_c^2 (1+k)} \frac{c_0 V_i}{1 + \kappa(T_h - 1)} \qquad (16.33)$$

where ε_t is the total column porosity, L its length, d_c its diameter, T_h is the peak tailing measured at a given peak height fraction h, and κ is a constant depending on h. It is about 4 when T_h is measured at 10% of the peak height.

The strong dependence of c_{max}, and thus the sensitivity, on the peak shape is often overlooked. For a peak with a tailing of 1.5 (measured at 10% of peak height), the response is about one-third of that of a perfectly symmetric peak, while a peak with a tailing of 2 is 5 times smaller than a symmetric peak. Tailing is a function of the nature of the solute and its interaction with the stationary phase. Strongly basic compounds are known to give rise to tailing peaks, and the column choice is very important in order to achieve low detection limits.

As defined, LOD should not depend on the matrix to be analyzed. However, in practice, the compound of interest is in a given matrix and the overall separation performances greatly affect the practical detection and quantitation of a given peak in the matrix. Unfortunately, there is no clear distinction in the literature between *absolute* LOD, which may be defined as the lowest solute concentration that can be detected, under given chromatographic conditions, in a sample free of matrix interference; and a *specific* LOD, which would be the lowest solute concentration that can be detected in a given matrix under those same chromatographic conditions. In this case, the peak's own shape, but also the asymmetry of other peaks present in the matrix may limit its *specific* LOD. This is clearly demonstrated in the example in Figure 16.18. Because resolution loss occurs when larger amounts are injected, there is a critical ratio of the largest peak to that of its neighboring peak above which the smaller peak 3 can no longer be detected or quantitated. See also the discussion of the discrimination factor d_0 in Section 16.2.4.3.

In general, there is a compromise between sensitivity and injection volume for a given column. This stems from the fact that both column overloading and band-spreading effects associated with the injection volume may affect separation performance (as expressed by resolution between adjacent peaks or

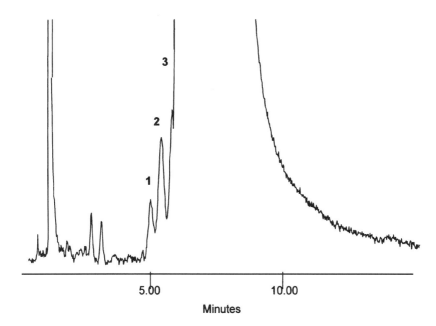

Figure 16.18 Loss of resolution due to the presence of neighboring peaks limits the sensitivity for the minor peaks in this separation of tamoxifen and its impurities. (*Chromatographic conditions*: Injection: 10 μL of 600-μg/mL solution of tamoxifen. Column: Zorbax Rx C_{18}, 4.6 mm × 150 mm. Mobile phase: 50 mM potassium phosphate, pH 3/acetonitrile 55:45. Flow rate: 1.4 mL/min. Detector: UV, 240 nm. (Chromatograms courtesy of D. J. Phillips, Waters Corp.)

efficiency for a specific peak). Mass overloading is easily recognized by the shark-fin-like peak shape and the decrease of peak retention time, but let us examine here the effect of the injection volume. If one assumes a given loss in efficiency, θ^2, associated with the injection of a volume V_i, the resulting efficiency, N_{inf}, is then given by:

$$N_{inf} = (1 - \theta^2)N_0 \qquad (16.34)$$

where N_0 is the efficiency obtained for an infinitely small injection volume. The maximum injection volume $V_{i,max}$ leading to a loss in resolution θ is then given by the following relationship:

$$V_{i,max} = \frac{\pi}{4}\,\varepsilon\theta\xi Ld_c(1 + k)\,\frac{1}{\sqrt{N_0}} \qquad (16.35)$$

where ξ is a constant depending on the injection profile, and is typically around 2 (66). It is worth noting that the column efficiency is in the denominator. This means that more efficient columns are more sensitive to the increase in injection volume and will show a steeper decrease in efficiency as injection volume increases, but the absolute value of the efficiency will still lead to a better resolution for higher-efficiency columns. At the same time, the relationship in Equation (16.35) shows that columns with a larger volume can tolerate much larger injection volumes for the same loss in efficiency. Therefore, a reduction of the column diameter as a means for increasing sensitivity is misleading. If a sufficient sample volume is available, a judicious column diameter choice would be to select the smallest possible column diameter that accomplishes the requirements for sensitivity and still maintains the resolution of the separation. This fact is clearly shown in Figure 16.19, where the relative sensitivity is plotted as a function of the injection volume for different column diameters. The upper limit of the curve corresponds to the injection volume for which 5% loss in resolution is observed, and below this injection volume a proportional increase in peak height with increasing injection volume was assumed. For this specific case ($k = 8$, $L = 15$ cm, and $N_0 = 10,000$), one can inject about 20 μL on the smallest diameter column before the quality of the separation starts to deteriorate. However, the same sensitivity can be obtained on a larger-diameter column (4.6 mm i.d.) when about 100 μL is injected. Note that these two conditions are equivalent only, if we assume that band-spreading contributions are primarily due to the injection volume. This is not entirely true unless all HPLC system volumes (tubing, connections, and detector cell) are reduced proportionally to the column volume. To achieve

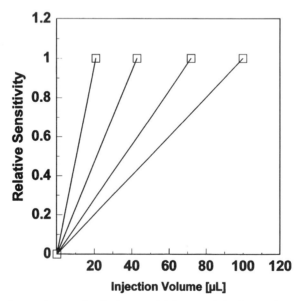

Figure 16.19 Dependence of relative sensitivity on injection volume for different column diameters.

this, a microbore HPLC system must be used with smaller columns. Generally, if the sample volume is not limited, the choice of column diameter is irrelevant for sensitivity. The only benefit of smaller column diameters is a reduction in solvent consumption, which is not without value in today's world, where the cost of waste disposal is becoming exorbitant, and an environmentally responsible behavior has become part of the culture of the chemical industry. In our experience, we found that with current HPLC systems, the 3-mm-i.d. column can replace larger columns in the majority of applications, leading to a reduction of almost 60% in solvent consumption. In our view, the 3-mm column is a wise choice to start with for any new method development, especially since this size is now widely available.

Retention also plays a role in sensitivity through the combined effects of Equations (16.33) and (16.35). Equation (16.33) shows that the signal is much higher when the retention factor k is smaller. But at the same time, Equation (16.35) suggests that for peaks with higher retention times it is possible to inject a larger volume without a significantly decrease in resolution. This is due to the fact that band broadening stemming from the injection itself becomes less significant for late-eluting peaks as the process of dilution becomes the major contributor to bandwidth. In fact, all extracolumn band-broadening contributions become less severe as retention increases. Therefore, there is an optimum retention factor for which sensitivity exhibits a maximum. However, it is difficult to change the retention of all peaks in the mixture without impacting the specificity of the method.

16.5 TEMPERATURE EFFECTS

Recently, temperature optimization in reversed-phase HPLC has become the topic of many investigations [67–69]. Temperature is known to play a significant role in biomolecule and chiral separations, but its influence on the separation of small molecules in reversed-phase HPLC is much less important. Its major effect is a reduction in retention and an improvement in peak efficiency as temperature increases. For complex samples, this may be of great utility, as the column peak capacity is increased significantly. The increase in efficiency is the result of a decrease in solvent viscosity as temperature increases. This enhances the mass-transfer rate between mobile and stationary phase. For monomeric stationary phases, selectivity is not very much affected by temperature changes, but polymeric stationary phases may show significant selectivity changes (70) due to a stationary-phase transition that occurs above a certain temperature. A packing material based on the modification of silica with N-isopropyl acrylamide (NIPAM), a thermally reversible polymer, is quite sensitive to temperature changes (71). Below 32°C, it shows a hydrophilic behavior as NIPAM chains hydrate and swell, while above that temperature they shrink and the material acts as a hydrophobic stationary phase. Figure

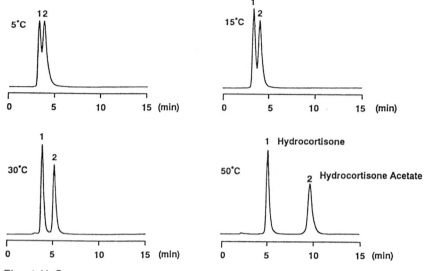

Eluent: H₂O.
Flow Rate: 1.0ml/min.
Detection: 254nm(UV).
Sample: Hydrocortisone, Hydrocortisone Acetate

Figure 16.20   Change of selectivity as a function of temperature for hydrocortisone and hydrocortisone acetate. (*Chromatographic conditions*: Eluent: water. Column: NIPAM.)

16.20 shows the effect of temperature on the separation of hydrocortisone and hydrocortisone acetate.

More recently, temperature programming has become commercially available, making it possible to accurately modify the temperature during analysis. One benefit of this technique may be that some methods that use solvent gradients may now be replaced by methods in which the temperature is gradually increased, without any change in solvent composition.

Besides the constraints on instrumental design and the requirement of solvent preheating, one major disadvantage of operating at high temperatures is that faster dissolution of the silica matrix occurs, and column lifetime will inevitably become shorter (72).

REFERENCES

1. L. R. Snyder, *High Performance Liquid Chromatography. Advances and Perspectives*, Vol. 1, Cs. Horváth, ed., Academic Press, New York, 1980, p. 207

2. L. R. Snyder, J. W. Dolan, and J. R. Grant, *J. Chromatogr.* **165**, 3 (1979).

3. L. R. Snyder, J. L. Glajch, and J. J. Kirkland, *Practical HPLC Method Development*, Wiley-Interscience, New York, 1988.

4. L. R. Snyder, in *Chromatography*, E. Heftmann, ed., Elsevier, Amsterdam, 1992, Chapter 1.

5. L. R. Snyder, M. A. Stadalius, and M. A. Quarry, *Anal. Chem.* **55**, 1412A (1983).

6. P. J. Schoenmakers, H. A. H. Billiet, R. Tijssen, and L. De Galan, *J. Chromatogr.* **149**, 519 (1978).

7. D. Ambrose, A. T. James, A. I. M. Keulsmans, E. Kovats, H. Rock, C. Rouit, and F. H. Stross, in *Gas Chromatography*, R. P. W. Scott, ed., Butterworths, London, 1960, p. 423.

8. E. Grushka, *Anal. Chem.* **44**, 1733 (1972).

9. W. L. Jones and R. Kieselbach, *Anal. Chem.* **30**, 1590 (1958).

10. L. R. Snyder, *J. Chromatogr. Sci.* **10**, 200 (1972).

11. M. Martin and G. Guiochon, *Anal. Chem.* **57**, 289 (1985).

12. E. Glueckauf, *Trans. Faraday Soc.* **51**, 34 (1955).

13. T. R. C. Boyde, *Sep. Sci.* **6**, 771 (1971).

14. P. R. Rony, *Sep. Sci.* **3**, 239 (1968).

15. P. R. Rony, *Sep. Sci.* **3**, 357 (1968).

16. R. Kaiser, *Chromatographie in der Gas Phase*, Vol. I, Gas-Chromatographie, Bibliographisches Institut, Mannheim, 1960, pp. 16, 32.

17. R. Kaiser, *Gas Chromatographie*, Akademie Verlagsgesellschaft Geest & Portig K.-G., Leipzig, 1960, p. 33.

18. W. Wegscheider, E. P. Langmayr, and M. Otto, *Anal. Chim. Acta* **150**, 87 (1983).

19. A. B. Christophe, *Chromatographia* **8**(4), 455 (1971).

20. H. J. G. Debets, B. L. Bajema, and D. A. Doornbos, *Anal. Chim. Acta* **151**, 131 (1983).

21. M. Z. El Fallah and M. Martin, *Chromatographia* **24**, 115 (1987).

22. M. Z. El Fallah and M. Martin, *Analusis* **16**, 241 (1988).

23. P. J. Schoenmakers, *Optimization of Chromatographic Selectivity*, *Journal of Chromatography Library*, Vol. 35, Elsevier, Amsterdam, 1986, pp. 121–123.

24. M. Z. El Fallah, Ph.D. thesis, University of Paris (1989).

25. P. J. Schoenmakers, Ph.D. thesis, Technical University of Delft, The Netherlands (1981).

26. R. Tijssen, H. A. H. Billiet, and P. J. Schoenmakers, *J. Chromatogr.* **122**, 185 (1976).

27. L. R. Snyder, *J. Chromatogr. Sci.* **16**, 223 (1978).

28. S. C. Rutan, P. W. Carr, W. J. Cheong, J. H. Park, and L. R. Snyder, *J. Chromatogr.* **463**, 21 (1989).

29. J. H. Park, A. J. Dallas, P. Chau, and P. W. Carr, *J. Phys. Org. Chem.* **7**, 757 (1994).

30. J. H. Park, A. J. Dallas, P. Chau, and P. W. Carr, *J. Chromatogr.* **A677**, 1 (1994).

31. L. Rohrschneider, *Anal. Chem.* **45**, 1241 (1973).

32. S. D. West, *J. Chromtogr. Sci.* **27**, 2 (1989).

33. V. R. Meyer, *Practical High-Performance Liquid Chromatography*, Wiley, New York, 1988, p. 136.

34. B. P. Johnson, M. G. Khaledi, and J. G. Dorsey, *Anal. Chem.* **58**, 2354 (1986).

35. J. G. Dorsey and B. P. Johnson, *J. Liq. Chromatogr.* **10**, 2695 (1987).

36. B. P. Johnson, M. G. Khaledi, and J. G. Dorsey, *J. Chromatogr.* **384**, 221 (1987).

37. M. J. Kamlet, J. L. M. Abboud, and R. W. Taft, *Prog. Phys. Org. Chem.* **13**, 485 (1983).

38. M. J. Kamlet, J. L. M. Abboud, and R. W. Taft, *Prog. Phys. Org. Chem.* **13**, 591 (1981).

39. P. J. Schoenmakers, H. A. H. Billiet, and L. de Galan, *J. Chromatogr.* **205**, 13 (1981).

40. J. C. Giddings, *Anal. Chem.* **39**, 1072 (1967).

41. E. Grushka, *Anal. Chem.* **42**, 1142 (1970).

42. G. Guiochon, *J. Chromatogr.* **14**, 378 (1964).

43. J. M. Davis and J. C. Giddings, *Anal. Chem.* **1983**(55), 418 (1983).

44. A. Tchapla, P. Allard, and S. Heron, *LC-GC Intl.* **6**, 86 (1993).

45. M. J. Kamlet, J. M. L. Abboud, M. H. Abraham, and R. W. Taft, *J. Org. Chem.* **48**, 2877 (1983).

46. J. Barbosa and V. Sanz-Nebot, *Anal. Chim. Acta* **283**, 320 (1993).

47. M. Martin, *J. Liq. Chromatogr.* **11**, 1809 (1988).

48. M. Z. El Fallah and G. Guiochon, *Anal. Chem.* **63**, 859 (1991).

49. D. V. McCalley, *J. Chromatogr. A* **1995**(708), 185 (1995).

50. J. C. Berridge, *Techniques for the Automated Optimization of HPLC Separations*, Wiley, New York, 1985.

51. J. L. Glajch and L. R. Snyder, eds., *Computer-Assisted Method Development for High-Performance Liquid Chromatography*, Elsevier, Amsterdam, 1990.

52. DryLab Chromatography Modeling, LC Resources, Walnut Creek, CA, Spring 1994.

53. S. M. Hitchen, HiPac Chromatography Optimisation Software, Phase Separations, Deeside, UK (1992).

54. S. V. Galushko and G. Wieland, paper presented at The 20th International Symposium on High performance Liquid Phase Separations and Related Techniques, June 16–21, 1996, San Francisco, CA.

55. L. De Galan and H. A. H. Billiet, in *Advances in Chromatography*, Vol. 25, J. C. Giddings, E. Grushka, J. Cazes, and P. R. Brown eds., Marcel Dekker, New York, 1987, p. 63.

56. T. Hamoir and D. L. Massart, in *Advances in Chromatography*, Vol. 33, P. R. Brown and E. Grushka, eds., Marcel Dekker, New York, 1993, p. 97.

57. J. W. Dolan and L. R. Snyder, *J. Chromatogr. Sci.* **28**, 379 (1990).

58. Q.-S. Wang, R.-Y. Gao, B.-W. Yan, and D.-P. Fan, *Chromatographia* **38**, 187 (1994).

59. M. Rozbeh and R. J. Hurtubise, *J. Liq. Chromatogr.* **17**, 3351 (1994).

60. J. L. Martinez-Vidal, P. Parillo, A. R. Fernandez-Alba, R. Carreno, and F. Herrera, *J. Liq. Chrom.* **18**, 2969 (1995).

61. P. Kaufman, B. R. Kowalski, and J. Alander, *J. Chemom. Intell. Lab. Syst.* **23**, 331 (1994).

62. P. Kaufman, *J. Chemom. Intell. Lab. Syst.* **27**, 105 (1995).

63. W. G. Lan, K. K. Chee, M. K. Wong, H. K. Lee, and Y. M. Sin, *Analyst* **120**, 281 (1995).

64. H. B. Wan, W. G. Lan, M. K. Wong, and C. Y. Mok, *Anal. Chim. Acta* **289**, 371 (1994).

65. K. K. Chee, W. G. Lan, M. K. Wong, and H. K. Lee, *Anal. Chim. Acta* **312**, 271 (1995).

66. M. Martin, doctoral thesis, University of Paris (1974).

67. B. Ooms, *LC-GC* **14**, 306 (1996).

68. N. M. Djordjevic and F. Houdiere, paper presented at The 20th International Symposium on High Performance Liquid Phase Separations and Related Techniques, June 16–21, 1996, San Francisco, CA.

69. P. L. Zhu, L. R. Snyder, J. W. Dolan, N. M. Djordjevic, D. W. Hill, J.-T. Lin, L. C. Sander, L. Van Heukelem, and T. J. Waeghe, paper presented at The 20th International Symposium on High Performance Liquid Phase Separations and Related Techniques, June 16–21, 1996, San Francisco, CA.

70. L. C. Sander and S. A. Wise, *Anal. Chem.* **61**, 1749 (1989).

71. H. Kanazawa, K. Yamamoto, Y. Kashiwase, Y. Matsushima, N. Takai, T. Okano, and Y. Sakurai, in *International Symposium on Chromatography*, H. Hatano and T. Hanai, eds., World Scientific Publishing, Singapore, 1995, p. 587.

72. B. E. Boyes and J. J. Kirkland, *Peptide Res.* **6**, 249 (1993).

17 Column Maintenance

In die Ecke,
Besen! Besen!
Seid's gewesen!
Denn als Geister
Ruft euch nur zu seinem Zwecke
Erst hervor der alte Meister.

        —J. W. von Goethe, *Der Zauberlehrling*

In this chapter, we will cover the care and use of HPLC columns. Then we will discuss various chromatography problems and ways to find and eliminate the source of these problems. The topics covered are not necessarily "column" problems in the strictest sense, but are often viewed as such in the eyes of the user. The troubleshooting section is organized by the observed phenomena, such as peak-shape problems or retention-time problems. This allows the reader to quickly go to the chapter that is closest to the phenomenon that he is trying to troubleshoot. We also included many visual examples to help the reader find the subject that appears to be closest to the observed problem.

17.1 CARE AND USE OF CHROMATOGRAPHIC COLUMNS

Chromatographic columns require some care, but the required effort is rather small. However, we have to keep in mind that they are high-performance parts of a high-performance instrument, and as such should be treated with a reasonable amount of respect. For example, while a column might survive a 1-m drop onto a concrete floor, it is something that one really should try to avoid. The same applies to other topics, such as column freezing or drying. In this section we will cover just a few of these common-sense do's and don'ts of column care.

17.1.1 Installation and Equilibration

Before you connect a column to your HPLC instrument, you must check the compatibility of the solvents in the column and the column itself with the solvent(s) in the instrument. Has the instrument been completely purged to the intended starting solvent? This can be a problem, when the instrument has been used for reversed-phase chromatography and will next be used for normal-phase chromatography. As we have pointed out in the chapter on normal-phase chromatography, silica or alumina columns are extremely sensitive to even small amounts of water in the mobile phase. Thus residual water in the instrument can result in lengthy equilibration times and distorted peak shapes for normal-phase columns. Also, the storage solvent of a reversed-phase column is frequently 100% of a polar solvent like methanol or acetonitrile. If we were to start this column with a mobile phase that contains a buffer or a salt, it is possible that the buffer precipitates in the column on exposure to the storage solvent. This, in turn, can result in a local region of low permeability and damage to the packed bed. Third, we have to make sure that the storage solvent in the column and the startup mobile phase are completely miscible, but this is rarely a problem.

Once we identified the solvent used to store the column, we need to plan the equilibration protocol for the column. Assume that we are dealing with a reversed-phase column that was stored in acetonitrile, and that we want to use a mobile phase consisting of a 1:1 mixture of a buffer and methanol. In order to avoid a precipitation of the buffer in the column, we should first purge the column with a few column volumes (1–3 is enough) of a 1:1 mixture of methanol and water before we switch to the methanol–buffer mobile phase. You should also be aware that the column backpressure will increase roughly

5-fold during this conversion, due to the viscosity differences between aceto-nitrile and the methanol–water mixture.

As another example, assume that we would like to use a propylamino column stored in hexane in ion-exchange mode using an aqueous buffer as mobile phase. This conversion needs to be planned carefully. Hexane is only partially miscible with methanol and acetonitrile, so we need another solvent between hexane and the polar solvent. Also, on the aqueous side, we need to use water as an intermediate between the buffer and the polar solvent. Therefore, the complete conversion would require 1–3 column volumes each of ethylacetate (or methylene chloride), methanol (or acetonitrile), water, and finally the buffer that was intended as mobile phase.

When installing the column, make sure that the mobile-phase flow is in the direction indicated on the column. Most modern HPLC columns, especially those based on hard spherical particles, can be used in both directions without detrimental effect, but it is nevertheless good practice to have a standardized direction of flow.

Also, as you install the column, make sure that the seat of the ferrule on the tubing is completely compatible with the column fitting. For different manu-facturers, the distance between the tip of the tubing and the tip of the ferrule is different (see Table 17.1). This can lead to two problems (see Fig. 17.1): in the case that this distance is too long for the new column, the ferrule needs to be completely reseated and you have some difficulty sealing the column. In the opposite case, you end up with some dead volume between the end of the tubing and the seat of the tubing inside the fitting, which results in band spreading and lowered column performance. You can avoid this issue in several ways. The best way is to use a universal connector, which uses a plastic, typically a fluorocarbon ferrule, which can slide up and down the tubing. Or you can only use columns that use the same standard, such as those from only one manufacturer. Or you make yourself a set of connectors that have been prepared to convert from your standard instrument end fitting to the different

TABLE 17.1   Critical Distance of Tubing Length Pro-truding Beyond the Tip of the Ferrule for Several Fitting Types

Fitting Type	Critical Distance (mm)
Parker	2.3
Rheodyne	4.3
Swagelok	2.3
Upchurch	2.3
Valco	2.0
Waters	3.3

Source: Adapted from Reference 1.

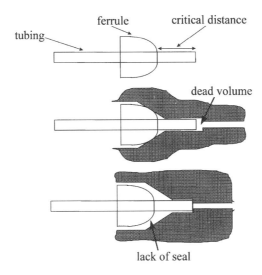

Figure 17.1 Effect of the critical distance between the tip of the ferrule and the tip of the tubing. If the tip is too short for the type of fitting used, a dead volume results that causes band spreading. If the tip is too long, the ferrule will not seat and the connection will leak.

column end fittings of the columns that you commonly use. Also, some column packers provide column fittings according to your specifications. Of course, you can also remove the old ferrule and prepare a new one for the new column, but this is a lot of work if you frequently change columns.

Most HPLC columns do not require a specific startup procedure. You can simply switch to the desired flow rate, and you are up and running. However, some soft polymeric packings require a slow incremental increase of the flow rate until the final flow is reached. Look for the manufacturer's recommendations for these columns.

Before you start to use the column for an analysis, you have to ensure that the column is properly equilibrated. This is best done by running a few standard analyses and observing any changes in retention or quantitation. If stable results are obtained, you can proceed to use the column for the analysis of unknowns.

Equilibration times vary widely with the type of chromatography and the history of the column. A column that had previously been equilibrated and was stored in the mobile phase should be up and running within a few, maybe 10, column volumes, provided the new mobile phase is really identical to the old mobile phase. If the column was stored in a solvent other than the mobile phase, equilibration times vary with the chromatographic mode. Normal-phase chromatography using silica or alumina columns may take a long time for equilibration; several days at a flow rate of 1 column volume per minute have been observed. The equilibration with normal-phase bonded phases is faster.

Also, if mobile phases with the same level of water saturation are used, equilibration can be fast.

Reversed-phase columns, ion-exchange columns, or columns for hydrophilic or hydrophobic interaction chromatography usually equilibrate quickly. Normally, equilibration is obtained after some 10–20 column volumes of mobile phase. The only reason for a slow equilibration is a low concentration of a strongly retained mobile-phase constituent. An example of this is paired-ion chromatography on a reversed-phase column. The concentration of an ion-pair reagent in the mobile phase is typically 5 mmol/L. After complete equilibration, the concentration of the ion-pair reagent on the surface of the packing is typically about $1 \, \mu mol/m^2$ or about $150 \, \mu mol/mL$. This means that we need already about 30 column volumes of mobile phase to deliver to the column the amount of ion-pair reagent necessary to saturate the surface, but in reality we need an additional 10–20 column volumes of mobile phase for a complete equilibration. If the concentration of the reagent in the mobile phase is lower, or the concentration on the surface is higher, proportionally higher amounts of mobile phase are needed for equilibration. Thus we have to be careful about any situation, where an active ingredient of the mobile phase is present in only minute molar quantities. Lengthy equilibration times may be the consequence.

17.1.2   Column Use

Most of the constraints on column use have been discussed in previous chapters, so we will recapitulate just a few items in this section. Column protection is covered in a later section. The key limitation of silica-based columns is the hydrolysis of the silica at high pH values and the instability of the bonded phase at low pH values. This constraint is the same for all silica-based columns, independent of the type of silica or the type of bonded phase, but a variation of the stability exists. Most manufacturers state that silica-based columns can be used only in the pH 2–8 range. But this should be taken only as a rough guide. At elevated temperature, a short column life can be observed within this pH range for polar bonded phases, such as a cyanopropyl packing. But a C_{18} bonded phase is dissolving at a 1000-fold lower rate than a cyanopropyl bonded phase, and at room temperature one may still get a reasonable column life outside this pH range. Also, according to studies by Kirkland (2), the nature of the buffer plays a role, and phosphate buffers at neutral pH are more aggressive than alternatives, for example, Tris buffers. Furthermore, the type of silica can influence the stability of a packing. Silicas with a small specific pore volume are more rugged than silicas with a high specific pore volume. To complicate things further, the ligand density influences the stability of a packing. High ligand densities protect the silica better than low ligand densities.

Any bonded phase will exhibit some level of hydrolysis in an aqueous mobile phase. Under most circumstances, the hydrolyzed bonded-phase silanol self-adsorbs to the packing. This results in a local hydrolysis equilibrium, which actually suppresses further hydrolysis. However, the hydrolyzed bonded phase is only bound by adsorption, and it can be removed from the surface of the packing by washing with an organic solvent. This disturbs the local equilibrium and is actually potentially detrimental to the stability of the bonded phase. We will discuss this further in the discussion of column storage.

Polymeric packings swell and shrink depending on the nature of the mobile phase. Column manufacturers adjust the swelling of the resin during column packing to match the use of the column. In many cases this simply means that the column is packed in the same solvent in which it will be typically used. The manufacturer's literature usually contains information about the mobile phases that are not compatible with the packing or the packing technique.

When using polymeric packings, flow-rate changes should be made gradually. This includes the startup of the column. Gradually increase the flow to the desired flow rate! Some HPLC instruments have built-in programs for this.

Some columns require special care. Amino columns, for example, should not be exposed to solvents that contain aldehyde or keto groups. Amino columns also are only partially stable in an aqueous environment. When exposed to water, the high concentration of bonded phase in the pores creates a strongly basic environment. Under these circumstances, the silica surface dissolves slowly, creating acid silanols. At some point, the acidic silanols neutralize the local basic environment, and the dissolution of the phase is slowed down or even halted. Since amino columns are often used in an aqueous environment as ion exchangers or for the analysis of carbohydrates, you should be aware of this effect.

17.1.3 Performance Monitoring during Use

Columns do have a finite lifetime. They may deteriorate following contamination with strongly adsorbed sample constituents, debris from instrument seals or chemical degradation, for example, due to hydrolysis. This makes it desirable to monitor column performance in regular time intervals.

In order to monitor the performance of a column during use, it is worthwhile to set up standard conditions and a group of reference parameters. Column manufacturers test the column plate count and peak asymmetry under a well defined set of conditions using samples that are optimized for this test. You may choose to adopt this test as a routine measure of column performance. The advantage of this kind of test is primarily that it uses the same baseline as the column manufacturer, thus making a comparison with the column test results at the manufacturer's site possible. In this way, you can be assured that the column did not get damaged during transport. This test is also a good reference point, if the column is used for many different assays.

However, this test has several disadvantages. For one, the test typically only checks column efficiency and asymmetry, but does not give you any indication of the state of the surface chemistry. This is especially true for reversed-phase columns. The test compounds commonly used for efficiency tests interact only with the hydrocarbon layer of the surface, but not with surface silanols. Thus they are insensitive to a deterioration of column performance due to hydrolysis of the stationary phase. To capture this aspect of column performance, you need to use or create a test that measures the activity of surface silanols. See Chapter 10 for tests designed to do this for RPLC columns. The other disadvantage of a standard column efficiency test is that it requires its own mobile phase. Thus you have to change the mobile phase from your usual operating conditions to the test mobile phase, equilibrate the column in the test mobile phase, and run the test. Then, after the test, you need to change back to the mobile phase of your assay, reequilibrate the column, run a standard, to make sure that your assay works, and only then you can start running the assay. This is a fair amount of work for just the information on the status of the column.

Alternatively, if the column is used primarily for one particular assay, you may use the assay itself as your reference point. The main advantage of this approach is that it creates a minimal amount of additional work. The second advantage is that the test will capture all performance characteristics of the column that are relevant for your assay, be it plate count, peak symmetry, retention, or selectivity. So for many practical situations, the latter approach is preferred.

Whether you use a standardized test or your assay, it is worthwhile to check column performance on a regular basis and keep a log of it. With today's computerized HPLC instruments this is fairly easy to do, and you can easily generate control charts of the important column characteristics. I recommend monitoring for at least one peak plate count, peak symmetry, and retention time, and relative retention for a critical pair of analytes. Resolution is not as instructive a parameter, since it is affected by both plate count and relative retention. Thus it does not tell you anything about which of the underlying parameters is changing.

You should also monitor the backpressure of your column in regular intervals. Increases in backpressure are early warning signs that tell you that the column top or the column filter is clogging. Clogged filters or column tops result in a decreased column performance and peak distortion.

Another important reference parameter is the band spreading of the instrument, also called *extracolumn band spreading*. It is defined as the accumulated effect of all noncolumn hydrodynamic influences on the standard deviation of a chromatographic peak, thus including the influence of the injection volume, the injector, the connection tubing from injector to column and from column to detector, the end fittings and column frits, and the detector. In practice, the influence of the column fittings and the frits are not measurable by the user. To perform a measurement of the extracolumn band spreading, the column is disconnected from the system, and the injector is connected directly to the detector. Then, a predetermined small amount of analyte is injected at a given flow rate, and the standard deviation of the peak is measured. Figure 17.2 demonstrates the measurement of the extracolumn band spreading based on

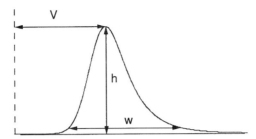

Figure 17.2 Measurement of extracolumn band spreading. A sample is injected without a column in place. The width of the peak is measured at 4.4% of the height of the peak. This width w is then equal to 5 standard deviations.

Waters 5σ convention, which assumes that the width of the peak at 4.4% of the peak height is 5 times the standard deviation of the peak. This convention holds well for even fairly asymmetric peaks.

When carrying out this measurement, a high sampling rate should be chosen and the detector time constant should be chosen to be negligible (0.1 s), unless one intentionally wants to include the influence of the detector time constant into the measurement. The measurement can either be carried out under standardized conditions (e.g., using methanol at 1 mL/min with a 1% solution of acetone in methanol as sample), or better with the actual mobile phase and a typical sample. The result is reported as the standard deviation in volume units, (e.g., $15\,\mu L$). It subsequently can be used to check the performance of the instrument, especially if any part of the fluid path has been changed.

17.1.4 Column Protection

When properly cared for, a column can last for a long time. Since the major cause for column deterioration is contamination, the protection of the column from contamination is a primary task of column care. Column lifetimes of over 10,000 analyses have been achieved with columns protected by guard columns. Under the identical set of conditions, but without this protection, column life was limited to about 2000 injections.

Column contamination may come from either particulates in sample or mobile-phase stream or from strongly adsorbed sample constituents that build up on the column. Both problems can be eliminated effectively through the use of a guard column, while in-line filters or off-line sample filtration solve only part of the problem. It should be pointed out though that guard columns have a limited lifetime themselves, and need to be replaced on a regular basis. Otherwise, the contaminants might break through onto the analytical column.

A guard column should be regarded as a sacrificial extension of the analytical column. Therefore, in order to perform its job well, it should contain the identical packing as the analytical column, both with respect to the type of packing and the particle size. On one hand, this guarantees maximum protection of the separation column. On the other hand, this minimizes the chance of a deterioration of the separation due to the influence of the guard column. For example, if we were to use a different C_{18} with different retention characteristics in the guard column than we use in the separation column, the guard column could actually destroy the separation achieved in the analytical column. On the other hand, with a well-designed and well-packed guard column, we can actually observe a gain in separation efficiency due to the guard column.

Other, less efficient ways to protect the column are the use of an in-line filter between injector and column, the filtering of the sample or sample cleanup through solid-phase extraction. Filtering removes only debris above the effective size of the filter. Smaller particles still break through and may accumulate on the top of the column through a chemical interaction. Many contaminants can be removed through solid-phase extraction, which is, in essence, a chromatographic process akin to analytical chromatography. However, the separations are quite crude, and often span a fairly large elution window. Thus only a part of the potential contaminants are captured.

17.1.5 Storage

If the column is not used for a short period of time, for example, overnight or over the weekend, the column is preferentially stored in the mobile phase, in which it is used. This minimizes equilibration times on startup. However, for longer storage, the column should be stored in the solvent that the manufacturer recommends for a particular column. For bonded phases used in reversed-phase chromatography, the most common solvent is an organic solvent such as methanol or acetonitrile. Nonaqueous solvents minimize the hydrolysis of silica-based bonded phases. When the column is converted to such a solvent, care should be taken to remove any salt prior to the conversion. It is highly recommended to put a label on the column that describes the storage solvent.

The conversion of a reversed-phase column to an organic solvent is also a way to wash off material that is adsorbed on the column. However, this is a two-edged sword. During regular use of the packing, a small amount of hydrolysis occurs. Usually, the hydrolyzed bonded-phase ligands adsorb to the surface, and a local equilibrium develops between hydrolysis and reformation of the bonded phase. When we convert a column to an organic solvent, this self-adsorption is eliminated and the hydrolyzed ligands are washed out of the column. This can actually result in an increase in column aging. This is another reason for the recommendation that the column should be converted to an organic solvent only for long-term storage.

Some polar bonded phases such as cyanopropyl phases tend to become mechanically unstable in polar organic solvents. In such a case, a storage in an aqueous mobile phase might be preferred. To reduce hydrolysis, the column can be stored at low temperature.

Generally, storage at low temperature is a good idea, but in most cases not necessary. When a column is stored at low temperature, make sure that the mobile phase does not freeze. This could damage the packed bed. Glass columns can actually burst on freezing.

For ion exchangers or aqueous SEC packings based on organic polymers, the typical storage solvent is an aqueous solution. To prevent the growth of bacteria, sodium azide is commonly added to the storage solvent at a concentration of about 0.05%. Also, most of these packings tolerate a fair amount of organic solvent. If they do, mixtures of water with acetonitrile or methanol are good storage solvents. Consult the manufacturer's recommendations!

For packings used in normal-phase chromatography or organic SEC, by far the best storage solvent is the mobile phase. Normal-phase mobile phases do not damage the packings, and reequilibration is faster. For SEC packings used at elevated temperature with high-viscosity mobile phases, the columns should be brought up to temperature first before flow is started.

17.2   TROUBLESHOOTING OF COLUMN PROBLEMS

In this section we will learn to diagnose and eliminate common chromatography problems. They are not necessarily column problems; quite often the cause of a supposed column problem might be an incorrect setting of an instrumental parameter, or it may be related to the sample. For example, most peak-shape problems are not column problems.

We divided this section into two major subsections: peak-shape problems and retention-time problems. The last section then covers problems that do not fit into the first two categories, such as pressure problems. This structure should allow you to quickly move to the chapter that covers the problem that you are trying to solve. In the section of peak-shape problems, example chromatograms should also help to point you to the most appropriate section. While this structure is good for rapid troubleshooting, it also makes repetition unavoidable. Therefore, we would like to remind the reader that this part of the book is not designed to be read as a consecutive text.

17.2.1   Peak-Shape Problems

Peak-shape problems represent the most common problems in HPLC. In many cases, the chromatographer has come to accept less-than-perfect peak shape, as long as the peak distortion is not excessive. But distorted peaks are causes for integration problems, and are a sign that the best column performance is not reached.

One can learn a lot from a careful examination of the peak shape. If you see double peaks or fronting peaks, especially if the same peak distortion is observed throughout the chromatogram independent of the retention of the peak, it is highly likely that the packed bed is damaged. You should then consult Section 17.2.1.6 or 17.2.1.7. If broad and tailing peaks are observed at low retention, but you get symmetric peaks and high plate count at large retention, then the cause of the problem is most likely some extracolumn effect. Extracolumn effects are covered in Section 17.2.1.1. If only the larger peaks are distorted, but small peaks give good peak shape, we might expect that we are dealing with an overload phenomenon, discussed in Sections 17.2.1.2–17.2.1.4. Finally, if only specific peaks in the chromatogram show a problem, while other peaks give good peak shape and good plate count, we would suspect specific chemical interactions as the cause of the problem, covered in Section 17.2.1.5.

17.2.1.1 Extracolumn Effects   Extracolumn effects are a more significant source of peak-shape problems than most people think. As a matter of fact, there is barely a chromatogram in the world that is free from extracolumn effects, but we have come to accept them as part of the reality of chromatography. Only when these effects become excessive do we attempt to reduce them. Also, extracolumn effects may cause problems when methods are transferred from one instrument to another or from a standard column to a microbore column.

All extracolumn effects are easily recognized in isocratic chromatography, if we measure the plate count for all peaks in the chromatogram. If we see that the plate count tends to increase with retention, there is a high likelihood that extracolumn effects play a significant role. A visual inspection of chromatogram shows tailing for early-eluting peaks, which decreases with retention (Fig. 17.3).

There are two types of extracolumn effects: extracolumn band spreading and the influence of the detector time constant (and related issues). Both will not gradually change as time progresses, but one might see a sudden change in either. If we are able to trace, what we have changed or what has changed, we should be able to quickly identify the source of the problem.

17.2.1.1.1 Extracolumn Band Spreading   This category includes all contributions of extra-column volume to band spreading. The injection volume, the volume of the injector tubing, the performance of the injector, the connection tubing from injector to column and from column to detector, the quality of the connections inside the fittings, precolumn filters, and the volume of the detector itself all contribute to extracolumn band spreading. With the possible exception of injector performance, none of these causes of extracolumn band spreading will change gradually. They may be present from the beginning of the development of a method and may change, as the method is transferred to another instrument. When they occur suddenly with an established method, the

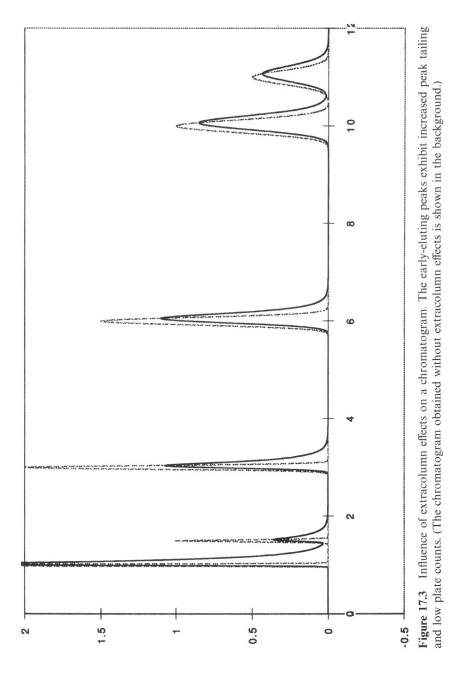

Figure 17.3 Influence of extracolumn effects on a chromatogram. The early-eluting peaks exhibit increased peak tailing and low plate counts. (The chromatogram obtained without extracolumn effects is shown in the background.)

first question that you want to ask yourself is which of all the possible causes has changed just prior to the observation of the deteriorated peak shape. The most common answer is that the column was reconnected to the system, after it had been disconnected, either because the system was not used or was used for something else. In this case the very first thing to do is to examine the connections to the column. Is the ferrule seated in the correct position? The fittings of different manufacturers require a different position where the ferrule is seated relative to the end of the tubing. If this position is incorrect, then a dead volume is formed between the end of the tubing and its seat, which is the cause of the band spreading. Table 17.1 (in Section 17.1.1) contains information about the ferrule position for different manufacturers of column hardware. If you do not find the problem there, think about what other things might have changed. Has any of the connection tubings been replaced recently? Is the detector equipped with the correct cell? Some detectors are available with different detector cells for preparative, analytical or microbore work. Preparative detector cells are often supplied with connection tubing with a large internal diameter. Is the detector connected correctly? Usually the outlet tubing of a detector cell has a much larger internal diameter than the inlet tubing. So if both tubings are interchanged accidentally, a large extracolumn band spreading can result.

In some injector types, worn seals or partially clogged tubing can cause excessive extracolumn band spreading. In other injectors, the material of construction can adsorb the sample. A measurement of the extracolumn band spreading and a comparison to previous results should help in the diagnosis of this cause. Consult the manual of the manufacturer of the injector for specific effects and troubleshooting recommendations.

The effect of too large an injection volume can easily be identified and remedied. A more subtle effect can be caused by changes in the makeup of the sample, including the solvent or the pH of the sample. If the sample is prepared in a solvent or at a pH that reduces the retention of the analytes at the column top, a distortion of the peaks can result. A change of ingredients in a formulation can cause similar disturbances. See also Section 17.2.1.4!

17.2.1.1.2 Detector Time Constant   This is a common cause of peak distortion, especially with fast chromatography on short columns. It occurs during the development of a method, but it is rarely the cause of trouble with an existing method. A peak with a retention time of one minute can easily have a standard deviation of under one second with today's high-performance columns. A setting of the detector time constant to one second, as is a commonly found default value, will definitely distort this peak. From the standpoint of column performance, you should select a detector time constant of about 0.1 s for all but the slowest HPLC applications.

In older equipment, the detector time constant is usually selectable through a switch on the detector. For newer equipment, the detector time constant is programmed. Check the setup parameters of your detector for a programmable time constant! This is a setting that is rarely adjusted; therefore it is rarely a source of trouble with existing methods on previously used equipment. But it can be the source of problems when an existing method is converted from one instrument to another or from a standard column to a shorter column packed with a smaller particle size, where a significant decrease in analysis time is expected.

17.2.1.2 Column Overload

Quite often, distorted peaks and low plate counts are due to column overload. One may encounter a situation like this when the concentrations of the analytes of interest span several orders of magnitude, for example, in the assays for impurities and degradants in pharmaceutical compounds. In this example, the minor compounds need to be quantitated down to a level of 0.1% of the parent compound. This means that the concentration of the parent compound is 3 orders of magnitudes higher than the concentration of the minor compounds. In order to achieve an acceptable signal for the compounds present in small concentration, the major compound often shows overload. Another possible case is volume overload, when the analytes are present only in small concentrations, forcing a large injection volume. Other related overload phenomena, such as detector overload or mobile phase overload, are dealt with in the subsequent sections.

17.2.1.2.1 Volume Overload

Volume overload is caused by too large an injection volume. It is characterized by broad, possibly flat-topped peaks at the beginning of the chromatogram, while later-eluting peaks are more normal (Fig. 17.4). Volume overload is encountered normally only during method development. In an established analysis, one would know whether the injection volume had changed significantly. A simple remedy is to reduce the injection volume, if the sensitivity of the analysis allows this. If this is not possible, then one should consider sample enrichment prior to HPLC analysis. Alternatively, one might consider diluting the sample with a weaker-eluting solvent than the mobile phase and increase the injection volume in the correct proportion. Under the right circumstances, this can improve the situation, since the presence of the weaker solvent forces the sample to become enriched on the top of the column (3). The following example illustrates this situation. Let us assume that the mobile phase in a reversed-phase separation contains 50% organic solvent, and that we have dissolved the sample in the mobile phase. Assume also that an injection volume of $20\,\mu L$ overloads the column. If we dilute the sample 1:1 with water, the sample now contains only 25% organic solvent. If we now inject $400\,\mu L$ of the diluted sample, we have injected the same mass as with $200\,\mu L$ of the undiluted sample, but because of the enrichment of the sample on the column top, the distortion of the peaks from volume overload is eliminated. In fact, in the situation just described, you are likely to be able to inject an even larger volume ($>1\,mL$) of the diluted sample without peak distortion.

17.2.1.2.2 Mass Overload Mass overload is encountered, when the mass injected onto the column exceeds a certain limit. For most HPLC packings and for low-molecular-weight analytes, this limit is somewhere around 10^{-4} to 10^{-3} g/mL column volume. If you use a UV detector, this translates for a compound with a typical extinction coefficient to a peak height of about 0.1 AUFS. Thus a quick glance at the y axis of the chromatogram can tell you immediately whether the observed peak distortion may be due to mass overload. Also, simply injecting a smaller amount will allow you to pinpoint the problem. See Figure 17.5 for an example of mass overload!

If there is no good reason to inject such a large mass of sample, all you need to do is inject less sample. However, in some situations an overload may be unavoidable, for example, when you need to quantitate trace analytes in the presence of major components. In such a case, a change to a more sensitive detection mode may be the only solution.

17.2.1.3 Detector Overload In some cases, the concentration of an analyte peak may be outside the range of a linear detector response. In a mild case of nonlinearity, the distortion of the peak is insignificant, and the fact that we are not in the linear response range of the detector is found only through a calibration curve. In some cases, however, it can be so severe that the top of the peaks become flattened, not unlike the case of volume overload. However, while the phenomenon of volume overload is strongest for early-eluting peaks and declines for later-eluting peaks, the detector overload will affect only large peaks.

If the peaks that exhibit detector overload do not need to be quantitated, then we can just ignore this overload phenomenon. Otherwise we should inject a smaller amount or use a detector (or detector setting) with a larger linear dynamic range.

17.2.1.4 Mobile-Phase Overload There are two basic types of mobile-phase overload. One is due to the solubility limits of the analyte in the mobile phase; the other is due to the presence of the sample solvent or sample matrix components, which may interfere with the adsorption of the sample on the column top. Both can result in significant peak distortions. In the case of analyte insolubility, only the analyte peak will be distorted. In the case of a disturbance of the loading of the sample onto the top of the column, it is possible that only some peaks are affected or that all peaks are affected equally. For example, if the pH of the sample is different from the pH of the mobile phase, only those analytes whose retention changes with pH could be affected, while other analytes remain unchanged.

17.2.1.4.1 pH-Control If we are dealing with ionic or ionizable analytes, it is important to control the pH of the mobile phase. The concentration of the mobile-phase buffer should be high enough to control the ionization of the analytes. We will encounter peak distortions due to this effect at high analyte concentration, at low buffer concentration, and/or at the limits of the buffering capacity of a buffer. A buffer has its maximum buffer capacity at the pK_a of the buffer. We need to keep this in mind when we develop a method with a buffered mobile phase.

Similarly, if the pH of the sample is significantly different from the pH of the mobile phase, and if we inject a large sample volume, the pH environment at the column top, when the analyte is supposed to adsorb onto the packing, may be affected. This can interfere with the adsorption of the analyte at the column top and result in peak distortion. One may encounter effects of this nature in dissolution testing, where the sample pH is dictated by the dissolution environment. If we encounter such a situation, we should consider increasing the buffer concentration of the mobile phase, injecting a smaller amount of sample or adjusting the pH of the sample.

An extreme example of peak distortion due to mobile-phase overload is shown in Figure 17.5. A sample of tricyclic antidepressants was dissolved at acidic pH, while the mobile phase was a phosphate buffer at neutral pH. The amount of acid injected overloaded the buffer, creating an acidic environment in the column. At acidic pH, the analytes are retained less than at neutral pH. Therefore, broad and less retained peaks are obtained for the acidified sample (solid line) compared to the standard sample, which was dissolved in mobile phase (dotted line). Usually, the effects of mobile-phase overload are not as extreme as in this example, and the peak distortions are more subtle.

17.2.1.4.2 Incorrect Sample Solvent If the solvent in which the sample is dissolved is a stronger-eluting solvent than the mobile phase, it can interfere with the adsorption of the sample onto the column top, especially when large injection volumes are used. The result are distorted peaks, often throughout the chromatogram. Fronting, tailing, and other peak forms can be observed. We can test for this phenomenon by preparing a sample in mobile phase instead of the usual preparation. Also, the best solution to the problem is the preparation of the sample in mobile phase. If this is not possible, a dilution of the sample with the weaker-eluting solvent together with a higher injection volume can solve the problem.

17.2.1.4.3 Solubility With the exception of the case of hydrophobic interaction chromatography, it is rare that the insolubility of the sample in the mobile phase causes peak distortion problems. In most cases the response factor of the detector is large enough to generate a sufficient signal height long before any sample solubility problems are encountered. Only in preparative chromatography are we likely to encounter solubility problems, but not in analytical chromatography. In many cases the peak distortion is similar to that caused by the mass-overload phenomena described above, but other forms of peak distortion can also be encountered. The solution to the problem is a reduction in the amount of sample injected.

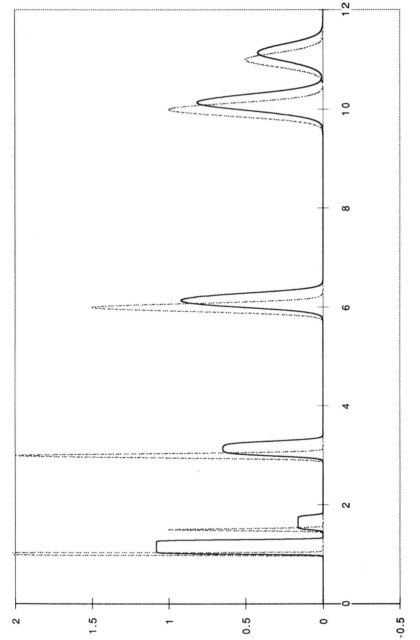

Figure 17.4 Broadening of the peaks due to excessive injection volume. Peaks early in the chromatogram exhibit a flat top, which becomes less pronounced at higher retention.

17.2.1.5 Tailing Due to Secondary Interactions Peak distortion, usually tailing, due to multiple interactions, is quite common in HPLC. A typical example is the tailing of basic analytes in reversed-phase chromatography due to the combined interaction of the analytes with the reversed-phase ligands and surface silanols. It is recognized easily if only some analytes show tailing at low concentrations while other analytes give symmetric peaks. This tailing phenomenon is discussed in Chapter 10. Appropriate additives to the mobile phase prevent this occurrence. For example, the addition of amine modifiers to the mobile phase will eliminate the silanol-dependent tailing in reversed-phase chromatography.

In reversed-phase chromatography, tailing may show up or increase as the column ages. This is due to a partial and slow hydrolysis of the stationary phase. This process exposes additional silanols, which cause or increase peak tailing. This is a natural part of the column aging process. If the column lifetime is too short as a result of this effect, you may want to incorporate amine modifiers into the mobile phase, even if they are not needed with a brand-new column.

17.2.1.6 Sample Distribution All HPLC columns are terminated by filter assemblies, which often have incorporated some means to ensure that the sample is distributed uniformly over the cross section of the column. The proper functioning of these devices can be impaired by blockage. If this is the case, the sample band at the column top is distorted, which can result in a reduced column performance. The phenomenon is indistinguishable from a collapsed bed: all peaks exhibit a similarly distorted profile independent of the retention (Fig. 17.6). Open the column and inspect the filter and the packed bed. If the packed bed appears intact, it is worthwhile to try to clean the filter assembly or replace it with a new one. But the appearance of an undisturbed bed does not guarantee that the column has not collapsed. Often, a bed collapse has occurred somewhere in the middle of the packed bed and is not visible at the ends.

The proper distribution of the sample over the column cross section can also be destroyed by sample constituents that are strongly adsorbed on the packing at the column top. In such a case, a washing of the column with a strong solvent can repair the damage. If you have an idea of the nature of the contaminant, you can design a wash sequence that is specific for the contaminant. Otherwise, solvents of increasing elution strength should be tried. For example, in reversed-phase chromatography one would use a sequence of water–THF–methylene chloride–THF–water. For normal-phase chromatography, the sequence methylene chloride–THF–water–methanol–methylene chloride should be tried.

If a replacement of the filter assembly and a washing of the column do not remedy the problem, it is highly likely that the packed bed is destroyed.

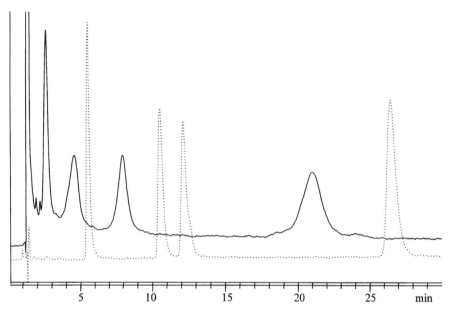

Figure 17.5 Peak distortion due to mobile-phase overload. Comparison of the chromatograms of an acidified sample of tricyclic antidepressants (solid line) and a sample dissolved in mobile phase at pH 7 (dotted line). (Chromatograms courtesy of Y. F. Cheng, Waters Corp.)

17.2.1.7 Column Collapse Despite the efforts of column manufacturers, column collapse is still a common mode of failure of standard HPLC columns. Its signal is the same peak distortion throughout the chromatogram (Fig. 17.6). However, as we have seen, other phenomena can have the same signature. If we can exclude the other causes, we can conclude that the packed bed has collapsed.

There is no true remedy for a collapsed bed, only prevention. Some authors suggest the "repair" of a column by filling the void with additional packing or glass beads. I do not recommend this, since in nearly all cases the problem is only partially remedied, and the performance of the column after such a repair is nowhere close to the performance of a new column. This is especially undesirable in a validated environment.

Bed collapse may occur if the column has been shocked mechanically, for example, by an accidental drop on a hard floor. If a column has accidentally dried out, it can collapse on restart. Immiscible solvents or the precipitation of a constituent of the sample or the mobile phase can cause a shift of the bed due to localized high-pressure drops. Bed collapse can be caused by a continued pulsation of a malfunctioning pump, or by storing the column in the wrong solvent. For example, the bed of silica-based cyanopropyl packings can collapse in solvents of intermediate polarity, such as acetonitrile or THF,

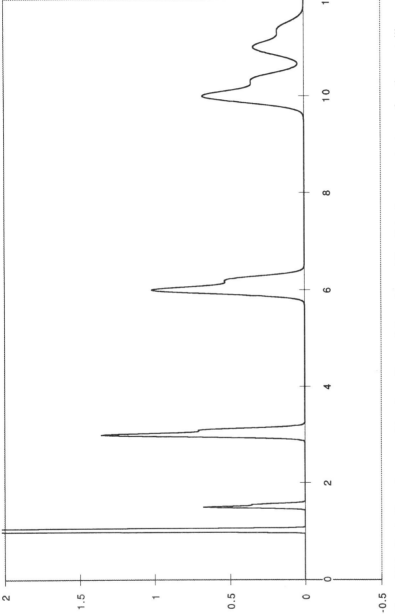

Figure 17.6 Peak distortion caused by the collapse of the column. The distortion may start as a fronting or tailing, and end up in the form of double peaks. It is characteristic for column voiding that the same peak distortion is observed for all peaks throughout the chromatogram. The same pattern can be caused by a blocked column inlet filter assembly.

especially if the column is stored in these solvents. Packings based on organic polymers can shrink in the incorrect solvent, which, in turn, results in a collapse of the bed. Some polymeric packings are also sensitive to rapid flow-rate changes. Finally, a bed may collapse because of the hydrolysis of the packing. To avoid bed collapse in the future, you need to investigate the possible causes of the problem carefully and then take steps to eliminate them.

17.2.1.8 Secondary Equilibria   Peak-shape problems due to secondary equilibria are fairly rare, and they can present quite a puzzle to the investigator trying to solve the problem. In this section we discuss the case in which the analyte itself exists in one or more forms or conformations that can be distinguished chromatographically and that are in an equilibrium with each other. The conversion rate between the different forms is fast enough that during the chromatographic run a substantial conversion can occur, but too slow to make the analyte appear to be chromatographically uniform.

The best known example is the phenomenon encountered in the separation of carbohydrates. Sugars exist in two anomeric forms, which can easily be separated chromatographically if the conversion between both forms can be slowed down enough. This happens, for example, in neutral to acidic mobile phases. In these mobile phases, two cleanly separated peaks are observed for each sugar. The conversion rate between the anomers can be increased by adjusting the pH to higher values. As the conversion rate increases, a step develops between the anomer peaks. As the conversion rate increases further, the anomer peaks themselves become smaller and smaller and the step between the peaks becomes larger and larger. Ultimately, a peak develops at intermediate retention between the anomer peaks. This peak is usually very broad and can be highly distorted. Ultimately, at very fast conversion rates, a single, narrow, symmetric peak develops.

Sugars are not the only compounds exhibiting this problem. Compounds containing an amide bond to the nitrogen of proline show a similar behavior. They exist as two conformers, which interconvert slowly at chromatographic speeds (4). The drugs captopril and enalapril are examples for this class of compounds. Figure 17.7 shows an example of the peak shapes obtained for captopril at various pH values. If a compound contains more than one of these centers, even more complicated peak shapes can arise, as shown in Figure 17.8 for the dimer of captopril.

When you encounter this kind of problem, steps should be taken to increase the rate of interconversion between the different forms. This can always be accomplished by raising the temperature. In many cases a change in pH can be used to manipulate the interconversion rate. Usually, one desires to obtain only a single peak for every compound, but if we understand the problem, there is also nothing wrong with slowing down the interconversion rate until we obtain two clean peaks. The intermediate peak forms are seldom desirable, since the peak recognition algorithms of the integration software cannot deal with odd peak shapes.

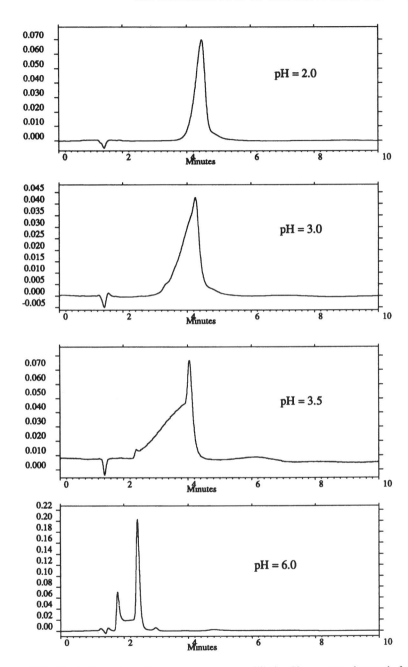

Figure 17.7   Peak distortion due to secondary equilibria. Shown are the peak forms obtained for captopril at various pH values, from top to bottom: pH 2.0, pH 3.0, pH 3.5, pH 6.0. (*Chromatographic conditions:* Column: 5 μm Symmetry C18, 3.9 mm × 150 mm, Waters Corp. Mobile phase: 35/65 methanol/50 mM potassium phosphate adjusted to the indicated pH values. Temperature: 25°C. Flow rate: 1 mL/min.)

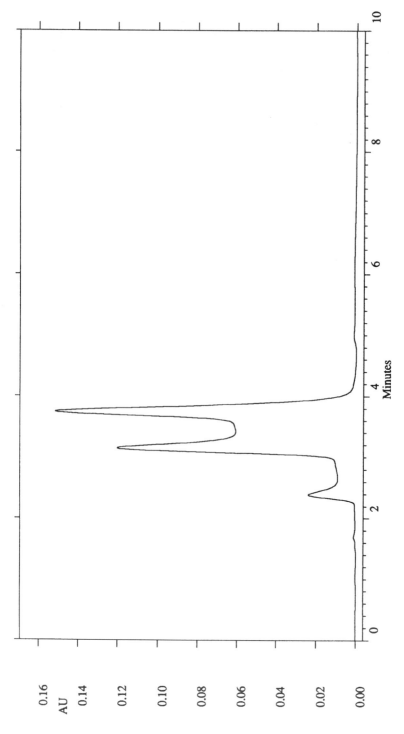

Figure 17.8 Peak distortion due to secondary equilibria when two interconverting centers are involved. The sample is the dimer of captopril: captopril disulfide. (*Chromatographic conditions:* Column: 5 µm Symmetry C18, 3.9 mm × 150 mm, Waters Corp. Mobile phase: 35/65 methanol/5 mM potassium phosphate pH 7.0. Temperature: 32°C. Flow rate: 1 mL/min.)

17.2.2   Retention-Time Problems

In this section, we will discuss various causes of retention-time problems and examine their causes. We will discuss the influence of temperature, mobile-phase composition, column contamination, column aging, and various other topics. We will also briefly discuss again column-to-column reproducibility.

The chapter is divided into two sections. In the first section we deal with retention time problems that are random in nature. The second section then covers the cases where the retention times are drifting in one direction only. If we can differentiate between these cases, we can immediately exclude some causes of the observed problem. For example, random fluctuations of retention times are not a symptom of column aging.

17.2.2.1   Retention-Time Reproducibility   Let us first discuss fluctuations in retention time. When we observe these fluctuations, we should try to correlate them with other known events. Was a fresh mobile phase prepared just before the observed change? Does the change correlate with the cycles of the air-conditioning or heating system of the lab? Or with the time of the day? The following sections will discuss specific effects in more detail and also give you some rules of thumb with respect to the magnitude of various effects.

It should be noted that retention-time reproducibility is not necessarily the most important parameter of an assay, but that the selectivity of a separation is more important. The influence of retention time shifts on quantitative results can be prevented by using peak area integration instead of peak-height measurements.

17.2.2.1.1   Temperature   The most common cause of retention-time fluctuations is a temperature shift. Retention times change typically by about 2% per degree Celsius. This effect is rather small, which is why many HPLC applications are still run without temperature control. However, the nighttime or weekend temperatures in many labs can be significantly different from the temperature maintained during working hours, and this can cause differences in retention during automated analyses over night or over the weekend. The solution to the problem of temperature fluctuations is simple: provide a controlled temperature environment for the column. Both column ovens and column thermostats with heating and cooling capability are commercially available. In a well-designed column oven or thermostat, the temperature of the mobile phase as it enters the column is controlled as well. This is usually accomplished through a piece of coiled tubing, typically 30–50 cm long, that is located in front of the column and in the same temperature environment as the column. Of course, there is a tradeoff, since this tubing causes extracolumn band spreading. Most of the time, retention decreases with increasing temperature. But this is not necessarily always the case. It is not impossible that a secondary equilibrium shifts with the temperature change, and that this shift is overcoming the reduction in retention that is expected. An example could be an equilibrium between a protonated and a deprotonated form of an analyte. Typically, the retention times between both forms can be drastically different. If the temperature change causes a change in this secondary equilibrium that increases the concentration of the more retained form of the analyte, then an increase in retention can be observed. Also, if the equilibrium shifts the concentration toward the less retained analyte, the change of retention with temperature can be several times the normal effect. Retention time shifts as large as 10% or 20% per degree Celsius have been observed. In such a case, retention can be affected even by the heat generated by the friction of the mobile phase in the packed bed, which normally falls within the noise of other experimental parameters. In this case, retention times slowly decrease over the first 1–1.5 h after column startup, after which they stabilize. Unfortunately, this phenomenon cannot be prevented by the use of a column thermostat.

17.2.2.1.2 Mobile-Phase Composition   The preparation of the mobile phase gives rise to random experimental errors. One needs to pay close attention to the reproducible preparation of the mobile phase. In the following sections we will provide some rules as to the magnitude of the effects caused by fluctuations in the percent modifier, the concentration of additives or the pH of the mobile phase. From these rules you can then decide to what level of precision the mobile phase should be prepared to obtain a certain retention-time reproducibility.

17.2.2.1.2.1 MODIFIER CONCENTRATION In reversed-phase and hydrophilic interaction chromatography, a change in the modifier concentration (concentration of the organic solvent) has a fairly uniform effect on retention. A reasonable rule is that for every 1% change in the modifier concentration, the retention factor changes by approximately 10%. This is not completely true over the entire range of solvent compositions, and it also depends on the organic solvent and on the molecular weight of the analyte, but we can use this rule of thumb as a good approximation for many applications. In practice this means that we need to prepare the solvent to an accuracy of 0.1%, if we would like to maintain the variation of retention times under 1%. This is not possible with graduated cylinders, therefore it is recommended to prepare mobile phases by weighing out the different components. This may be a little more work, but it pays off in the reproducibility of a method.

In methods for normal-phase chromatography, the solvent composition effects are similar in magnitude to those for mixtures of solvents of intermediate polarity. For mixtures of nonpolar solvents such as hexane with polar solvents such as isopropanol, significant changes in retention are observed with small relative changes in the concentration of the polar solvent. Also, the polar solvent is used in only low concentrations. Thus one needs to make sure that it is well mixed with the nonpolar solvent. The changes in solvent strength, which determine the relative changes in retention times, can be observed in Figure 9.1 (in the chapter covering normal-phase chromatography).

17.2.2.1.2.2 pH AND IONIC STRENGTH   The retention of an ionizable analyte can vary strongly with the pH of the mobile phase, especially around the pK_a of the analyte. It is not unusual to observe changes of as much as 10% in retention with a pH change of as little as 0.1 units. This applies to all types of chromatography carried out in an aqueous medium: reversed-phase, ion-exchange, hydrophilic interaction, and hydrophobic interaction chromatography. In this case the pH of the mobile phase needs to be tightly controlled. It is, of course, also possible to develop a method in a pH range where the dependence of retention on pH is less, specifically, far away from the pK_a of the analyte, but this is realistic only in special cases. In many real separation problems the sensitivity of the retention of analytes stretches over a broad pH range. Also, a separation may only be possible at a particular pH value. Therefore there is fundamentally nothing wrong with a method that exhibits a high level of pH sensitivity, as long as we are aware of this sensitivity and control the pH appropriately. However, we need to choose the buffer carefully. The buffering capacity of a buffer is a measure of the degree to which a buffer can maintain a pH when its concentration is perturbed by a small amount of acid or base. The optimal buffering capacity of a buffer occurs at its pK_a, and it becomes vanishingly small at about 1.5 pH units way from the pK_a. Therefore we should choose the buffer carefully to yield a good buffering capacity for the target pH. Table 17.2 shows the pK_a values of commonly used buffers.

The buffering capacity also depends directly on the concentration of the buffer. In all applications the maximum buffer concentration is constrained. In reversed-phase and hydrophilic interaction chromatography, the buffer concentration is constrained by the solubility in the presence of organic solvents, and in ion-exchange and hydrophobic interaction chromatography retention is directly influenced by the buffer concentration.

In reversed-phase applications, the concentration of a buffer is rarely a critical parameter. The buffer constituents are commonly weighed, and this procedures provides adequate accuracy. The dependence of retention on ionic strength is small, and only significant weighing errors could cause a measurable shift in retention.

The question is frequently asked how the pH of a reversed-phase mobile phase should be measured: in the presence of the organic solvent or without the organic solvent. For the purpose of a reproducible measurement, it really does not matter, as long as it is always done in the same way. But for the purpose of a consistent lab practice, and to put different data on the same basis, it is preferred to measure the pH *before* the addition of the organic solvent. In some cases, this may not be possible. For example, when octylamine is used as an ion-pairing reagent or as a buffer component to suppress the tailing of basic compounds, we need to take into consideration that its solubility in water is limited. In this case, we are forced to add the organic solvent before the pH measurement.

TABLE 17.2   pK_a Values of Commonly Used Buffers (at 25°C)

Buffer	pK_a
Acetate	4.75
Ammonium	9.24
Borate 1	9.24
Borate 2	12.74
Borate 3	13.80
Citrate 1	3.13
Citrate 2	4.76
Citrate 3	6.40
Ethylenediamine 1[a]	7.56
Ethylenediamine 2[a]	10.71
Glycine 1	2.35
Glycine 2	9.78
2.(N-Morpholino)ethanesulfonic acid (MES)	6.15
Oxalate 1	1.27
Oxalate 2	4.27
Phosphate 1	2.15
Phosphate 2	7.20
Phosphate 3	12.38
Trichloroacetic acid	0.52
Triethanolamine	7.76
Triethylamine	10.72
Trifluoroacetic acid	0.50
Tris(hydroxymethyl)aminomethane(Tris)	8.08

[a]Measured at 0°C.

17.2.2.1.2.3   Other Additives   Additives present in small concentrations in the mobile phase can have a strong influence on retention. Sometimes these low-concentration additives are added on purpose to influence a separation. An example of this could be an ion-pair reagent, which typically is added in concentrations around 5–10 mM. At high concentration levels of the ion-pairing reagent, retention is barely influenced by a change in concentration, but at lower concentrations it is possible to observe a change in retention of about 10% for a change of 1 mM in the concentration of the pairing reagent.

Retention can also be influenced by other components of the mobile phase present in small concentrations. The user may not be aware of the presence of these components. In such a case, troubleshooting can be quite difficult. A common example of this problem is water in normal-phase chromatography. Water is always present in all solvents (see Table 9.3) and can shift retention substantially. However, this is not the only example. For instance, a contamination of methanol with amines that influences the retention of basic analytes in reversed-phase chromatography has been observed. To avoid this situation, the use of HPLC-grade solvents is generally recommended for HPLC applications.

17.2.2.2   Retention-Time Drift   We distinguish retention time drift from random fluctuations of retention time and define drift as the steady change of retention time in one direction only. It helps in the diagnosis of the problem to clearly identify whether retention times are drifting. For example, a drift in retention-time may indicate column aging, while fluctuating retention times cannot be caused by column aging. Actually, the most common causes of retention time drifts are the different mechanism of column aging, such as stationary-phase deterioration (e.g., by hydrolysis) or column contamination (e.g., from sample or mobile phase).

17.2.2.2.1   Column Equilibration   When we observe retention time shifts, we should first consider whether the column is completely equilibrated with the mobile phase. Generally, equilibration with the mobile phase is very fast and is complete within 10–20 column volumes. However, if additives to the mobile phase are present in only very low concentrations (e.g., $\leqslant 10\,mM$) lengthy equilibration times can be observed. A common example is ion-pair chromatography. Another example is the presence of water in normal-phase mobile phases. The latter can be alleviated by the use of mobile phases that are half-saturated with water. See Chapter 9 for more detail.

The mobile phase may be contaminated with something that is present at very low levels and slowly accumulates on the column. Check the quality of the solvents! The use of HPLC-grade solvents is highly recommended. Water can easily be contaminated. Water that is deionized using an ion-exchange process should be treated subsequently with activated charcoal to remove small amounts of the ion-exchange resin, which otherwise would accumulate on a reversed-phase column and cause retention-time drift for compounds that can interact with the resin. Other column contamination mechanisms are discussed in more detail below.

17.2.2.2.2   Stationary-Phase Stability   The chemical stability of all stationary phases is limited. Even within the stability ranges commonly quoted for packings a slow deterioration can occur. For example, silica-based bonded phases are supposed to be stable between pH 2 and 8. But a slow hydrolysis is occurring even within these limits. Optimal resistance to hydrolysis is obtained at intermediate pH values, around pH 4. The speed of hydrolysis is a function of the mobile phase and the ligand. Di- and trifunctional ligands are more stable than monofunctional ligands. Bonded phases with longer chains are more stable than bonded phases with shorter chains. If the ligand contains a functional group, such as a cyano packing, a still lower stability is observed. The difference in stability between a C18 ligand and a cyano ligand may span 3 orders of magnitude; while on a C_{18} column, a shift in retention due to hydrolysis may be significant only after 3 months of use, the same shift may be observable on a cyano packing within an hour.

Stationary-phase hydrolysis can actually be accelerated by frequent cleaning of the column (see also Section 17.2.2.2.4). For example, after hydrolysis, the hydrocarbon ligands of reversed-phase packings adsorb to neighboring ligands under typical mobile-phase conditions. This effectively reduces the loss of ligand. However, this self-adsorption is eliminated when the column is converted to 100% organic solvent, which is recommended for column cleaning or for long-term storage of these columns. Therefore, the treatment of the column with organic solvent involves a compromise between column protection from buildup of contaminant, reduction of hydrolysis during storage, and the acceleration of the loss of ligand. For overnight storage or storage over a weekend, columns are best left in mobile phase. For longer storage, they should be converted to an organic solvent.

Other silica-based bonded phases also hydrolyze in an aqueous environment. Propylamino bonded phases are especially prone to bleed. The mechanism is discussed in more detail in Chapter 11. After initial rapid hydrolysis and retention-time drift, the hydrolysis slows down and the retention time stabilizes. In this state, the columns can be used for an extended period of time.

Methacrylate-based and polyvinylalcohol-based packings can also hydrolyze. As with silica-based packings, this is a function of the pH, and optimal stability is obtained at intermediate pH values.

17.2.2.2.3 Column Contamination   Another common cause of retention-time drift is column contamination. An HPLC column is a very effective adsorptive filter, and it will filter out and adsorb anything that is carried toward it by the mobile phase. Sources of contamination are manifold: the mobile phase, mobile-phase containers, connection tubing, seals in the pumps and injectors of the instrument, and — last but not least — the sample.

If retention-time drifts follow a consistent trend, it is often possible to discriminate between different causes of this drift. For example, one can design an experiment in which the column is run for alternating periods of time with injection of sample and without injection. We then plot the retention time against both time and the number of injections. The smoother trend line points toward the cause of the retention-time drift, in this case the injections or simply the pumping of mobile phase. Similarly, one˙ can determine whether the injection per se or the sample itself is causing the drift by interspersing blank runs with mobile-phase injections between the analyses. In many cases, suitable data are already available, since analyses are usually embedded between calibration runs using standards. A plot of retention time versus the total number of runs and the number of samples can reveal whether an ingredient of the sample is the culprit. The technique of plotting an observation versus the number of events that might cause the observation is a good general tool for troubleshooting.

Any constituent of the sample that might be strongly retained on the column is a potential cause of retention-time drifts. The source is usually the sample matrix: excipients in dosage forms, proteins and lipids in biological samples such as serum, starch in food samples, humic acids in environmental water samples, and similar. Often, the constituents of the sample that are strongly retained by the column have a high molecular weight. In this case, shifts in retention time are either accompanied or followed by increases in backpressure. But it also is possible that the molecular weight of the contaminants is low, such as in the case of lipids.

Problems with the sample matrix can often be avoided or alleviated by an improved sample preparation procedure, for example, through solid-phase extraction. If the contaminant is polymeric in nature, such as starch or protein, it can be separated from the analytes of interest by a simple precipitation. For instance, a large fraction of the proteins in serum can be removed by precipitation with an organic solvent, for example, acetonitrile. For a protein precipitation with acetonitrile, dilute the sample 2:1 with acetonitrile and centrifuge it.

The mobile phase itself can be the source of column contamination. Vessels used to prepare or contain the mobile phase may not have been clean, and whatever was in these vessels can now accumulate on the column. Generally, HPLC-grade solvents should be used. Reagent-grade solvents are less expensive, but not as pure as HPLC-grade solvents, and small amounts of contaminants present in these solvents can accumulate on the column and cause a retention-time drift. Also, the mobile phase carries with it debris from pump seals and injector seals, which may accumulate on the column frit or on the packing itself.

If an instrument is used for both normal-phase and reversed-phase chromatography, special care should be taken when converting the instrument from a reversed-phase mobile phase to a normal-phase mobile phase. If any water remains in any corner of the fluid path, it can later bleed slowly into the normal-phase eluent and contaminate the column. Dead-ended T-connections in the fluid path, used, for example, to attach pressure gauges, are a common source of this problem.

Often, the column contamination is accompanied with an increase in backpressure. This indicates the accumulation of either a precipitate or a polymeric compound on the top of the column. The rise in backpressure can cause an increase in mobile-phase temperature due to frictional heating, which, in turn, leads to a decrease in retention. A column contamination severe enough to cause a rise in backpressure also often leads to a significant peak distortion.

The simplest solution of all column contamination problems is prevention. It is much more difficult to pinpoint the source of the problem and then design effective washing procedures to remove the contaminant. Many washing procedures are of marginal value and remove only part of the contaminant. Typical washing procedures are given for each type of chromatography at the end of the respective chapter. They usually employ a mobile phase that is a strong eluent for the given chromatographic mode. But not all contaminants may be soluble in this mobile phase. For example, THF is a good solvent for removing many contaminants from a reversed-phase column, but proteins are rather insoluble in THF, and injections of DMSO (dimethylsulfoxide) are commonly used to remove protein from reversed-phase columns.

A frequently used procedure of column cleaning is column backflushing. In this case a solvent is pumped through the column in the direction opposite to the mobile phase flow. I generally do not recommend this procedure. It should be used as a last resort only. Although many modern HPLC columns tolerate a reversal of the flow direction, this procedure does have the potential of weakening the bed and causing a collapse of the packed bed. That's why it should be used only if nothing else seems to help.

The best remedy is prevention. Well-designed guard columns protect the analytical column quite effectively without a deterioration of column performance. They take care of all column contaminants, independent of the source. This way you do not have to sort out exactly what the source of the column problem is. You simply replace the guard column any time the problem is observed, or even better, if the occurrence of the problem is reproducible, just before the problem occurs.

Guard columns should generally be placed between the injector and the column. However, if the source of the contamination is the mobile-phase stream, they can be effective also when placed between the pump and the injector, but a situation that would require this position is quite rare.

17.2.2.2.4   Mobile-Phase Composition   Slow changes in mobile-phase composition are also a common cause of retention-time drifts. The most common cause is the evaporation of a mobile-phase constituent. In reversed-phase chromatography, the organic modifier is more volatile than water, and an increase in retention will occur over time. The evaporation of a part of the mobile phase can be amplified by "sparging" with helium, which is commonly used to remove air from the mobile phase.

Also, the mobile phase itself can become contaminated and cause retention-time drifts. This is especially the case when the mobile phase is recirculated to reduce solvent consumption. In this case, the injected samples accumulate in the mobile phase and can cause a drift in retention and other problems, such as baseline drift, ghost peaks, and quantitation problems. Recirculating the mobile phase is generally not recommended.

The composition of a mobile phase can change as a result of chemical reactions. For example, THF oxidizes with time. Metals can leach into the mobile phase from metal frits or from brown solvent bottles. But in both cases the amount of reaction product is so small that it scarcely causes a problem. In practice, a change in mobile-phase composition due to chemical reaction is extremely rare.

17.2.2.2.5   Other Column Aging Mechanisms   Here, we will discuss a few miscellaneous column aging mechanisms that have not been covered in other sections. These mechanisms are specific to certain columns, but the principles are generally instructive.

Ion exchangers used for the separation of biomolecules are commonly cleaned with a strongly alkaline solution, 0.1 M NaOH, pH 13. The principle is a hydrolysis of strongly retained contaminants into smaller fragments, which can be eluted more easily. Although the commonly used ion exchangers can survive this cleaning procedure, a small amount of deterioration of the exchanger happens during every cleaning step. Therefore small changes in the chromatographic properties of the ion exchangers accumulate and the number of regenerations is limited. The same holds true for regeneration procedures that use strong acids.

The primary amine in aminopropyl bonded phases reacts easily with aldehydes and ketones in both the mobile phase and samples. Therefore the exposure of these columns to aldehydes and ketones should be limited. A guard column can be used to protect the analytical column from small amounts of contaminants in the sample.

It is not unusual that the analytes or other sample constituents can react with the stationary phase or residual functional groups on the stationary phase. This can change the properties of the packing. An example is the SEC analysis of polymers that still contain reactive precursor molecules. These reactive molecules may bind to residual sites on the surface of a packing, thus modifying it and rendering it reactive to other analytes. In such a case it is best to dedicate the column for a single analysis.

Over extended periods of time, oxygen dissolved in the mobile phase can react with the surface of a packing. This is especially likely at high-temperature applications such as high-temperature SEC. A gradual change in the surface properties of styrene–divinylbenzene packings used for this application is possible, and it is best to dedicate the columns to the high-temperature application.

17.2.2.2.6 Hydrophobic Collapse   An unusual phenomenon is observed when well-endcapped reversed-phase packings with a small pore size are used with mobile phases that contain close to 100% water. The commonly used equilibration procedure involves first a "wetting" of the column with a mobile phase containing a large percentage of organic modifier followed by equilibration of the packing with the highly aqueous mobile phase. Good and reproducible separations are obtained, but if the pressure on the column is released either by switching off the pump or inadvertently through an air bubble in the system, the separation is suddenly lost and the analytes are significantly less retained or even unretained. One can recover the separation by repeating the "wetting" procedure described above. This phenomenon is not observed on unendcapped or poorly endcapped reversed-phase packings. The cause of this phenomenon is the large wetting angle ($>90°$) of the highly aqueous mobile phase on the highly hydrophobic surface; as the pressure is released, the mobile phase is driven out of the pores. A related phenomenon occurs, when the column has inadvertently dried out during storage. These highly aqueous mobile phases are not capable of wetting the surface.

The remedy to the problem is prevention; it will not occur if the pressure on the column is maintained or if a drying of the column is avoided. If it does occur, it is simply solved by "wetting" the column again with an organic solvent. Alternatively, columns that do not exhibit this phenomenon can be used in highly aqueous mobile phases. These include unendcapped columns and columns with an embedded polar functional group (see Section 10.2).

17.2.2.3   Column-to-Column Reproducibility   A common complaint in HPLC is the difficulty to reproduce a method on different columns. There are several elements that can contribute to this, and a good understanding of the factors involved can help to minimize this problem.

When the chromatograms obtained on columns prepared from the same identical batch of packing material are compared to each other, the differences in retention times should be small and differences in selectivity should be nonexistent. Retention-time differences may stem from differences in the column volume or the packing density and should not exceed a few percentage points. If this is not the case, and if we can exclude noncolumn influences on retention, then the difference between the columns is due to either lack of equilibration or aging-process differences of one of the columns. Equilibration differences can be caused by mobile-phase additives present in small concentrations (e.g., ion-pairing reagents), contaminants (e.g., water on a silica column), or hydrophobic collapse of a reversed-phase column (see Section 17.2.2.2.6). Differences in column aging are especially likely when an older column is compared to a new column or when one or both columns have been used with many different mobile phases and/or for multiple assays. In either case, one should not rely on the long-term reproducibility of the method until the origin of the column-to-column difference has been determined and the method has been verified on a properly equilibrated column prepared from the same batch of the packing. This precaution should be a common practice in method development.

If columns prepared from different batches of a packing are compared to each other, differences in retention times and selectivity of the separation can stem from differences in the surface chemistry of the different batches. One can discriminate between batch-to-batch reproducibility problems and column aging by requesting from the manufacturer a brand-new column packed with the same batch of packing as the original column. Many manufacturers are capable of supplying such a column. If the reproducibility of the method between the original column and the new column from the same batch of packing is satisfactory, we can eliminate column aging or equilibration as the source of the reproducibility problems and conclude that the issue is the batch-to-batch reproducibility of the packing. In this case, there are two solutions to the problem. Either a sufficient amount of packing from the qualified batch can be set aside by the manufacturer to ensure a supply of columns for as long as the method is needed, or we need to modify the method to make it less sensitive to the variation between different batches of the packing. In the first case we should reserve the packing itself rather than packed columns to prevent an aging of the columns in the presence of the storage solvent. In the latter case we should obtain additional batches from the manufacturer to obtain a good statistical sampling of the reproducibility of the packing.

If we attempt to modify the method to make it less sensitive to batch-to-batch differences, it is worthwhile to try to understand the source of the differences. In the case of silica-based bonded phases, differences in surface coverage or silanol activity are the most likely culprit. Often, the silanol activity can be masked by the addition of mobile-phase modifiers. See the respective chapters for more information.

17.2.3   Miscellaneous Problems

In this section we will cover three problem areas: the occurrence of extra peaks in the chromatogram, the source of excessive backpressure, and the causes of baseline drift. All three subjects are not necessarily column problems, but can be caused by other components of the HPLC system.

17.2.3.1   Extra Peaks   Extra peaks are very rarely due to a column problem. It is actually fairly difficult to envision a mechanism, by which the chromatographic column proper generates distinct peaks. Decomposition of the stationary phase can cause a drifting baseline, but a distinct peak can be obtained only if the injection of the sample is causing the decomposition.

A decomposition of the sample due to interaction with the stationary phase does not result in distinct sharp peaks, but rather broad and distorted profiles. However, the decomposition of an analyte on the surface of the column frit can result in the occurrence of a sharp additional peak in the chromatogram, but this condition is fairly rare. If this is truly the problem, it can be remedied by using non-metallic frits at the column inlet.

The most common cause of extra peaks in the chromatogram is the sample itself. The source of the sample may be different, it may have decomposed or aged otherwise. It should be easy to verify whether a particular sample is the source of the extra peak by comparison with samples from other sources or the injection of standards or blanks. If the injection of standards or blanks results in extra peaks, we should next troubleshoot the injector and the mobile phase.

If we inject manually, the syringe, the syringe needle or the injector seal can be contaminated and cause the occurrence of the extra peak. Simply interchanging different syringes can help pinpoint the source of the problem. In an automated injector, the same parts can be causing the problem. Cleaning procedures for these parts should be investigated.

Often, the extra peaks are due to injector carryover from previous injections. This is especially likely, if the concentrations of the previous samples was much higher than the concentration of subsequent samples. The best way to troubleshoot this problem is to make blank injections. The height of the peak should decrease with continued blank injections.

Stagnant solvent in parts of the HPLC instrument can cause extra peaks. The pressure pulse from the injection may cause a small quantity of the stagnant solvent to enter the mobile-phase stream and ultimately create additional peaks. This is one of the more difficult conditions to trace and to eliminate.

The mobile phase itself can generate additional peaks. Since the composition of the sample and the mobile phase are by definition different, the injection of the sample disturbs the equilibrium of the column with the mobile phase. Especially in cases where mobile-phase additives are used at low concentration, extra peaks corresponding to mobile-phase constituents can occur. This phenomenon is commonly observed in ion-pair chromatography. Both negative and positive peaks are possible.

A related phenomenon occurs when the mobile phase is recirculated. In this case, sample constituents slowly accumulate in the mobile phase. An injection of blank, freshly prepared mobile phase results in negative peaks at the same positions in the chromatogram, where the sample peaks usually occur. This condition can create problems with quantitation if the concentration of analytes spans a wide range. It should be avoided.

If we observe a distinct but broad peak in a chromatogram (Fig. 17.9), it is quite common that this peak stems from a previous injection. The position of the peak should change if we change the time interval between injections. We can also verify this by just recording the chromatogram for a longer period of time. This condition is easily recognized. The remedy can be simple, too — increase the run time of the chromatogram until the extra peak has eluted. However, this can significantly increase the run time and decrease productivity. Another solution might be to time the injections such that the extra peak appears in a section of the chromatogram where it does not interfere with the analysis. Alternatively, a step gradient to a stronger-eluting mobile phase, an increase in flow rate after completion of the run, or an improved sample preparation procedure via solid-phase extraction can be used to solve the problem.

In gradient chromatography, extra peaks can be generated by the accumulation of constituents of the starting mobile phase on the column top. The size of these peaks is proportional to the equilibration time with the starting mobile phase. This gives us an easy tool to verify the source of these peaks. If the starting mobile phase is the problem, higher-quality solvents should be considered. If you have a dual-pump gradient system with high-pressure mixing, you can incorporate a guard column into the flow of the weaker-eluting mobile phase. The guard column will scavenge the peaks that otherwise would accumulate on the analytic column and be washed out during the gradient. Unfortunately, this technique cannot be used with single-pump gradient systems with low-pressure mixing.

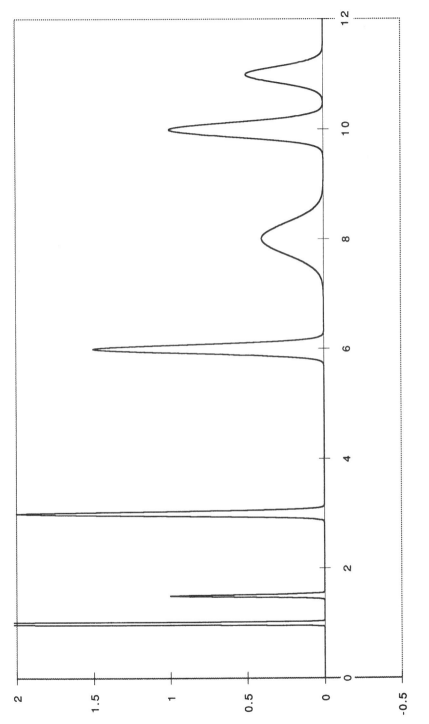

Figure 17.9 Broad extra peak in the chromatogram stems from the last injection.

Finally, in some rare cases, extra peaks can be generated by causes unrelated to chromatography. Temperature-sensitive detectors may respond to a brief draft caused by the opening of a door in the lab, and the resulting baseline disturbance can resemble a peak.

17.2.3.2 Excessive Backpressure

There can be several causes of excessive column backpressure, but before we start to troubleshoot them, it is worthwhile to verify that the pressure is indeed "excessive." To do this, we can estimate what the pressure should be by using the Kozeny–Carman equation discussed in detail in Section 2.2.5.

At a given flow rate F, viscosity η, column length L, column radius r, and particle size d_p, the pressure drop Δp is given by

$$\Delta p = 1000 \frac{F \eta L}{\pi r^2 d_p^2} \tag{17.1}$$

An example of the calculation including the conversion factors is given in section 2.2.5. The viscosities for various HPLC solvents are also listed in Section 2.2.5, including aqueous solvent mixtures. The influence of buffers and salts on the viscosity of an aqueous mixture is typically small, and we can use the viscosity of the solvent mixture without the buffer for an estimation of the pressure.

We will repeat the pressure calculation for another example here. The flow rate shall be 1 mL/min, which is $1/60\,cm^3/s$. The viscosity is 1 cP (water), which is 0.01 $(dyn \cdot s)/cm^2$. The column length is 15 cm, and the column diameter is 3.9 mm, which gives a column radius of 0.195 cm. The particle size is 5 μm, which is $5 \times 10^{-4}\,cm$. The equation yields a pressure of about 85 atm or 1200 psi.

The actual pressure measured on the HPLC instrument is usually somewhat higher, due to the flow resistance of the connection tubing. So we should add an allowance for this effect. Also, we should take into consideration that the particle-diameter designation given by the manufacturer may be be a nominal designation. Only if the column backpressure is significantly different from the calculation can we assume that the backpressure is excessive and that we need to troubleshoot the problem.

The most common problem is an accumulation of sample constituents on the column top. This could either be a precipitate or a high-molecular weight compound. Some possibilities have been discussed in Section 17.2.2.2.3. Other sources could be dust or debris from instrument seals. All of these problems can simply be eliminated by the use of a guard column that is replaced in regular intervals.

Occasionally, the cause of high backpressure is the precipitation of a buffer inside the column. This can occur either when the column has accidentally dried out, while containing a buffered mobile phase; or, more often, if the column contained a buffered mobile phase, but was flushed with an organic solvent that precipitated the buffer; or if the buffer precipitated during gradient elution. In all these cases, we need to attempt to redissolve the buffer by pumping water at a low flow rate into the column until the pressure decreases. This may take some time, but usually works without difficulty.

17.2.3.3 Baseline Drift   Baseline drift can also be produced by many different causes, such as the detector, the mobile phase, the column, or a combination of mobile phase and column. Typically, the same sources should be considered as the causes for retention time drift. This is especially true if baseline drift is accompanied by retention time drift. Consult Section 17.2.2.2 for a detailed discussion. If the drift cannot be pinpointed to the column, mobile phase, or temperature, other detector-specific causes should be considered.

REFERENCES

1. J. W. Dolan and P. Upchurch, *LC-GC* **6**(10), 886–892 (1988).
2. H. A. Claessens, M. A. van Straten, and J. J. Kirkland, *J. Chromatogr.* **728**, 259 (1996).
3. U. D. Neue and E. Serowik, *Waters Column* **VI**(2), 8–11 (1996).
4. U. D. Neue and M. Morand, *Waters Column*, 7–12 (Spring 1995).

Appendix

Teaching Column Chromatography

For chromatographers who are used to deal with peaks that emerge from metal tubes, it is always very instructive and exciting to observe the phenomena happening inside a column by direct visual inspection. NMR imaging has been used to look inside chromatographic columns, and at several recent conferences, tapes showing the migration of bands in a packed bed have found a large audience. However, megadollar equipment is not needed, and much simpler techniques that are accessible to everybody can be used to visualize the phenomena of column chromatography.

Two separate experiments will be described in this section. The first simply demonstrates a separation, using solvents of different elution strength. This experiment can be carried out anywhere, such as in a high-school classroom. The second experiment allows you to observe the development of a separation and the migration of bands inside a bed, but it requires more sophisticated equipment and should be carried out in a ventilated area, such as a hood.

Experiment 1: Separation of the Ingredients of the Kool-Aid Drink Mix

Principle: Separate the constituents of the Kool-Aid drink mix using stepwise elution from a solid-phase extraction cartridge.

This experiment is carried out using a Waters Sep-Pak Classic C18 solid-phase extraction cartridge. A kit containing two Sep-Pak Classic C18 cartridges, one envelope of Kool-Aid-brand unsweetened soft drink mix, and a detailed instruction sheet can be ordered from Waters Corporation under the part number WAT088253. In addition to the kit, you need a 10-mL syringe (with gradation and without needle), a container with a capacity of 2 L, two 20-mL beakers, at least four test tubes, and one bottle with 70% isopropanol and about 3 L water.

The Kool-Aid grape drink mix contains citric acid, ascorbic acid, calcium phosphate, food coloring (Red Dye #40 and Blue Dye #1) and artificial flavor. These ingredients are separated into four fractions, two of which — the dyes — are visible.

Prepare the Kool-Aid grape drink by dissolving one envelope of the drink mix in 2 L of water. Do not add any sugar.

Four eluents with increasing elution strength are needed. The weakest eluent is water, which elutes citric acid, ascorbic acid, and calcium phosphate. The next stronger eluent contains about 6.5% isopropanol, which is prepared by adding 2 mL of 70% isopropanol to 20 mL of water. This eluent elutes the red dye. The subsequent eluent contains 35% isopropanol and is prepared by mixing 10 mL of water with 10 mL of 70% isopropanol. This eluent elutes the blue dye. The final eluent is 70% isopropanol, which elutes the flavoring oils.

The Sep-Pak cartridge is prepared by pushing first 10 mL of 70% isopropanol through the cartridge, followed by 10 mL of water. Then 3 mL of the Kool-Aid grape drink are loaded onto the cartridge. The dyes will concentrate on the top of the bed.

Now use 10 mL of water to elute the acids and the calcium phosphate into the first test tube. The dyes will remain on the top of the cartridge. Next use 10 mL of the 6.5% isopropanol solution to elute the red dye into the second test tube. The blue dye will remain on the top of the cartridge. Then elute the blue dye with 10 mL 35% isopropanol into the third test tube. It is interesting to watch the dye band move through the bed. Finally elute the artificial flavor into the fourth test tube with 10 mL of 70% isopropanol. The fraction is colorless, but you can use your nose to detect the flavor.

The entire process can be repeated several times.

Experiment 2: Visualization of Band Migration in a Glass Column Using a Mobile Phase of Matched Refractive Index

Principle: A glass column is packed with silica particles. A mobile phase is prepared that has the same refractive index as the silica particles. The migration of a dye through the invisible bed is observed. Defects in the bed are easily observed.

The experiment is carried out using a glass column suitable for high-pressure work, equipped with a protective shield. Recommended is the Waters AP-1 10-mm × 100-mm glass column from Waters Corporation, part number WAT021901. The column is packed with a fully porous silica. A particle size of $\sim 30\,\mu$m is recommended. This keeps the packing procedure simple, maintains a low operating pressure, but still allows you to obtain narrow bands. The mobile phase is a mixture of 59% toluene and 41% methanol. An HPLC instrument is required for the demonstration.

The column can either be packed dry or wet. To pack the Waters AP-1 column, the inlet connector of the column is removed and replaced by the packing funnel. For other columns, see the instructions of the manufacturer. For dry packing, a spatula full of packing is added to the column. The column is then tapped gently on a wooden table or board until the packing stops settling. Then another spatula of packing is added and the tapping continues. This process is repeated until the desired height of the packed bed is reached.

TABLE A.1   Refractive Indices of Some Solvents

Solvent	Refractive Index at 20°C
Toluene	1.497
Methylene chloride	1.424
Methanol	1.328

Then the packing funnel is replaced with the inlet connector. The connector is then adjusted until the frit touches the top of the bed. For the Waters AP-1 column, about 3.5 g of silica is needed (the actual amount depends on the density of the silica).

For wet packing, suspend the estimated amount of silica in about 40 mL of methanol by vigorous stirring. Then pour the slurry into column (to which the packing funnel is attached). Let the packing settle into the column. Remove the excess slurry solvent by decantation. You can correct the height of the bed by either removing silica with a spatula or adding additional silica in a methanol slurry. As above, the preparation of the column is completed by replacing the packing funnel with the inlet connector and adjusting the plunger.

After preparation of the column, the bed is equilibrated with mobile phase, which consists of approximately 59% toluene and 41% methanol. The bed will rapidly become nearly completely transparent. Now, a suitable dye dissolved in the mobile phase can be injected, and the band can be observed moving through the transparent bed. One can also generate opaque bands by injecting methanol or other solvents.

One can also adjust the elution strength of the mobile phase and simultaneously balance the refractive index. This allows the separation of several dyes. This experiment is more tricky to carry out. Toluene has a refractive index that is higher than that of silica; methanol and heptane or isooctane have lower refractive indices (see Table A.1). Methanol has a high elution strength on silica, while the elution strength of the hydrocarbons is lowest. Thus one can adjust the elution strength by adjusting the ratio of methanol to heptane and toluene, while balancing the refractive index with the ratio of toluene to methanol and heptane. However, for most purposes the simple experiment described here suffices. It allows you to see how the band emerges from the inlet distributor, how it spreads while migrating down the column, and how it can be distorted by imperfections in the bed or radial nonuniformity of the bed. You can disturb the bed and observe the effect of the resulting irregularities on the band. Happy experimenting!

List of Symbols

A	packed-bed dispersion term
a	Mark–Houwink constant
α	relative retention factor
A_A	relative area of adsorbed analyte molecule
A_p	area of a peak
As	asymmetry factor
A_S	surface area of the stationary phase
A_{sp}	specific surface area
b_g	gradient steepness
b	bunching factor
B	axial dispersion coefficient
B_0	specific permeability
β	phase ratio
β_p	particle phase ratio
C	mass-transfer coefficient
C_{crit}	critical value of particle strength
c_0	solute concentration before injection
$c_{\max}$	solute concentration at the peak apex
c_M	concentration in the mobile phase
c_S	concentration in the stationary phase
c_v	heat capacity at constant volume
χ	surface coverage
d	diameter, thickness
d_0	discrimination factor
d_p	particle diameter
$d_{p,n}$	number-averaged particle diameter
$d_{p,v}$	volume-averaged particle diameter
d_c	diameter of the capillary
d_s	thickness of the stationary phase
d_t	tubing diameter
D	diffusion coefficient
D_M	diffusion coefficient in the mobile phase
D_S	diffusion coefficient in the stationary phase

Δ_{eas}	correction term for second-order contributions in Snyder's retention model
ΔG°	standard Gibbs free energy
Δp	pressure drop
E	separation impedance
E	velocity fluctuation parameter of the Berdichevsky equation
E_T	transition energy
ε°	solvent strength parameter
ε_f	fraction of the particle filled with liquid
ε_i	interstitial porosity of the column, interstitial fraction
ε_p	fraction of the column occupied by the pores of a porous packing
ε_{pp}	particle porosity
ε_{sk}	fraction of the particle occupied by the particle skeleton
ε_t	total porosity of the column
F	flow rate
F_a	fractional area
F_h	fractional height
f	factor
f_P	width of the peak front, subscript indicating the measurement technique
Φ	flow resistance parameter
g	gradient slope
G	peak compression factor in gradient chromatography
ϕ	volume fraction of particles in a slurry
γ	obstruction factor
γ_n	fraction of unused chromatographic space
γ_t	fraction of unused chromatographic time
h	reduced plate height
h_{rp}	relative height of two adjacent peaks
h_p	height of a peak
h_v	height of a valley between two peaks
H	height equivalent to a theoretical plate (HETP)
H_C	HETP derived from the coupling theory
H_D	contribution to the HETP from diffusion
H_F	contribution to the HETP from flow dispersion
η	viscosity
φ	volume fraction
φ_B	volume fraction of organic modifier B in reversed-phase chromatography
$[\eta]$	intrinsic viscosity
k	retention factor
k_a	retention factor at the beginning of a gradient
k_h	hypothetical retention factor
k_{ave}	average retention factor
k_e	retention factor in SEC

k_{max}	retention factor at the peak maximum
k_w	(extrapolated) retention factor in neat aqueous mobile phase in reversed-phase chromatography
K	partition coefficient or equilibrium constant
K_0	partition coefficient in size-exclusion chromatography
K_{iex}	ion-exchange equilibrium constant
K_M	Mark–Houwink constant
l	step length (in random-walk theory)
L	length (column length or migration distance)
λ	tortuosity factor
Λ	hydrophobic interaction constant
m	mass
m_S	molal salt concentration
m_{sat}	mass on the column at saturation of the stationary phase
M	molecular weight
n	number of steps (in random-walk theory)
n_B	relative molecular area of solvent B
n_c	number of columns in a bank
n_i	number of interfering compounds
n_m	number of compounds in the sample matrix
N	plate count, efficiency
N_{column}	plate count of a column
N_{inj}	plate count after taking into account the effect of the injection volume
N_M	number of molecules in the mobile phase
N_{mass}	plate count of an ideal column under conditions of mass overload
N_S	number of molecules in the stationary phase
N_T	total number of plates in a column bank
o	number
O	constant
p	number
P	constant
P'	solvent polarity parameter
P_c	peak capacity
π	performance index
qS	(surface) concentration on the stationary phase
Q	thermal energy
Q_g	peak width factor in gradient chromatography
QM	index of the quality of the separation medium
θ	factor describing loss of efficiency due to injection volume
$\bar{r}$	radius, column radius
r_m	molecular radius
R	universal gas constant
R_{po}	pore radius
Rs	resolution

R_{sp}	specific resolution (SEC)
ρ	density
ρ_{sk}	density of the particle skeleton
ρ_l	density of the liquid
s	slope
sn	signal-to-noise ratio
S	persistence-of-velocity span (in random-walk theory)
S_r	solvent strength parameter in reversed-phase chromatography
S^0	dimensionless interaction energy of analyte with sorbent
SA	specific surface area of a packing (m^2/g)
S_{SEC}	selectivity (SEC)
$\langle s^2 \rangle$	root-mean-square radius of gyration (of random polymer coil)
σ	standard deviation of a peak (length units)
σ_k	standard deviation of a peak in retention factor units
σ_t	standard deviation of a peak (time units)
σ_v	standard deviation of a peak (volume units)
σ^2	variance
T	temperature
T_c	compensation temperature
T	USP tailing factor
t	time
t_a	analysis time
$t_{A/D}$	sampling time
t_d	delay time (in a gradient)
t_g	gradient retention time
t_{gc}	corrected gradient retention time
t_G	gradient duration
t_P	width of the peak tail, subscript indicating the measurement technique
t_0	retention time of an unretained peak
t_M	time spent in mobile phase
t_R	retention time
t_S	time spent in stationary phase
τ	time constant
T_h	tailing factor measured at the height h of the peak
u	linear velocity
u_0	linear velocity based on unretained peak
u_i	linear velocity of the mobile phase between the particles, interstitial velocity
u_s	settling velocity
v	reduced velocity
V_C	column volume
V_d	gradient delay volume
V_e	excluded volume, interstitial volume
V_i	injection volume

V_M	volume of the mobile phase
V_p	particle volume
V_{prep}	volume of a preparative column
V_{pp}	pore volume
V_R	retention volume
V_S	volume of the stationary phase
V_{sc}	volume of a scaling column
V_{sk}	volume of the particle skeleton
V_{sp}	specific pore volume
w	velocity fluctuation factor (in random-walk theory)
w_g	peak width in a gradient
w_l	peak width (length units)
w_t	peak width (time units)
w_P	peak width, used in calculating plate count; subscript P indicates measurement technique (5σ method, tangent method, etc.)
Ψ	solvent association factor in the Wilke–Chang equation
x	mole fraction
X_e	proton acceptor parameter of a solvent
X_d	proton donor parameter of a solvent
X_n	dipole parameter of a solvent
ξ	factor for injection profile

INDEX